Algorithmic Probability

STOCHASTIC MODELING SERIES

Other titles in the series

Stable Non-Gaussian Processes: Stochastic models with infinite variance
G. Samorodnitsky and M.S. Taqqu

Stationary Marked Point Processes: An intuitive approach
K. Sigman

Large Deviations for Performance Analysis: Queues, communications, and computing
A. Shwartz and A. Weiss

Algorithmic Probability: A collection of problems
M. Neuts

Algorithmic Probability

A collection of problems

STOCHASTIC MODELING SERIES

Marcel F. Neuts

Department of Systems and Industrial Engineering,
University of Arizona, USA

CHAPMAN & HALL
London • Glasgow • Weinheim • New York • Tokyo • Melbourne • Madras

Published by
Chapman & Hall, 2–6 Boundary Row, London SE1 8HN, UK

Chapman & Hall, 2–6 Boundary Row, London SE1 8HN, UK

Blackie Academic & Professional, Wester Cleddens Road, Bishopbriggs, Glasgow, UK

Chapman & Hall GmbH, Pappelallee 3, 69469 Weinheim, Germany

Chapman & Hall USA, 115 Fifth Avenue, New York, NY 10003, USA

Chapman & Hall Japan, ITP-Japan, Kyowa Building, 3F, 2-2-1 Hirakawacho, Chiyoda-ku, Tokyo 102, Japan

Chapman & Hall Australia, 102 Dodds Street, South Melbourne, Victoria 3205, Australia

Chapman & Hall India, R. Seshadri, 32 Second Main Road, CIT East, Madras 600 035, India

First edition 1995

ISBN 0 412 99691 X

Library of Congress Cataloging-in-Publication Data

Neuts, Marcel F.
Algorithmic probability : a collection of problems / Marcel F. Neuts.
p. cm.
Includes bibliographical references and index.
ISBN 0-412-99691-X
1. Probabilities. 2. Algorithms. 3. Queuing. I. Title.
T57.35.N48 1995
519.2—dc20 94-36489
CIP

∞
Printed on acid-free text paper, manufactured in accordance with ANSI/NISO Z39.48-1992 (Permanence of Paper).

Table of Contents

PREFACE

Teaching is more than merely imparting information. It includes the structuring of a body of knowledge to guide the learner to an integrated understanding and to independent, critical thinking. From the teacher, it requires enthusiasm and the ability to involve the student in shared thought processes. In my experience, the purest, most delightful teaching has occurred in my interaction with talented doctoral students and research associates. From it, I have learned the importance of constantly formulating questions and of seeking new ways of looking at old problems. With those students, I can share deeper insights on why a particular result holds, why a certain method works, and what are its limitations. As soon as possible, I engage my students in individual work on a specific problem and urge them to explore questions of their own.

In my opinion, the talents and the independence of thought nurtured in graduate courses differ from those expected of the research student only in the breadth of their scope, not in their nature. In mathematical education, problems and exercises are so important because they require the integration and application of formal knowledge. At the elementary level, most exercises call only for direct implementations of a theorem, a technique, or an algorithm. To the more mature student, we give problems rather than exercises. In solving these, several steps of logic must be correctly carried out. They do not drill, but involve the student in thought processes. Good problems should not have obvious answers. Preferably, there should be alternative ways of looking at them and their solution should call for interpretation and generalization. Above all, they should be stimulating. Every problem solved should give the student a sense of accomplishment and a desire to tackle the next and harder ones.

As most persons who have taught many and diverse probability and statistics courses, I have accumulated a large collection of problems for class room use. I am familiar with the classics found in one version or another in all textbooks. Relatively few of these require complex thought processes. Their purpose, a valid one, is drill of material discussed in the predecing chapter. However, in teaching algorithmic thinking in probability or statistics, the routine problems are not adequate. No significant algorithmic thinking is learned from writing a program to generate some numerical examples for a simple formula. Algorithmically challenging problems are always more complex.

The probability problems in this book are elementary but rather complex.

The probability background needed for their solution is imparted in introductory courses for advanced undergraduate or beginning graduate students. However, few of the problems have simple explicit analytic solutions. Their emphasis is on the integration of the knowledge required to derive recurrence relations or matrix-analytic equations appropriate for numerical implementation. Precise probabilistic reasoning and a careful structural analysis of the model are essential to derive such equations. Recognizing and carrying out the formal steps leading to a valid computational procedure is what algorithmic thinking is about. It can be developed by doing the problems in this book.

The chapter headings indicate the general framework of the various large problem sets. In Chapter 1, we emphasize the recognition of recursive relations and schemes. The numerical solution of equations, particularly those that are only implicitly available, is the subject of Chapter 2. The broad topic of derived probabilities is addressed in Chapter 3. It deals with the computation of distributions of random variables relevant to models of some complexity. Markov chains are treated in Chapters 4 and 5. Because of the more delicate numerical solution of differential equations, we have separated discrete and continuous parameter chains into two chapters. In both, we stress the use of matrix formalism and of recursive methods for matrices.

In many problems, the student is asked for an interpretation of the numerical results obtained. The aim of analysis and computation is deeper understanding of the behavior of a physical probability model. The importance of an insightful discussion of the meaning of numerical results cannot be stressed enough. Our experience indicates that initially, many students find this quite difficult. Eventually, of course, that becomes an exciting endeavor. We encourage the students to see the algorithm as a tool for enquiry into the properties of a model. It is always a source of great satisfaction when they raise issues we had not anticipated or propose unexpected insights into the meaning of numerical results.

Many probability models of engineering or OR practice are so complex as to defy even an algorithmic analysis. For such problems, the computer can serve as a tool of experimental enquiry. A selection of computer experimental problems is given in Chapter 6. They require a basic knowledge of simulation methodology. Their emphasis is on the simulation of the correct formal structure and on the interpretation of computer experimental results. The visualization of experimental results, particularly for stochastic processes, is extremely important. To represent and interpret such results correctly can be challenging.

I began collecting and formulating the problems in this book in 1973. Many of them have been used the eight or so times that I have offered an introductory course in algorithmic probability. Their primary source is my occasionally fertile imagination. Others were suggested by material in journal articles on statistics, queueing theory, and Markov chains. In the formulation of the more technical problems, I have included the essential background material, but very few problems require specialized knowledge. The professional reader will recognize the broad spectrum of applications behind the problems. Discussions of their solutions can be serve as introductions to more specialized material such as tests of hypotheses, queueing and reliability models, and simulation studies. The book can only offer samples of the problems in these subjects; it by no means exhausts the rich variety of algorithmic problems of moderate difficulty found in applications of probability. I also hope to share with the student my enthusiasm for the exciting applications of our discipline. This collection of problems should do that better than the pervasive "balls in urns" or the calculations of probabilities of poker hands.

Quite a few of the students who have used drafts of this material have become professional users of probability models. Since completing their studies, some of them have made important research contributions to algorithmic methods in probability or to the application of these methods in biology, communications, or computer engineering. I like to think that some of what these problems taught them has helped in the success of their careers. I have benefited from their comments and been encouraged by their enthusiasm and progress. It is a pleasure to acknowledge their input. I do so, more or less in the chronological order of when our paths first met. From my years at Purdue University, I thank Charles C. Carson and David B. Wolfson; from those at the University of Delaware, S. Chakravarthy, S. Kumar, David M. Lucantoni, Kathleen Meier-Hellstern, Miriam Pagano, and V. Ramaswami. In addition to studying and solving many problems, my students at the University of Arizona have assisted with the onerous tasks of checking the mathematics and proofreading the text. For this, I owe a special debt of gratitude to Y. Chandramouli, Danielle Liu, David Rauschenberg, David Robinson, H. Sitaraman, and especially Jian-Min Li, whose enthusiasm for probability is boundless.

The significance and challenge of algorithmic thinking is now generally recognized. In earlier days, when that was not yet the case, I was inspired by the example and the writings of the late Richard Bellman and of

Professor Ulf Grenander. In education, there is much discussion on when and how to introduce algorithmic methods in various fields of mathematical study. The interested reader can find further expressions of my views on this in the articles:

"The Gambler's Ruin Problem with Markov Dependent Trials - An Example in Computational Probability." The Mathematical Scientist, 9, 25 - 36, 1984.

"An Algorithmic Probabilist's Apology." In *"The Craft of Probabilistic Modelling: A Collection of Personal Accounts."* J. Gani, Editor, Springer-Verlag, 213 - 221, 1986.

"Computer Experimentation in Applied Probability." in *"A Celebration of Applied Probability,"* A special volume (25A) of the Journal of Applied Probability, J. Gani, Editor, 31 - 43, 1988.

"Probabilistic Modelling Requires a Certain Imagination." Proceedings of the Third International Conference on Teaching Statistics (ICOTS 3), August 19-24, 1990, Dunedin, New Zealand, (Ed. David Vere-Jones), Vol. 2, 122 - 131, 1991.

"Algorithmic Probability: A Survey and a Forecast." Proceedings of the Second Conference of the Association of Asian-Pacific Operational Research Societies, Beijing, 1991, Cang-Pu Wu, Editor, Peking University Press, 13-25, 1992.

Tucson, Arizona
9 October 1994

CHAPTER 1

COMPUTATIONAL PROBABILITY: AN INTRODUCTION

1.1. AN HISTORICAL PERSPECTIVE

In its early stages mathematical education places much emphasis on analytic calculation. With continued schooling comes a gradual introduction into the rigor and structural insight of mathematical reasoning. A smaller group of students, those pursuing degrees in mathematics, the sciences, or engineering, are educated in the more powerful methods of modern mathematics. However, even in university education, courses and textbooks dealing with applied subjects devote major attention to analytic derivations that lead to relatively simple answers. Courses on specifically mathematical subjects, such as analysis, algebra, and abstract probability theory, deal exclusively with structural or asymptotic results with, perhaps, a few simple examples where particularly tractable solutions can be found.

In many ways and perhaps unconsciously, an individual's mathematical education retraces the historical development of the discipline, from an early stage of correct and purposeful manipulation of numbers, symbols, and simple geometric objects, through the great era of analytic mathematics stimulated by the development of the physical sciences in the 19-th century, to the structural, abstract mathematics of the 20-th century.

For obvious and valid reasons, classical applied mathematics puts a high premium on analytic and asymptotic methods that lead to useful results, while eliminating or at least alleviating the burden of numerical computation. With the advent of the modern computer, the capabilities of mathematical problem-solving have been greatly enhanced. In studying a mathematical model, we can now greatly relax some stringent assumptions needed only for the sake of analytic tractability. For many problems without a nice formula solution, we can now frequently offer a mathematically correct and implementable algorithmic solution.

The validity of that observation is now generally accepted, yet the integration of algorithmic analysis into university texts and courses has been rather slow. Already in the early 1970s, we noted that typical derivations and exercises in the books we used in teaching probability fell into two broad categories: simple models for which a few analytic operations lead to a nice explicit formula, or general problems of structure or asymptotics that do not involve the search for a specific solution. The latter require only theoretical insight and the writing of a valid proof. Except in courses on numerical analysis, there were essentially no exercises to develop the student's algorithmic competence.

When we started to offer a course on *Algorithmic Probability,* first at Purdue University, then at the University of Delaware, and currently at the University of Arizona, we made a number of common observations:

1. To offer exercises with a significant algorithmic challenge, it was necessary to develop an extensive collection of problems not commonly found in textbooks.

2. The audiences, comprised primarily of graduate students, found the courses refreshing and stimulating, although most came quite unprepared for the type of reasoning we wished to teach. Initially, most students would seek a formula solution and fail to recognize when analytic methods should yield to algorithms. Few of the problems required deep or advanced insights, but nearly all offered challenging complexity. The solutions requiring one or two steps of reasoning, separated by a few pages of calculation, so familiar from other courses were now absent. We devoted much time to learning how to break a complex model into simpler, manageable components to be processed by the algorithm.

3. Although we repeatedly stressed the importance of insight gained from numerical results, only a rare student could correctly interpret the significance of the computed numbers and formulate additional questions suggested by them.

It is testimony to the success of this course that, within one semester, most students acquired considerable algorithmic skills. Indeed, some of them went on to successful careers and became known for their use of integrated mathematical methodology. It is evident that to all areas of mathematical modelling, approaches leading to algorithmic solutions are now as important as those offered by classical analysis. In some disciplines, such as mathematical programming, the focus of research and implementation is almost exclusively algorithmic. Such disciplines are nurtured more by the structural insights of contemporary mathematics than by earlier calculational methods.

Our own area of special interest is ***probabilistic modelling***. In its research and advanced implementations, it now has a thriving and interesting algorithmic component. However, we believe that, in all areas of mathematical modelling, the life of an algorithmic subdivision will be transitory. As algorithmic thinking finds its way into standard textbooks and courses, the student will receive an integrated education in which structural insight will be used to evolve the most efficient and appropriate approaches to the problems at hand.

Although there are important books on algorithmic research methodology, we have repeatedly declined to write a textbook specifically on algorithmic probability. To bring a probability model of modest complexity to the stage where its solution can be studied numerically, one does not need much theoretical knowledge of probability beyond what is offered by the classical texts. What is essential is maturity of insight to use the assumptions and structure of the model to construct a mathematically correct algorithm. These are the same skills that underlie traditional analytic methods, except that a high premium on tractable solutions has led to avoiding problems with sufficiently complex structure.

Teaching algorithmic probability therefore proceeds mostly by examples. The intellectual skills to be nurtured are not new theoretical knowledge, but a mode of reasoning and a new approach to problem-solving. For that reason, we have written a *Book of Problems* containing a large set of exercises organized around common themes.

This first chapter deals with recognizing recurrence relations and their implementation in counting problems. The student is encouraged to review the chapter on combinatorics found in most introductory texts. That review should emphasize the combinatorial reasoning leading to the classical counting formulas, as much as the actual formulas themselves.

1.2. RECOGNIZING RECURRENCE RELATIONS

Combinatorial problems are common in probability. They are much more important than is suggested by the traditional balls-in-urns and card counting problems in elementary texts. Among other applications they are basic to the theory of codes, to nonparametric statistics, and to studying the complexity of algorithms. Counting problems have an extensive mathematical literature, some of which dates as far back as the 16th century.

A common approach to combinatorial problems is to find recurrence relations for the counts of interest. In some cases and often by using generating function methods, one can find analytic expressions for the counts in terms of familiar numbers such as factorials and binomial coefficients. If an analytic expression is complicated and particularly if it requires arithmetic operations with very large integers, the explicit formula may not be suitable for numerical implementation. Because of the computational limitations of the past, much clever mathematics was developed to obtain approximations and order-of-magnitude estimates for combinatorial quantities.

One of the earliest examples is *Stirling's Formula* (actually first proved by A. de Moivre):

$$s(n) = \frac{n!}{\sqrt{2\pi n}\, e^{-n}\, n^n} \to 1, \qquad \text{as } n \to \infty.$$

Apart from many theoretical uses, that formula is useful in calculating (or estimating) ratios involving large factorials. It has been historically fortunate that many early approximations also turned out to be numerically remarkably accurate. For Stirling's formula, that has been proved through many formal theoretical refinements. It is worthwhile to verify this through a direct numerical computation. That is the point of the following simple exercise.

Problem 1.2.1: By taking logarithms and also by using a recurrence relation for $s(n)$, compute $s(n)$ for $1 \le n \le 100$. Use double-precision arithmetic and print $s(n)$ to 15 decimal places. To appreciate the importance of Stirling's formula, determine (by using base-10 logarithms) the number of digits in the decimal representation of $n!$.

Using the computer and recurrence relations, many combinatorial quantities can be computed exactly over a wider range of parameter values than was previously feasible. Among other things, that is useful in examining the quality of limit theorems when they are used in approximations. We illustrate the derivation of a recurrence relation and examine the quality of an approximation by the following example:

Example 1.2.1: Ups in a Random Permutation: In a random permutation of $1, \cdots, n$ we count the number of times that a smaller integer *immediately precedes* a larger number. Call that an *up*. We are interested in the number $M(k;\, n)$ of permutations with exactly k ups.

Clearly, $M(0; 2) = M(1; 2) = 1$. These two values will serve as initial conditions for the recurrence relation to be established. The value of k ranges from 0 to $n - 1$, and there is exactly one permutation with 0 or with $n - 1$ ups. The counts $M(k; n)$ also satisfy the symmetry relation

$$M(k; n) = M(n - k - 1; n), \quad \text{for } 0 \leq k \leq n - 1.$$

To see that, we note that *reversing* a permutation with k ups yields a permutation with $n - k - 1$ ups. There is therefore a one-to-one correspondence between the sets of permutations with k and with $n - k - 1$ ups. Both sets have the same cardinality.

To find a recurrence relation between the arrays $\{M(k;n)\}$ and $\{M(k;n+1)\}$, we see what happens when the number $n + 1$ is inserted at one of the $n + 1$ available positions in a permutation of $1, \cdots, n$. No new up is created if we place $n + 1$ at the beginning or between two numbers where there already is an up. If we place $n + 1$ at the end or between two numbers having a down, exactly one new up is created.

Permutations of $1, \cdots, n + 1$ with k ups are therefore created by inserting $n + 1$ at one of the $k + 1$ positions of the $M(k; n)$ permutations with k ups, where no new up is created. That yields $(k + 1)M(k; n)$ permutations. The permutations of $1, \cdots, n$ with $k - 1$ ups have $n - k$ downs. They offer $n - k + 1$ places where we can put $n + 1$ to create the kth up. That yields $(n - k + 1)M(k - 1; n)$ additional permutations with k ups. As that exhausts all possibilities, we have the recurrence relation

$$M(k; n + 1) = (n - k + 1)M(k - 1; n) + (k + 1)M(k; n), \quad \text{for } 1 \leq k \leq n.$$

Notice that we carefully state the domain of validity of each formula. That is essential to numerical computation (as it should be to careful mathematical writing). In the computer code, we further specify that $M(0; n + 1) = M(n; n + 1) = 1$.

We can now write a program to evaluate the quantities $\{M(k; n)\}$ for a given n. We initialize an array with the values for $n = 2$. We compute those for $n = 3$ and store them in a second array. Successively switching back and forth between the two arrays, we calculate the quantities for $n = 3, 4, \cdots$ up to the desired value of n. In major computations such an economic use of storage space in the recursive computation of arrays is essential. Even for problems with small storage requirements we like to cultivate good programming habits.

The $M(k;\, n)$ are integers, so it seems natural to perform their computation in integer arithmetic. However, many combinatorial quantities grow rapidly and one soon runs into the limitation of word length. Although there are ways around this, such as storing large integers in several words of memory, for the computations in this book these are not needed. If the array containing the $M(k;\, n)$ is defined as double-precision real, it is possible to evaluate these quantities for somewhat higher values of n without loss of accuracy. To see the growth of combinatorial counts, the reader is asked to compute the $M(k;\, n)$ for n up to 40 in this manner. (In Fortran, use an f50.0 format for printout. For $n = 40$, you need to print integers with up to 50 digits.) In probability, large combinatorial integers usually occur in ratios, as in the related problem that follows. It is often possible, and indeed desirable, to derive recurrence relations for the probabilities themselves. The problem of the magnitude of the computed quantities is then avoided.

This example also illustrates other points. By a somewhat involved analysis using generating functions, one proves the explicit formula

$$M(k;\, n) = \sum_{j=0}^{k} (-1)^j \begin{bmatrix} n+1 \\ j \end{bmatrix} (k-j+1)^n, \quad \text{for } 0 \le k \le n-1.$$

However, as it expresses $M(k;\, n)$ as an algebraic sum of large integers, that formula is not well-suited for computation.

A Limit Law: $P(k;\, n) = (n!)^{-1} M(k;\, n)$ is the probability that a random permutation of $1, \cdots, n$ has k ups. This probabilistic interpretation and an appropriate use of a central limit theorem lead to estimates of $M(k;n)$ in the spirit of Stirling's formula. Via generating functions or by a counting argument, one can show that the mean $\mu(n)$ and the variance $\sigma^2(n)$ of the number X_n of ups in a random permutation are

$$\mu(n) = \frac{1}{2}(n-1), \quad \text{and} \quad \sigma^2(n) = \frac{1}{12}(n+1).$$

Furthermore,

$$\lim_{n \to \infty} P\left\{ \frac{X_n - \mu(n)}{\sigma(n)} \le x \right\} = \Phi(x),$$

where $\Phi(\cdot)$ is the standard normal distribution. In addition to its common uses in probability, this *central limit theorem* is also used to obtain an estimate of $M(k;n)$. In doing so, we use the *continuity correction* which is commonly used in approximating discrete probability densities. It is based on the following derivation. We write

$$P\{X_n = k\} = P\{k - \frac{1}{2} \le X_n \le k + \frac{1}{2}\},$$

which by the central limit theorem is approximately

$$\Phi\left[\frac{k + 0.5 - \mu(n)}{\sigma(n)}\right] - \Phi\left[\frac{k - 0.5 - \mu(n)}{\sigma(n)}\right] = B(k;n).$$

That leads to the estimate $M(k;n) \approx n!B(k;n)$, for the magnitude of $M(k;n)$. The symbol $\approx$ indicates that the ratio of the two sides tends to 1 as the parameter (in this case, n) tends to infinity.

The preceding discussion is indicative of the type of mathematical analysis we encourage the student to do for the problems in this book. With that discussion in place the numerical solution of the following problem is now straightforward. We repeat the complete statement of the problem to illustrate the style of problem formulation used in this book.

Problem 1.2.2: From among the $n!$ permutations of $1, \cdots, n$, a permutation is drawn at random. A permutation has an up whenever a smaller number immediately precedes a larger one. $P(k;n)$, $0 \le k \le n-1$, is the probability that the permutation has exactly k ups.

a. Show that for $0 \le k \le n$, the $P(k;n)$ satisfy the recurrence relation

$$P(k;n+1) = \frac{k+1}{n+1}P(k;n) + \left[1 - \frac{k}{n+1}\right]P(k-1;n).$$

An explicit formula for $P(k;n)$ is known, but its derivation is somewhat complicated. That formula reads

$$P(k;n) = \frac{1}{n!}\sum_{j=0}^{k}(-1)^j(k-j+1)^n\binom{n+1}{j}, \quad \text{for } 0 \le k \le n-1.$$

It is also known that for large n the number X_n of ups has approximately a normal distribution with mean and variance given by

$$E(X_n) = \frac{n-1}{2}, \quad \text{and} \quad \sigma^2(n) = \frac{n+1}{12}.$$

b. Note that $P(k; n) = P(n-k-1; n)$, (why?) and use that equality in conjunction with both the recurrence relation and the explicit formula to compute tables of $P(k; n)$ for n up to 30. For $n \geq 5$, compare the exact values of $P\{X_n \geq k\}$ to those obtained from the normal approximation *with continuity correction.* In terms of the standard normal distribution $\Phi(x)$, that approximation to $P\{X_n \geq k\}$ is given by

$$1 - \Phi\left[\left(\frac{n+1}{12}\right)^{-\frac{1}{2}}\left(k - \frac{n}{2}\right)\right].$$

Reference: David, F. N. and Barton, D. E., *"Combinatorial Chance,"* London: Ch. Griffin & Company Ltd., pp. 150 - 154, 1962.

There are many cases for which, even with an explicit formula available, there is clear advantage in the recursive computation of probability densities and other quantities. In the second example, we discuss a probability density important in nonparametric statistics. Its explicit formula is derived by a nice combinatorial argument.

Example 1.2.2: The Density of the Number of Runs: Consider a string of m zeros and n ones. A *run* is a maximal uninterrupted string of symbols of the same kind. For example, the string 0, 0, 1, 1, 1, 0, 1, 1, 0, 1, 0 of 5 zeros and 6 ones exhibits 7 runs, respectively of lengths 2, 3, 1, 2, 1, 1, 1. $X(m, n)$ is the number of runs in a *random* string of m zeros and n ones. A combinatorial argument found in most basic texts shows that the probability density $\{p(k)\}$ of $X(m, n)$ is given by

$$p(2k) = \frac{2\binom{m-1}{k-1}\binom{n-1}{k-1}}{\binom{m+n}{m}},$$

$$p(2k+1) = \frac{\binom{m-1}{k}\binom{n-1}{k-1} + \binom{m-1}{k-1}\binom{n-1}{k}}{\binom{m+n}{m}}.$$

We sketch the argument for the first formula; that for the second one is entirely similar. We can think of the $2k$ runs as $2k$ boxes. In these, the m zeros and the n ones must be placed so that no box is empty and the boxes alternatingly contain zeros or ones. There are

$$\binom{m-1}{k-1}$$

ways of placing the m zeros into the odd-numbered boxes (with no box empty) and similarly

$$\binom{n-1}{k-1}$$

ways of putting the n ones into the even-numbered boxes. The number of ways of placing the zeros and ones is therefore the product of these two integers. An equal number of allocations is obtained if the zeros go into the even and the ones into the odd-numbered boxes. That accounts for the factor 2 in the numerator of the formula for $p(2k)$. All

$$\binom{m+n}{m}$$

strings of m zeros and n ones are equally likely. The formula for $p(2k)$ is now evident.

We may always assume that $m \le n$, as the number of runs remains the same when the roles of zeros and ones are interchanged. In nontrivial cases m and n are both positive. There are then always at least two runs and, for $m < n$, there are at most $2m + 1$ runs. However, for $m = n$, there can be at most $2m$ runs. Now, give a precise specification of the values of k for which the stated formulas hold.

In computing the density $\{p(k)\}$ efficiently, it is advantageous to set

$$A(m, n; k) = \frac{\binom{m-1}{k-1}\binom{n-1}{k-1}}{\binom{m+n}{m}}, \quad \text{for } 1 \le k \le m.$$

Recurrence relations for the binomial coefficients then imply that

$$p(2k) = 2A(m, n; k), \quad \text{and} \quad p(2k+1) = \left[\frac{m+n}{k} - 2\right] A(m, n; k).$$

The quantities $A(m, n; k)$ can be computed recursively by the scheme

$$A(1, 1; 1) = 0.5,$$

$$A(r, n; 1) = \frac{r}{r+n} A(r-1, n; 1), \quad \text{for } 2 \le r \le n,$$

$$A(m, n; k) = \left[\frac{m}{k-1} - 1\right]\left[\frac{n}{k-1} - 1\right] A(m, n; k-1), \quad \text{for } 2 \le k \le m.$$

There is no need to store the $A(m, n; k)$. They can be evaluated by appropriately nested DO-loops and for each k, the corresponding terms $p(2k)$ and $p(2k+1)$ can be computed and stored.

We conclude Example 1.2.2 by a remark about the mean of nonnegative integer-valued random variables. Let N be such a random variable. Its mean $E(N)$ may be written in the equivalent forms

$$E(N) = \sum_{i=0}^{\infty} ip(i) = \sum_{i=0}^{\infty} [1 - F(i)],$$

where $p(i) = P\{N = i\}$, and $F(i) = P\{N \le i\}$. That equality is proved by a simple interchange of summations. The second form is useful in numerical computations. The second sum is easily accumulated while the terms $p(i)$ or $F(i)$ are computed. Furthermore, if we truncate a probability density at some index i^*, the second sum (truncated at that same index) gives a more accurate value of the mean than the first. (Show this!)

When an explicit formula for $E(N)$ is known, we usually prefer to truncate a calculation at an index i^* such that

$$\left| E(N) - \sum_{i=0}^{i^*} [1 - F(i)] \right| < \varepsilon,$$

rather than computing until the tail probability is less than ε. For the same value of ε, that usually requires computing a few more terms. However, it guarantees that also the mean of the truncated density is close to its theoretical value. For heavy-tailed probability densities, this may require significantly more terms. We suggest deriving and discussing the corresponding formulas for $E(N^2)$ and $E(N^3)$.

The next two problems deal with the density of the number $X(m, n)$ of runs.

Problem 1.2.3: Write a computer code to evaluate the probability density, the mean, and the variance of $X(m, n)$. In computing the variance, use a formula for the *second* moment, analogous to that given for the mean in the example. Your code should be able to handle values of m and n up to 100. Prove that when $m = n$, the terms of the density with an even index satisfy

$$p(2k) = p(2m - 2k + 2), \quad \text{for} \quad 1 \leq k \leq m,$$

and examine the symmetry properties (if any) of the terms with odd indices. Use these as accuracy checks in your program.

Remark: Provided that min(m, n) is sufficiently large, the probability density of the number of runs also has an excellent normal approximation. That approximation can be examined exactly as in Example 1.2.1.

Problem 1.2.4: By a careful examination of the recurrence relations for the probabilities $p(2k)$ and $p(2k + 1)$, one can prove (try to do so!) that the index r of the *largest* term of the density of the $X(m,n)$ satisfies

$$\frac{2mn}{m + n} < r < \frac{2mn}{m + n} + 3.$$

Use that inequality to write a program to compute $p(r)$ as efficiently as possible [that is, by computing as few terms of the density as needed to find r and $p(r)$.] Check the accuracy of this program by locating the maximum term $p(r)$ in a table of the density computed by the code for Problem 1.2.3.

The Modal Term: The index r (the mode) and the value of the largest term $p(r)$ are interesting, respectively, as the most likely value taken by the random variable and the corresponding probability. Knowing the index r without first computing the density is occasionally useful in numerical calculations. It often happens that the density concentrates nearly all probability mass in a number of terms on either side of r and that the other terms are negligibly small. In computing such a density, we first evaluate $p(r)$

and then alternatingly compute terms on either side of r until the total computed probability exceeds $1 - \varepsilon$ for some specified ε. That idea is explored in the following problems.

Problem 1.2.5: For the Poisson density with parameter $\lambda > 0$,

$$p(k) = e^{-\lambda}\frac{\lambda^k}{k!}, \quad \text{for } k \geq 0,$$

for the binomial density with parameters $n > 0$ and $0 < p < 1$,

$$p(k) = \binom{n}{k} p^k (1-p)^{n-k}, \quad \text{for } 0 \leq k \leq n,$$

and for the negative binomial density with parameters $n > 0$ and $0 < p < 1$,

$$p(k) = \binom{n+k-1}{k} p^n (1-p)^k, \quad \text{for } k \geq 0,$$

determine the modal index r. Write subroutines to evaluate $p(r)$ directly. In computing $p(r)$ use logarithms or careful grouping of factors to ensure high accuracy. Use recurrence relations for the $p(k)$ to evaluate terms on either side of r until the total computed probability exceeds $1 - 10^{-9}$. Implement your code for the Poisson density with $\lambda = 250$; the binomial density with $n = 300$, $p = 0.35$; and the negative binomial density with $n = 15$ and $p = 0.20$. What happens if, say, for the Poisson density with $\lambda = 250$, you implement the recurrence relation starting with $k = 0$?

Alternative Recurrence Relations: In many situations, the quantities of interest satisfy several recurrence relations. It is important to learn to recognize those that are most numerically stable, require the fewest arithmetic operations, get most quickly to the quantities we need, and so on. There is rarely a general theoretical basis for choosing the best recurrence relation. In this chapter and in the sequel, alternative recursion formulas are given for quite a few problems. We suggest that the reader discuss the merits of each *before* doing any computer programming. The following problem addresses such an issue.

Problem 1.2.6: Restricted Occupancy Numbers: The *restricted occupancy number* $N_a(n, m)$ is the number of distinct ways in which n identical items can be placed in m containers, so that each container contains *at most*

a items. Prove the following three recurrence relations:

$$N_a(n, m) = \sum_{j=1}^{m} \binom{m}{j} N_{a-1}(n - j, j),$$

$$N_a(n, m) = \sum_{r=0}^{m} \binom{m}{r} N_{a-1}(n - ar, m - r),$$

$$N_a(n, m) = \sum_{j=0}^{a} N_a(n - j, m - 1).$$

Write a program, using double-precision real arithmetic, to compute a table of the integers $N_a(n, m)$ for $1 \le m \le 25$, $1 \le n \le 25$, and $1 \le a \le 5$. Use each of the recurrence relations in turn and discuss which one you find most efficient for the task at hand. You will need to print your results using an f20.0 format, as some of the counts result in very large integers.

Without calculations prove that $N_a(n, m) = N_a(am - n, m)$, for $0 \le n \le am$. Check that this symmetry relation is reflected in all but the largest numbers computed by the three subroutines. The discrepancy in some of the final two or three digits stems from the fact that, even in double-precision arithmetic, these computations push the limits of full numerical accuracy.

Use the third recurrence relation to evaluate $N_a(n, m)$ for $a = 10$, $1 \le m \le 10$, and $0 \le n \le 100$. For this task, there is no loss of accuracy at all. Your numbers should have the stated symmetry.

Comments: To establish the first formula, let j be the number of nonempty boxes. These can be chosen in $\binom{m}{j}$ ways. If one item is placed in each of these boxes, the remaining $n - j$ items are to be distributed among the j boxes, so that at most $a - 1$ more items are added to a box. The arguments for the other two formulas are similar.

Effective Summation Limits: In writing recurrence relations, we have followed the common mathematical convention that it is clear from the context which terms in a sum are obviously zero. However, in numerical computations it is important to identify the summation limits that give only nonzero contributions. For instance, in the first formula the contribution to the sum is positive only if $j \le m$ and $n - j \le j(a - 1)$. By combining these inequalities, we see that the *effective* upper and lower summation limits are

$$j_L = \min\left\{1, \left[\frac{n}{a}\right]\right\}, \quad \text{and} \quad j_U = m,$$

where $[\cdot]$ denotes the integer part. If $j_L > j_U$, then $N_a(n, m) = 0$. The effective summation limits are similarly found for the other two formulas.

Embedding in Recurrence Relations: We must often evaluate quantities that apparently do not satisfy recurrence relations. One of the most useful skills in all numerical work is recognizing recursive schemes in which those quantities are embedded. At the expense of computing some numbers that are not germane to our problem, we can extract those we need from an implementation of the recurrence relation.

Embedding is usually accomplished by treating one or more constants of the given problem as a variable. Sometimes, that leads to a recursive scheme that is implemented until the given value of the given constant is reached and the needed solution can be picked off. Here also it is important to look for alternative approaches with possibly better algorithmic properties. The following problem offers a simple example of embedding.

Problem 1.2.7: Fáa di Bruno's Formula: In the formula, attributed to Fáa di Bruno, which expresses the nth derivative of the composite function $f[g(x)]$ in terms of the derivatives of $f(\cdot)$ and $g(\cdot)$, it is necessary to form the set $E_{n,r}$ of all n-tuples $j_1, \cdots, j_n$, of nonnegative integers that satisfy

$$\sum_{k=1}^{n} j_k = r, \quad \text{and} \quad \sum_{k=1}^{n} k j_k = n,$$

for given integers $n \geq r \geq 0$. Find an appropriate recurrence to compute the number of elements of $E_{n,r}$ for $0 \leq r \leq n \leq 50$.

Hint: There is no convenient *direct* recurrence relation for the cardinality of the set $E_{n,r}$. Consider a slightly more general problem, involving one more variable. Several recurrence relations for the counts in that problem may be derived. These are well-suited for numerical computation. The required counts are obtained for special choices of the indices.

Remark: Fáa di Bruno's formula states that, if all derivatives involved exist,

$$\left\{\frac{d^n}{dx^n} f[g(x)]\right\}_{x=a} = \sum_{r=1}^{n} \left\{\frac{d^r}{dy^r} f(y)\right\}_{y=g(a)}$$

$$\times \sum_{E_{n,r}} \frac{n!}{j_1! j_2! \cdots j_n!} \left[\frac{g^{(1)}(a)}{1!}\right]^{j_1} \left[\frac{g^{(2)}(a)}{2!}\right]^{j_2} \cdots \left[\frac{g^{(n)}(a)}{n!}\right]^{j_n}.$$

Reference: For an application related to queueing theory, see Klimko, E. M. and Neuts, M. F., "The single server queue in discrete time - Numerical analysis II," *Naval Research Logistics Quarterly,* 20, 297 - 304, 1973.

1.3. MAIN PROBLEM SET FOR CHAPTER 1

The following problems will aid in developing the skills of setting up recurrence relations, of the use of limit theorems as approximations, and of using the mode in the computation of probability densities. These essential skills are needed to tackle more complex problems in the next chapters.

EASIER PROBLEMS

Problem 1.3.1: With each subset of $\{1, \cdots, n\}$, associate the sum of its elements. Clearly, the possible sums are between 0 and $n(n+1)/2$. $A(n; k)$ is the number of different subsets with sum k.

a. Give a quick argument (no calculations!) to show that, for every n, the array $A(n; k)$ is symmetric in k, that is:

$$A(n; k) = A\left[n; \frac{n(n+1)}{2} - k\right], \quad \text{for all } n.$$

b. By considering whether or not the integer n is used in forming the sum k, derive a simple recurrence relation for the $A(n; k)$.

c. Write a code, using integer arithmetic only, to print a table of the counts $A(n; k)$ for any given value of n up to 38. That is the largest n that can (conveniently) be handled on most computers by using integer operations only.

d. Show how, by using double-precision floating point arithmetic, the $A(k; n)$ can be *exactly* evaluated for values of n much larger than 38.

Remark: For a use of the integers $A(n; k)$ in probability, see Problem 1.3.10.

Problem 1.3.2: As efficiently as possible, determine the number of points with integer coordinates (x, y) that satisfy the inequality

$$\frac{x^2}{a^2} + \frac{y^2}{b^2} \leq 1.$$

Assume that $a \geq b$ are integers. Your code should be able to handle all values of $b \leq 100$. Special care is needed not to miss points on the boundary of the ellipse. For these, the equality must be checked in integer arithmetic.

Problem 1.3.3: For a group of m people, all 365^m ways of choosing their birthdays are equally likely. Show that the probability $P(k; m)$ that there are exactly k pairs of people with matching birthdays (falling on k distinct dates) and that all remaining persons have distinct birthdays is

$$P(k; m) = 365^{-m} \binom{365}{k} \binom{m}{2k} \frac{(2k)!}{2^k} (365 - k)(365 - k - 1) \cdots (365 - m + k + 1)$$

$$= \frac{365!\, m!}{365^m 2^k (m - 2k)! k! (365 - m + k)!}, \quad \text{for } 2k \leq m \leq 365 + k.$$

Compute the probabilities $P(k; m)$ for $0 \leq k \leq 10$ and for all m for which they exceed 10^{-5}.

Problem 1.3.4: A class has m students and an instructor. It was found that one student had the same birthday as the instructor and that k more pairs of students had common birthdays. The remaining $m - 2k - 1$ individuals all had different birthdays. All the pairs had distinct birthdays. Assume that all 365 days of the year are equally likely to be birthdays and that the birthdays of all $m + 1$ individuals are selected independently.

a. Show that the probability $P(k; m)$ of that event is given by

$$P(k; m) = \frac{m}{365^m} \binom{m-1}{2k} \binom{364}{k} \frac{(2k)!}{2^k}$$

$$\times (364 - k)(364 - k - 1) \cdots (364 - m + k + 2)$$

$$= \frac{m!\, 364!}{365^m 2^k (m - 2k - 1)! k! (364 - m + k + 1)!},$$

for all nonnegative integers m and k satisfying $2k + 1 \le m \le 365 + k$.

b. Write an efficient computer program to evaluate $P(k; m)$ for $0 \le k \le 5$ and $2k + 1 \le m \le 2k + 100$.

Problem 1.3.5: Table Tennis: We consider a game, such as table tennis, between two players, I and II. The players compete for successive points and the first player to accumulate n points wins provided that she leads by at least two points at that time. If her opponent has $n - 1$ points when her score reaches n, the game continues until one player leads the other by two points. For table tennis, $n = 21$.

We assume that the competitions over successive points can be treated as independent Bernoulli trials with probability p of success for Player I and with probability $q = 1 - p$ for Player II. We thereby ignore the effect of whether the one who serves has a better chance of winning the point. For given values of p and n, we wish to compute the probability $P^*(n)$ that Player I wins. Let $P(i, j)$ be the conditional probability that Player I wins, given that the scores have reached i for her and j for her opponent.

a. Show that the $P(i, j)$ satisfy the recurrence relations

$$P(i, j) = pP(i + 1, j) + qP(i, j + 1), \quad \text{for} \quad i \ge 0, j \ge 0,$$

with the boundary conditions

$$P(n, j) = 1, \quad \text{for } 0 \le j \le n - 2, \quad \text{and} \quad P(i, n) = 0, \quad \text{for } 0 \le i \le n - 2,$$

$$P(n, n-1) = P(n+k, n+k-1),$$

$$P(n-1, n) = P(n+k-1, n+k),$$

$$P(n-1, n-1) = P(n+k-1, n+k-1), \quad \text{for } k \geq 0.$$

From these equations derive the explicit formulas

$$P(n, n) = p^2(p^2+q^2)^{-1},$$

$$P(n-1, n) = p^3(p^2+q^2)^{-1},$$

$$P(n-1, n-1) = (p^2+2pq)(p^2+q^2)^{-1}.$$

b. By a direct combinatorial argument, prove the explicit formula

$$P^*(n) = P(0,0) = \sum_{r=0}^{n-2} \binom{n+r-1}{n-1} p^n q^r + \binom{2n-2}{n-1} p^n q^{n-1} P(n, n-1)$$

$$= \sum_{r=0}^{n-2} \binom{n+r-1}{n-1} p^n q^r + \binom{2n-2}{n-1} p^{n+1} q^{n-1} (p^2+q^2)^{-1}.$$

c. Set $A(i, j) = P(n-i, n-j)$ in the recurrence relations and write an efficient program to compute the array $A(i, j)$. Note that for all $\nu \geq 2$, $A(\nu, \nu) = P^*(\nu)$. A single computation of the array therefore yields $P^*(\nu)$ for all values up to the given n.

d. Use formulas for the absorption probabilities in the classical Gambler's Ruin problem to obtain an explicit formula for $P^*(n)$ for the game, in which to win, Player I must lead her opponent by k points. First consider the case $k \geq n$, which is essentially the Gambler's Ruin problem. Use the results obtained to handle the case $k < n$ by considering the cases where Player I wins in at most $2n-k$ steps and where $2n-k$ or more steps are needed.

e. Compute the probability $P^*(n)$ for table tennis and for $p = 0.40 + 0.01j$,

$0 \leq j \leq 10$, by using the recurrence relations and the explicit formula.

Problem 1.3.6: Table Tennis Generalized: The model for table tennis is generalized as follows: The successive points are still treated as independent Bernoulli trials. The probability that Player I wins a point is p_1 if she serves and p_2 when her opponent serves. Player I gets to serve the first five balls of the set so that she also gets to serve the 11th through the 15th, the 21st through the 25th, and so on. However, after the score (20, 20) is reached, the players serve alternatingly.

Write a computer program to evaluate the probability $P^*(n)$ for given values of p_1 and p_2. Interpret your numerical results to see the importance of a strong serve.

Problem 1.3.7: Tennis: In tennis a player needs to win at least four points but must lead by two points to win a *game*. In order to win a *set,* she must win at least six games but should lead her opponent by two games. Finally, to win the *match,* three out of five sets must be won. Assuming that the successive points in a game of tennis may be treated as independent Bernoulli trials in which the player has a probability p of winning a point, compute the probability $P(p)$ that the player wins the match. Tabulate $P(p)$ for $p = k/100, \; 1 \leq k \leq 50$.

AVERAGE PROBLEMS

Problem 1.3.8: A permutation of $1, \cdots, n$ has a *peak,* whenever a number is larger than *both* its neighbors. A peak cannot occur at either end. For example, the permutation 7 2 $\underline{4}$ 3 $\underline{6}$ 1 5 has peaks at 4 and at 6. Let $N(k; n)$ be the number of permutations of $1, \cdots, n$ with k peaks. Verify that $N(0; 2) = 2$, $N(0; 3) = 4$, and $N(1; 3) = 2$. Also, give an easy argument to show that $N(0; n) = 2^{n-1}$ for $n \geq 2$.

a. Show that the integers $N(k; n)$ obey the recurrence relation

$$N(k; n+1) = 2(k+1)N(k; n) + (n - 2k + 1)N(k-1; n),$$

and compute a table of the positive $N(k; n)$ for $2 \leq n \leq 15$.

b. Derive a recurrence relation for the probabilities $p(k; n) = (n!)^{-1}N(k; n)$, that the permutation has k peaks. For $n = 10r, \; 1 \leq r \leq 100$, compute and plot the probability density of the number of peaks. Also print a table of the means and standard deviations of these densities.

Problem 1.3.9: A box contains m red and n white balls. Balls are drawn out at random, without replacement. The random variable $T(r)$ is the number of drawings required until r, $1 \le r \le m$, red balls have been drawn.

a. Show that the probability $P(k) = P\{T(r) = k\}$ is given by

$$P(k) = \frac{\binom{k-1}{r-1}\binom{m+n-k}{m-r}}{\binom{m+n}{m}}, \quad \text{for } r \le k \le n+r.$$

b. Show that $P(k)$ is largest for

$$k^* = \max\left\{r, \left[\frac{r(m+n)-n-1}{m-1}\right]\right\},$$

where $[\cdot]$ denotes the integer part. Further show that $P(k)$ is strictly decreasing in k for $r \le k \le n+r$, if and only if

$$r > \frac{m+n}{n+1}.$$

c. Write a computer program to evaluate the probabilities $P(k)$ for $r \le k \le n+r$, as well as the mean and variance of $T(r)$. Execute your program for $m = 50$, $n = 100$, and $r = 20$. Compare the numerical values so obtained to the corresponding terms of the negative binomial probability density $P_1(k)$ with parameters $p = m(m+n)^{-1}$ and r. For reference,

$$P_1(k) = \binom{k-1}{r-1} p^r q^{k-r}, \quad \text{for } k \ge r.$$

d. Write a subroutine to compute the index k^* and the corresponding values of $P(k^*)$ and $P_1(k^*)$ separately.

Problem 1.3.10: The combinatorial integers $A(n;k)$ of Problem 1.3.1 arise in the following model. The random variables U_i, $i \ge 1$, are independent, identically distributed with $P\{U_i = 1\} = P\{U_i = 0\} = 1/2$.

a. Show that $P\{\sum_{i=1}^{n} iU_i = k\} = 2^{-n}A(n;k)$, for all values of k such that $0 \le k \le n(n+1)/2$.

b. Writing $U_1 + 2U_2 + \cdots + nU_n = X(n)$, derive formulas for the mean $\mu(n)$ and the variance $\sigma^2(n)$ of $X(n)$.

c. Show that the probability generating function $P_n(z)$ of $X(n)$ is given by

$$P_n(z) = 2^{-n}\prod_{r=1}^{n}(1+z^r).$$

d. By performing series expansions in the moment generating function show that, as $n \to \infty$, the normalized random variable $[\sigma(n)]^{-1}[X(n) - \mu(n)]$ has an asymptotic $N(0, 1)$ distribution.

e. From this central limit theorem, we obtain the following estimates for the magnitude of the integers $A(n;k)$:

$$A(n;k) \approx 2^n \frac{1}{\sigma(n)\sqrt{2\pi}}\exp\left\{-\frac{1}{2}\left[\frac{k-\mu(n)}{\sigma(n)}\right]^2\right\} = A_1(n;k),$$

and with continuity correction,

$$A(n;k) \approx 2^n\left\{\Phi\left[\frac{k+0.5-\mu(n)}{\sigma(n)}\right] - \Phi\left[\frac{k-0.5-\mu(n)}{\sigma(n)}\right]\right\} = A_2(n;k).$$

Compute and discuss a table of the relative errors

$$\frac{A(n;k)-A_1(n;k)}{A(n;k)} \quad \text{and} \quad \frac{A(n;k)-A_2(n;k)}{A(n;k)},$$

for $n = 40$ and all values of k.

Problem 1.3.11: In a random string of m zeros and n ones, the symbols are examined by pairs from left to right. A pair is said to be *heterogeneous* if it is either 01 or 10. X is the number of heterogeneous pairs *leading* the string, i.e., $X = k$, if and only if the first k pairs are heterogeneous. Show that

$$P\{X \geq k\} = 2^k \frac{\binom{m+n-2k}{m-k}}{\binom{m+n}{m}}, \quad \text{for } 0 \leq k \leq \min(m, n).$$

For given values of m and n, compute and print the probabilities $P\{X \geq k\}$ and $P\{X = k\}$ for all k satisfying $0 \leq k \leq \min(m, n)$.

Problem 1.3.12: In a random string of m zeros and n ones, the ith zero is preceded by exactly j ones. Show that this occurs with probability

$$\theta(i; j) = \frac{\binom{m+n-i-j}{n-j}\binom{i+j-1}{j}}{\binom{m+n}{m}}, \quad \text{for } 1 \leq i \leq m, \quad 0 \leq j \leq n.$$

For given m, n, and i, compute and print the probabilities $\theta(i; j)$ for all j for which they exceed 10^{-7}.

Problem 1.3.13: At time $n = 0$, you are seated in the $(k + 1)$st car of a row of $m + 1$ cars, $0 \leq k \leq m$, waiting to drive off. However, you cannot leave unless the cars numbered $1, \cdots, k$, ahead of you, have left. At every next time point, one driver arrives with probability p, $0 < p \leq 1$, and none arrive with probability $q = 1 - p$. Arrivals are independent events. Each arriving driver enters an unoccupied car chosen at random and drives off immediately if all cars ahead have left. If not, that driver also waits. The random variable T is the time of your departure.

a. Show that for all $n \geq k$,

$$P\{T = n\} = P(k; n) = \sum_{\nu=k}^{m} \binom{n-1}{\nu-1} p^{\nu} q^{n-\nu} \frac{k}{\nu} \binom{\nu}{k} \binom{m}{k}^{-1}.$$

b. Using an efficient recursive scheme, write a program to compute the densities $\{P(k; n)\}$ and the corresponding distribution functions for all k with $0 \leq k \leq m$. Compute terms until the remaining tail probabilities are less than 10^{-7}. Execute your program for $m = 19$ and $p = 0.60$. Plot the graph of the density for $k = 9$.

Problem 1.3.14: Suppose n identical items are distributed at random into m containers, numbered $1, \cdots, m$. The random variable R is defined as the difference between the highest and lowest numbers of the boxes that contain at least one item. For example, if the contents of six boxes are respectively 0, 4, 1, 0, 5, 0, then R is 3 as the highest and lowest numbers of nonempty boxes are 5 and 2. The possible values k of R satisfy $0 \le k \le m - 1$. Show that

$$P\{R = k\} = p(k) = (m - k)\frac{\binom{n + k - 2}{n - 2}}{\binom{n + m - 1}{n}}, \quad \text{for } 0 \le k \le m - 1.$$

Examine the ratio $p(k)/p(k - 1)$ to prove that $p(k)$ is largest at

$$k^* = \left[\frac{n - 2}{n - 1}m\right]$$

For all but exceptional cases, $k^* = m - 1$.

Write a program to evaluate the density $\{p(k)\}$, starting from the highest index $m - 1$ and computing terms recursively downwards. Stop as soon as the sum of the computed terms exceeds $1 - 10^{-8}$, and print the computed terms.

Implement your code for $m = 10, 20, 40, 60, 100, 150, 200, 400, 500$. For each m, set the corresponding n first equal to $3m/2$ and next to $2m$. Notice that, in each string of numerical results, the higher terms of the density are only slightly different. That suggests that the random variable $m - R$ has a limiting density if m and n tend to infinity in such a manner that $m/n \to \alpha$, where α is a positive number. Now examine the expressions for the probabilities $p(m - k)$ for constant values of k. Determine the limits of these probabilities as m and n tend to infinity in the prescribed manner. State and prove the suggested limit theorem for the density of $m - R$.

Hint: The limiting density is negative binomial with parameters $\nu = 2$ and $\theta = (1 + \alpha)^{-1}$ on the positive integers.

Problem 1.3.15: The World Championship match in chess consists of $2n$ games between the title holder (Player I) and the challenger (Player II). To retain the title, Player I must accumulate at least n points. For each game won, a player gets 1 point, 0 for a game lost, and 1/2 point for a draw. Assume that successive games can be treated as independent trinomial trials and that the players are of equal ability i.e., the probabilities of a win and a loss are equal and $(1 - x)/2$, where x is the probability of a draw. X is the total number of points accumulated by Player I.

a. Show that the probability that the champion retains the title satisfies

$$P\{X \geq n\} = \frac{1}{2} + \frac{1}{2}P\{X = n\},$$

where

$$P\{X = n\} = \sum_{\nu=0}^{n} \frac{(2n)!}{(\nu!)^2(2n-2\nu)!}\left[\frac{1-x}{2}\right]^{2\nu} x^{2n-2\nu}.$$

b. Write a program to compute $P\{X \geq n\}$ for a given n and for $x = k/100$, $0 \leq k \leq 100$. Execute your program for $n = 12$. Plot the resulting probabilities as a function of x.

c. Using properties of Bessel functions, one can prove that $P\{X \geq n\}$ is a strictly convex function of x on $[0, 1]$. The minimum of that function is attained in the interval $(0, 1/2)$. Write an efficient subroutine to compute, to within 10^{-4}, the abscissa x^* of the point where the minimum is attained.

d. Find the probability $P_1(k; n; x)$ that the chess championship is decided after exactly k games with Player I retaining the title. Also find the corresponding probability $P_2(k; n; x)$ that the match ends after k games with the challenger winning. Write a program to compute $P_1(k; n; x)$ and $P_2(k; n; x)$ for a given n and all k satisfying $n \leq k \leq 2n$. Tabulate these quantities for $x = k/100$, $0 \leq k \leq 100$. Execute your program for $n = 12$, but use the cases $n = 1$ and $n = 2$ to check your analytic and numerical solutions.

d. Compute the expected number of games played in the match.

e. By $P_1(\nu; k; n; x)$ and $P_2(\nu; k; n; x)$ denote the probabilities that the championship is won, respectively, by Player I and Player II, that the match ends after exactly k games, ν of which end in a draw. Tabulate these probabilities for a given n and for all relevant values of ν and k.

Problem 1.3.16: $p(r; m, n)$ is the probability that a random string of m zeros and n ones has at least one substring of r or more consecutive ones.

a. By considering the place of the first zero, if there is one, among the first r symbols, show that

$$p(r; m, n) = \binom{m+n}{m}^{-1} L(r; m, n),$$

where $L(r; m, n)$, the number of strings in which there are at least r consecutive ones, satisfies the recurrence relation

$$L(r; m, n) = \binom{m+n-r}{m} + \sum_{\nu=0}^{K(r)} L(r; m-1, n-\nu).$$

$K(r) = \min(r-1, n-r)$. State boundary conditions for this recurrence relation.

b. Derive the following recurrence relation for the $p(r; m, n)$:

$$p(r; m, n) = \frac{\binom{m+n-r}{m}}{\binom{m+n}{m}} + \sum_{\nu=0}^{K(r)} \frac{m}{m+n-\nu} \cdot \frac{\binom{n}{\nu}}{\binom{m+n}{\nu}} p(r; m-1, n-\nu).$$

Use it to compute $p(5; 20, 20)$, $p(5; 30, 20)$ and $p(5; 40, 20)$.

HARDER PROBLEMS

Remark: If you cannot find the combinatorial argument leading to the recurrence relations for the following problems on random permutations, it may help to compare the cases $n = 5$ and $n = 6$ by enumeration. The algorithm of Problem 1.3.33 can be used to get the computer to do some of the tedious work for you.

Problem 1.3.17: From among the $n!$ permutations of the numbers $1, \cdots, n$, a permutation is drawn at random. An upward run is a maximal uninterrupted string of increasing numbers and a downward run is such a string of decreasing numbers. For instance, the permutation 3 5 6 2 1 8 7 4 has the upward runs 3 5 6 and 1 8 and the downward runs 6 2 1 and 8 7 4. By X_n we denote the total number of runs (both upward and downward) in a random permutation of $1, \cdots, n$, and $T(k; n)$ denotes the number of permutations with k runs.

a. Give a simple argument to show that $T(k;\,n)$ is even.

b. Show that

$$T(k;\,n+1) = kT(k;\,n) + 2T(k-1;\,n) + (n-k+1)T(k-2;\,n),$$

for $1 \le k \le n$, with the boundary conditions $T(0;\,n) = 0$, $T(1;\,n) = 2$, for $n \ge 2$ and $T(k;\,2) = 0$, for $k \ge 2$.

c. Write a program in integer arithmetic to generate a table of $T(k;\,n)$ for $2 \le n \le 15$.

d. Let $P(k;\,n)$ be the probability that a random permutation of $\{1, \cdots, n\}$ has k runs. Write a program in floating point arithmetic to evaluate the probabilities $P(k;\,n)$ for a given value of n. Execute your program for $n = 30$. Compute numerically the mean and variance of X_n. Compare them to the values given by the formulas

$$E(X_n) = \frac{2n-1}{3}, \quad \text{and} \quad \mathrm{Var}(X_n) = \frac{16n-29}{90}.$$

e. As in Example 1.2.1, compare the $P\{X_n \ge k\}$ to the values obtained from the normal approximation *with continuity correction.*

Hint: By reversing a permutation, it is easily seen that $T(k;\,n)$ is always even. Also, insertion of $n+1$ in the permutations of $1, \cdots, n$ can create at most two new runs. The count $T(k;\,n+1)$ is therefore obtained from the sets of permutations involved in the counts $T(k;\,n)$, $T(k-1;\,n)$ and $T(k-2;\,n)$. There are three cases to be considered. $n+1$ can either be placed inside a permutation of $1, \cdots, n$; between an element at the end and its neighbor, or at either end.

Moreover, if k is even, half of the permutations in the count $T(k;\,n)$ have $k/2$ peaks with a certain pattern at the end points and the other half have $k/2 - 1$ peaks, again with a certain pattern at the end points. There is a similar partition when k is odd. Consider all cases and examine what happens if $n+1$ is inserted in all possible ways. A careful count yields the stated recurrence relation.

Reference: David, F. N. and Barton, D. E., *"Combinatorial Chance,"* London: Ch. Griffin & Company Ltd., pp. 158 - 162, 1962.

Problem 1.3.18: The numbers $1, \cdots, n$ are written in random order. The position of the number i in the random permutation is denoted by $\pi(i)$. The random variable $W_n(m)$ is defined by $W_n(m) = \sum_{i=1}^{m} \pi(i)$, for $1 \leq m \leq n$. Show that $W_n(m)$ takes values in the set of integers k satisfying

$$\frac{1}{2}m(m+1) \leq k \leq \frac{1}{2}m(2n-m+1).$$

Develop recurrence relations for the probability density of $W_n(m)$. Efficiently code these recurrence relations and print out the densities of $W_{20}(m)$ for all allowable values of m.

Problem 1.3.19: A string of $m + n$ binary digits is given. It is the binary representation of some integer C. Next, a random string of m zeros and n ones is formed. It is considered as the binary representation of a random integer X. Write a program to compute the probability $P\{X \leq C\}$.

Hint: By scanning the given string from left to right, the event $\{X \leq C\}$ can be written as the finite union of disjoint events. Their probabilities can be computed and accumulated to yield $P\{X \leq C\}$. When $m + n$ is large, some caution is needed in computing the probabilities of some of the disjoint events accurately.

Problem 1.3.20: Write a general purpose subroutine to evaluate expressions of the form

$$u = \frac{m_1!\, m_2! \ \cdots \ m_r!}{n_1!\, n_2! \ \cdots \ n_k!},$$

for given integers $m_1, \cdots, m_r$ and $n_1, \cdots, n_k$ less than 1,000. One suggested approach is as follows: Write u as

$$u = 2^{c_2}\, 3^{c_3} \ \cdots \ M^{c_M},$$

where M is the maximum of the data $\{m_i\}$ and $\{n_j\}$. Find the exponents $\{c_\nu\}$ by a straightforward counting routine.

Next, using a table of the primes smaller than 1,000, test each of the factors for divisibility and determine the exponents in the form

$$u = p_1^{a_1} p_2^{a_2} \cdots p_K^{a_K},$$

where the integers p_v are the prime factors of the numerator and denominator of the reduced fraction for u. The positive and negative exponents correspond to the prime factors of the numerator and denominator, respectively. Now use logarithms to evaluate u. Note that u can have very large or very small values. Therefore use care in taking the antilogarithm of u. Write u in the form $u = U \cdot 10^R$, and calculate R and U separately.

Reference: For a discussion of this problem, see Neuts, M. F., "On computing ratios of factorials." *The Mathematical Scientist,* 18, 64 - 66, 1993.

Problem 1.3.21: Write a general purpose subroutine that calls the routine developed in Problem 1.3.20, to compute expressions of the form

$$u = \frac{\binom{m_1}{i_1}\binom{m_2}{i_2} \cdots \binom{m_r}{i_r}}{\binom{n_1}{j_1}\binom{n_2}{j_2} \cdots \binom{n_k}{j_k}}.$$

Test your program on problems of the following type: From a set of balls containing c_v balls of color v, $1 \leq v \leq k$, a sample of N balls is drawn without replacement. What is the probability that the sample contains d_v balls of color v? Obviously $d_1 + \cdots + d_k = N$. Report your numerical answers in the form specified in Problem 1.3.20.

Problem 1.3.22: For a random string of m zeros and n ones, $p(k; m, n)$ is the probability that there are k zeros between the leftmost and the rightmost of the symbols 1.

a. Show that

$$p(k; m, n) = (m - k + 1)\frac{\binom{n+k-2}{k}}{\binom{m+n}{m}},$$

for $0 \leq k \leq m$, and that $p(k; m, n)$ is largest for

$$k^*(m,n) = \left[\frac{(n-2)(m+1)}{n-1}\right],$$

where $[\cdot]$ denotes the integer part.

b. $X(m,n)$ is the number of zeros between the outermost symbols one in a random string of m zeros and n ones. For a fixed value of n, study *as efficiently as possible* the numerical behavior of the probabilities

$$P\{X(m,n) = k^*(m,n)\} \quad \text{and} \quad P\{k^*(m,n) - 5 \le X(m,n) \le k^*(m,n) + 5\},$$

as a function of m and n. Successively increase m and print both probabilities. Stop at the first m for which the second probability is smaller than 0.1. Implement your code for $n = 10$ and discuss the significance of your numerical results.

Problem 1.3.23: Two independent random strings, each of m zeros and n ones, are drawn and compared starting from the left. X is the number of successive initial characters that agree in both strings. For example, for the strings 0 1 1 0 0 1 1 and 0 1 1 1 0 1 1, X has the value 3.

a. Show that for $0 \le k \le m+n$,

$$P\{X \ge k\} = \sum_{r=K_L(k)}^{K_U(k)} \frac{\binom{k}{r}\binom{m+n-k}{m-r}^2}{\binom{m+n}{m}^2},$$

where $K_L(k) = \max(0, k-n)$, and $K_U(k) = \min(k, m)$.

b. Develop an algorithm to compute the probability density $\{p_k\}$ of X. Use appropriate recurrence relations to evaluate the terms in the stated sum.

c. If m and n tend to infinity in such a manner that $m/n \to \alpha$, a positive constant, show that for every fixed k,

$$P\{X \ge k\} \to \left[\frac{1+\alpha^2}{(1+\alpha)^2}\right]^k, \quad \text{for } k \ge 0.$$

In the limit, X therefore has a geometric density with $p = 2\alpha(1 + \alpha)^{-2}$, on the nonnegative integers.

d. Examine how well the exact density is approximated by its limit by computing the exact densities of X for $m = 2\nu$, $n = 3\nu$, for $\nu = 1, 2, 5, 10, 25$, and 50.

Problem 1.3.24: Two independent random strings, each of m zeros and n ones, are drawn and compared starting from the left. X is the index k for which the first coincidence, either of a symbol 0 or 1 occurs.

a. Show that the probability $P\{X = k\} = p_k$, is given by

$$p_k = \sum_r \frac{\binom{k-1}{r}\binom{m+n-k}{n-r-1}\binom{m+n-k}{n-k+r}}{\binom{m+n}{m}^2}$$

$$+ \sum_r \frac{\binom{k-1}{r}\binom{m+n-k}{n-r}\binom{m+n-k}{n-k+r+1}}{\binom{m+n}{m}^2}.$$

The summations run over all r between 0 and $k - 1$, for which the binomial coefficients are non-zero. Find the effective summation limits for both sums. Note that the first sum is the probability that the first coincident symbols are ones; the second is the probability that the first coincident symbols are zeros.

b. Show that if m and n tend to infinity in such a manner that $m/n \to \alpha$, a positive constant, then for every fixed k, p_k tends to the geometric probability with $p = (1 + \alpha^2)(1 + \alpha)^{-2}$ for $k \geq 1$.

c. Find the probability that a 1–coincidence occurs before a 0–coincidence. Also, the probability that a 0–coincidence occurs before a 1–coincidence. Why don't these probabilities always sum to one? What is the remaining probability?

d. Examine how well the exact density is approximated by its limit by computing the exact densities of X for $m = 2\nu$, $n = 3\nu$, for $\nu = 1, 2, 5, 10, 25$, and 50.

Problem 1.3.25: In a football pool, there initially are m persons trying to predict the winners in N games. Each person bets one dollar. The person who correctly predicts most games wins the entire pot. If there are several players with the maximum score, the pot is evenly divided among them.

A new player, Joe, joins the group and asks to be allowed to place two bets. Joe has noticed that the other players just guess which teams will win. Each player's score therefore has a binomial density with parameters $p = 1/2$ and N. To come up with his predictions, Joe also performs N Bernoulli trials with $p = 1/2$. For his second bet, he just predicts the opposite outcomes for each game. That form of betting has been called the *evil twin strategy.*

a. Show that if $N = 2n + 1$, Joe's expected return E for the evil twin strategy is

$$E = \frac{2(m+2)}{m+1}\left[1 - \frac{1}{2^{m+1}}\right].$$

Notice that when the number of games is odd, the expected return does not depend on n.

b. Obtain an expression for E for the case $N = 2n$. Now, E depends on both m and n and the formula is a bit more involved.

c. For $3 \le m \le 10$ and $1 \le N \le 20$, examine the expected return E of the evil twin strategy. For which values of m and n is Joe's expected profit $E - 2$ largest?

d. Suppose that the other m players can predict winning teams with probability p. Joe and his twin use the same strategy as before. By computing the expected return $E(p;m,N)$ for $0.45 \le p \le 0.55$, and for selected values of m and N, examine the sensitivity of the evil twin strategy to p.

Reference: DeStefano, J., Doyle, P. and Snell, J. L. "The evil twin strategy for a football pool," *The American Mathematical Monthly,* 100, 341 - 343, 1993.

CHALLENGING PROBLEMS

Problem 1.3.26: Review the derivation of the formulas for the density of the number of runs in a random string of m zeros and n ones. You will see that, with minor changes, the same reasoning yields an explicit formula for the probability $p(\nu;i,j)$ that the random string has ν runs, that the shortest string of zeros has length i and the shortest string of ones is of length j. The formulas in Example 1.2.2 correspond to $i = j = 1$. As in the example, one gets a different expression for $\nu = 2k$ and for $\nu = 2k+1$. Carefully specify the domains of validity of both formulas.

The array $\{p(\nu;\, i,\, j)\}$ contains the joint probability density of the number of runs N and the lengths L_0 and L_1 of the shortest runs of zeros and ones respectively. Write a computer program capable of handling values of m and n up to forty to evaluate that joint density. Also print the marginal densities of L_0, L_1 and the marginal joint density of L_0 and L_1. Compute the means, variances and the correlation coefficient of L_0 and L_1. Implement your code for three sets of pairs (m,n) of your choice and discuss the qualitative features of your numerical results.

Remark: Problems involving the longest rather than the shortest run lengths are considerably more involved. They require use of the restricted occupancy numbers defined in Problem 1.2.6.

Problem 1.3.27: The combinatorial argument for the density of the number of runs in a random string of m zeros and n ones, consists in thinking of the $2k$ or $2k + 1$ runs as boxes alternatingly filled with zeros or ones so that no box remains empty. By using the classical occupancy numbers, we readily obtain the formulas in Example 1.2.2. The probability that there are ν runs and that the run lengths of zeros and ones are *at most* i and j respectively, can be expressed using the *restricted occupancy numbers,* defined in Problem 1.2.6. Use the properties of those numbers to write a code to generate, for m and n up to 15, the probabilities $p(2k;\, i,\, j)$ and $p(2k + 1;\, i,\, j)$ of these events.

The next problem deals with a case where the index for which a certain combinatorial probability is largest, arose naturally in an application.

Problem 1.3.28: A lot of foil packets was impounded during a drug raid. k of the packets, chosen at random from the lot, were tested and all contained a prohibited substance. Later, r of the remaining untested packets were sold to an individual during a sting operation. That individual was arrested but had been able to dispose of the packets. During the trial, the defense argued that there was a possibility that some packets in the original lot contained only an inoffensive substance; the defendant may only have purchased such packets. Therefore, the plea went, reasonable doubt about his guilt remains.

a. If we assume that the original lot contained m offensive and n inoffensive packets, what is the probability that the first $k \le m$ drawn are offensive and the next $r \le n$ inoffensive?

b. With $m + n$ fixed (and much larger than $k + r$,) how would you determine the n for which that probability is maximum?

Using specific numbers, it was shown during the trial that this maximum probability is sufficiently large to support the argument of reasonable doubt. However, at that time, the prosecutor revealed that the remaining $m + n - k - r$ packets were still available. It was proposed to test s additional packets. s was to be chosen such that, if all these were found to be offensive, the maximum probability that the r packets were inoffensive would be less than 0.01. It was agreed that such a test, if positive, would remove all reasonable doubt.

c. Show that the probability $P(m, n; k, r, s)$ that the first k packets are offensive, the next r inoffensive and the next s again offensive, is

$$P(m, n; k, r, s) = \frac{\binom{m}{k}\binom{n}{r}\binom{m-k}{s}}{\binom{m+n}{k}\binom{m+n-k}{r}\binom{m+n-k-r}{s}}.$$

For successive values of s, determine the value of n for which the largest value $P^*(m, n; k, r, s)$ of $P(m, n; k, r, s)$ is attained. Stop at the smallest s for which $P^*(m, n; k, r, s) \le 0.01$. Search for the maximizing indices as cleverly as possible. Implement your algorithm for $m + n = 496$, $k = 4$ and $r = 2$, the actual data of the case on which this problem is based.

Reference: This problem is based on Shuster, J. J., "The statistician in a reverse cocaine sting," *The American Statistician,* 45, 123 - 124, 1991.

Problem 1.3.29: n identical items are distributed at random into m containers, numbered $1, \cdots, m$. We mark the successive boxes with a 1 or a 0 depending to whether the box is non-empty or empty. R is defined as the number of runs in the string of length m that is so obtained. R can assume all integer values between 1 (2 if $m > n$) and the smaller of m and $2n + 1$. Give a careful combinatorial argument to show that for allowable k, the probability $p(2k) = P\{R = 2k\}$ is given by

$$p(2k) = 2\sum_{r=k}^{m-k} \frac{\binom{m-r-1}{k-1}\binom{r-1}{k-1}\binom{n-1}{r-1}}{\binom{m+n-1}{n}},$$

and for $k > 0$,

$$p(2k+1) = \sum_{r=k}^{m-k-1} \frac{\binom{m-r-1}{k}\binom{r-1}{k-1}\left[\binom{n-1}{r-1} + \binom{n-1}{m-r-1}\right]}{\binom{m+n-1}{n}}.$$

Finally, the probability of a single run (all m boxes are non-empty) is

$$p(1) = \frac{\binom{n-1}{m-1}}{\binom{m+n-1}{n}}.$$

By an insightful argument, verify the remarkable equality

$$p(2k+1) = \frac{m-2k}{2k} p(2k).$$

Write a computer program to evaluate the density of R, capable of handling accurately all values of m up to 75 and n up to 125.

Hint: Denote the term in the first sum by $a(k, r)$. Examine the ratio $a(k+1, r)/a(k, r)$ for fixed r to find the index $k^*(r)$ for which $a(k, r)$ is largest. Determine carefully the ranges of the indices k and r and compute the $a(k, r)$ over the allowable indices. Compute the largest values of $a(k, r)$ first, by using logarithms. Next use recurrence relations to fill in the other values. Once the quantities $a(k, r)$ have been evaluated, the density $\{p(\nu)\}$ is easily computed. Use double precision, avoid underflow and check that the terms of the computed density $\{p(\nu)\}$ sum to 1.

Problem 1.3.30: Two independent random strings, each of m zeros and n ones, are drawn and scanned from left to right. By $p(k; r)$, we denote the probability that the k–th time ones are found in the same places in both strings, occurs at the location $r + k$.

a. Show that for $0 \le k \le n$,

$$p(k; r) = \sum \frac{\binom{r + k - 1}{k - 1}\binom{r}{j_1}\binom{r - j_1}{j_2}\binom{m + n - k - r}{n - k - j_1}\binom{m + n - k - r}{n - k - j_2}}{\binom{m + n}{m}^2}.$$

The summation extends over all the pairs of integers (j_1, j_2) satisfying

$$0 \le j_1 + j_2 \le r, \quad \max(0, r - m) \le j_1 \le n - k, \quad \max(0, r - m) \le j_2 \le n - k.$$

b. If m and n tend to infinity in such a manner that $m/n \to \alpha$, a positive constant, use Stirling's formula and the classical limit

$$\left[1 - \frac{x}{n}\right]^n \to e^{-x},$$

to show that for fixed values of k, r and j,

$$\frac{\binom{m + n - k - r}{n - k - j}}{\binom{m + n}{m}^2} \to \frac{\alpha^{r - j}}{(1 + \alpha)^{k + r}}.$$

Use that result to show that for fixed values of $k \ge 0$ and $r \ge 0$,

$$p(k; r) \to \binom{r + k - 1}{k - 1}\left[\frac{1}{(1 + \alpha)^2}\right]^k \left[1 - \frac{1}{(1 + \alpha)^2}\right]^r,$$

the negative binomial density with $p = (1 + \alpha)^{-2}$, on the nonnegative integers. Give an intuitive explanation for that result.

c. Write a carefully organized program to compute the exact probabilities

$p(k; r)$ for given m, n and k. Compute terms of the density $\{p(k; r)\}$ until their sum exceeds $1 - 10^{-5}$. Execute your program for several choices of m, n and k. Compare the numerical results to those obtained using the limiting density as an approximation.

Hint: Denote the terms in the exact formula for $p(k; r)$ by $\theta(j_1, j_2)$ and obtain recurrence relations for these quantities. Sketch the set of allowable pairs (j_1, j_2) and carefully evaluate and accumulate all needed $\theta(j_1, j_2)$. Exercise caution to avoid starting any recurrence at a term that could cause underflow.

SOME SPECIAL PROBLEMS

The following problems require general algorithmic thinking. They are not particularly difficult, but require good organizational skills.

Problem 1.3.31: Fáa di Bruno's Formula: Develop an efficient way to generate the sets $E_{n,r}$ in Problem 1.2.7 for r and n with $0 \le r \le n \le 20$. Do not print the elements of the sets, but write a table of the *largest number of positive* j_k, found in the n–tuples of each set. (It is advisable to print out the sets of n–tuples for some small value of n to ascertain that your algorithm generates all n–tuples and without duplication.

Problem 1.3.32: You are asked to distribute the 52 cards of a standard deck at random to four players. For each player the hand should appear ranked by suit and by value; for example, if the player has cards of all four suits, the spades should be listed first, the hearts next, followed by the diamonds and the clubs. Cards within each suit should be ranked from lowest (ace) to highest (King). That should be accomplished without sorting. The solution in which the deck of cards is first permuted, then assigned to the four players, followed by a sorting of each hand, is lacking in elegance and efficiency. That is not an acceptable solution.

Remark: To do this problem, you need to know how to generate a random permutation of the elements in a list.

Problem 1.3.33: Write a program to form all permutations of the symbols $1, \cdots, n$, without storing the permutations in an array. Your code should be able to handle all values of n up to ten. As each permutation is generated, the elements should be stored in the integer identifiers $I(1), \cdots, I(n)$. Give an argument to show that your code indeed generates all permutations without duplication. Print out the solution for $n = 5$.

Problem 1.3.34: Consider a tournament with n teams in which each team plays every other team exactly once. Each match ends in one of the teams winning; there are no tied games. In a match between teams i and j, team i wins with probability $P(i, j)$. Clearly the matrix $P = \{P(i, j)\}$ satisfies $P(i, j) = 1 - P(j, i)$, and for convenience, we set $P(i, i) = 1$, for $i = 1, \cdots, n$. The outcomes of successive matches are independent.

We shall say that there is a *perfect progression* in the final scores, if the best team wins exactly n games, the next one $n - 1$, and so on, with the last team losing all n games. Show that the probability θ of a perfect progression is given by

$$\theta = \sum \prod_{\pi(i) <\cdot\, \pi(j)} P(i, j),$$

where the sum is taken over all permutations $\{\pi(1), \cdots, \pi(n)\}$ of $1, \cdots, n$. $\pi(i) <\cdot\, \pi(j)$ signifies that $\pi(i)$ strictly precedes $\pi(j)$ in the permutation.

By a combinatorial argument, show that if $P(i, j) = 1/2$ for all $i \neq j$, then $\theta = n!2^{-n(n-1)/2}$. Write a computer program to evaluate θ for any given set of parameters with $n \leq 10$. Use the particular case as a test of your program. You will need the code for Problem 1.3.33 as a subroutine, since you neither need nor want to store all permutations of $1, \cdots, n$.

Reference: Madsen, R., "On the probability of a perfect progression," *The American Statistician,* 45, 214 - 216, 1991.

Problem 1.3.35: Use the procedure developed in Problem 1.3.33 to generate successively the permutations of $1, \cdots, 10$. With each permutation associate a binary string of length 9 by writing a 1 if a smaller number immediately precedes a larger one. Otherwise write a 0. For instance, to the permutation 2, 5, 6, 3, 4, 7, 1, 9, 10, 8 this associates the string 1, 1, 0, 1, 1, 0, 1, 1, 0. By interpreting that string as the binary digits of an integer, we so assign to each permutation of $1, \cdots, 10$, a value k between 0 and 511. The given permutation is assigned $k = 438$. For all k, $0 \leq k \leq 511$, how many permutations are assigned the value k?

CHAPTER 2

SOLVING EQUATIONS

2.1. COMMON EQUATIONS

Solving equations pervades all of applied mathematics. It is therefore a major topic in numerical analysis. For equations with a single unknown, there are many classical methods such as Newton's method, bisection, the secant method, and others. We assume that the reader is familiar with these. They often appear as modules to be implemented in the more involved situations common in probability.

An essential step is to ascertain mathematically that the equation $f(x) = 0$, to be solved has at least one zero in an interval $[a, b]$ of interest. Under quite general conditions, which are usually fulfilled in probability problems, one can construct successive approximants x_k that converge to a root of the equation.

No single method is entirely foolproof. Techniques for locating zeros should always be used with caution and in conjunction with accuracy checks. The soundest basis for a correct solution of an equation is good understanding of the qualitative behavior of the function $f(\cdot)$ in a neighborhood of the possible zero. In probability, such an understanding usually comes from good insight into the specific probability model.

Techniques used to locate zeros can be adapted to find local maxima or minima of the function $f(\cdot)$. The correct identification of critical points requires similar caution. We shall discuss some general examples with indications on where special caution is needed.

Example 2.1.1: Finding a Percentile of a Probability Distribution: The αth percentile of a probability distribution is the smallest value $x(\alpha)$ for which

$$F[x(\alpha)] \geq \frac{\alpha}{100}.$$

Since $F(\cdot)$ is nondecreasing, such a value $x(\alpha)$ always exists and is unique.

For a continuous probability distribution, the inequality is satisfied with equality. Solving that equation numerically depends on how easy or difficult it is to compute $F(x)$ for given values of x. For the time being, let us assume that a function subroutine is available to compute $F(x)$.

The first step is to identify an interval $[x', x'']$ that brackets $x(\alpha)$. If the support of $F(\cdot)$ is a finite interval $[a, b]$, such an interval is immediately available. If the mean μ'_1 and the variance σ^2 of $F(\cdot)$ are known, we can use the intervals with endpoints $x' = \mu'_1 + 2\sigma$ and $x'' = \mu'_1 + 3\sigma$. Generally, these yield good trial values for α in the 60-80 range. Similar guesses can be made for other values of α.

A somewhat different approach is used if we want to compute several percentiles, say, for $\alpha = 5k$, $1 \le k \le 19$. Using a small x' and a large x'', we first find the 5–percentile. Next, to find the 10–percentile, we replace x' by $x(5)$. Continuing in this manner, we use information obtained earlier to bracket the next root to be computed by an interval that becomes progessively shorter. Within a bracketing interval, we preferably use the bisection or secant methods. Newton's method usually converges with fewer iterations, but requires computation of the derivative. Also, near high percentiles the derivative of $F(\cdot)$ is usully very small. Its accurate computation can then be delicate.

Equations rarely arise in isolation. Prior analysis based on the application at hand yields information on their solution. The time to do that analysis is well spent. That is illustrated by the next example.

Example 2.1.2: The Mode of a Mixture of Gamma Distributions: We wish to develop a general purpose algorithm to find the local maxima of the probability density

$$\phi(u) = pe^{-\lambda_1 u}\frac{(\lambda_1 u)^{\beta - 1}}{\Gamma(\beta)}\lambda_1 + (1-p)e^{-\lambda_2 u}\frac{(\lambda_2 u)^{\beta - 1}}{\Gamma(\beta)}\lambda_2, \quad \text{for } u \ge 0.$$

$\phi(\cdot)$ is a *mixture* of two gamma densities. These gamma densities are the *components* of the mixture. λ_1, λ_2, and β are positive parameters and $0 < p < 1$. To limit the amount of preliminary analysis that is needed, we focus attention only on the case where both components have a common shape parameters β. If $0 < \beta \le 1$, both densities are strictly decreasing. When $\beta < 1$, they tend to infinity at 0. For that case, the answer is obvious. We only need discuss the case $\beta > 1$.

Then, the means of the components are $(\lambda_1)^{-1}\beta$ and $(\lambda_2)^{-1}\beta$ respectively. The case $\lambda_1 = \lambda_2$ is trivial, so we confine our attention to the case $\lambda_1 > \lambda_2$. Without loss of generality, we let the second component have the larger mean.

In equations where the effect of some parameters is obvious, it is advisable to eliminate or simplify these whenever we can. λ_1 and λ_2 are *scale* parameters. Setting $\lambda_1 = \beta$, and $\lambda_2 = \beta\gamma^{-1}$, $\gamma > 1$, therefore amounts to choosing the unit on the u–axis equal to the mean of the first component. γ is then the mean of the second component. Such a redefinition of the parameters is easily done immediately after reading in the data to a problem. However, we must remember that, after solving the necessary equations, our answers should be reported in terms of the original parameter values.

The density is now rewritten as

$$\phi(u) = pe^{-\beta u}\frac{(\beta u)^{\beta - 1}}{\Gamma(\beta)}\beta + (1-p)e^{-\beta\gamma^{-1}u}\frac{(\beta\gamma^{-1}u)^{\beta - 1}}{\Gamma(\beta)}\beta\gamma^{-1}.$$

Since $\phi(u)$ and $K\phi(u)$ have the same critical points, it suffices to find the local maxima of the function

$$f(u) = u^{\beta - 1}\left[pe^{-\beta u} + (1-p)\gamma^{-\beta}e^{-\beta\gamma^{-1}u}\right].$$

Setting the derivative $f'(u)$ equal to zero yields the following equation for the locations of the possible critical points in $(0, \infty)$:

$$(\beta - 1)\left[pe^{-\beta u} + (1-p)\gamma^{-\beta}e^{-\beta\gamma^{-1}u}\right] = \beta u\left[pe^{-\beta u} + (1-p)\gamma^{-\beta-1}e^{-\beta\gamma^{-1}u}\right].$$

To express that equation in a transparent form, it is successively rewritten as

$$u = \frac{\beta - 1}{\beta}\,\frac{pe^{-\beta u} + (1-p)\gamma^{-\beta}e^{-\beta\gamma^{-1}u}}{pe^{-\beta u} + (1-p)\gamma^{-\beta-1}e^{-\beta\gamma^{-1}u}},$$

and

$$u = \frac{\beta - 1}{\beta}\,\frac{1 + c\gamma e^{du}}{1 + ce^{du}}$$

$$= \frac{\beta - 1}{\beta}\left\{1 + (\gamma - 1)\left[1 - \frac{1}{1 + ce^{du}}\right]\right\}, \tag{1}$$

where the constants c and d are defined by

$$c = (1 - p)p^{-1}\gamma^{-\beta-1}, \quad \text{and} \quad d = \beta\gamma^{-1}(\gamma - 1).$$

Since $\gamma > 1$, c and d are positive. Furthermore, the right hand side of (1) is clearly increasing in u. From the value $(\beta - 1)\beta^{-1}$ at $u = 0$, it increases to $(\beta - 1)\beta^{-1}\gamma > 1$ at infinity. The graph of the right hand side intersects the line $y = u$ in a single point. For $\beta > 1$, the density $\phi(u)$ has a unique mode.

We can obtain further information on the location of the mode. For $u = (\beta - 1)\beta^{-1}$, the right hand side of (1) exceeds $(\beta - 1)\beta^{-1}$, the location of the mode of the first component. The desired root must therefore be to the left of that value. Knowing that the mode is unique and lies in the interval $(0, (\beta - 1)\beta^{-1})$ greatly simplifies the search for its value. Problem 2.3.27 calls for the implementation of one of several classical procedures to compute its location and the corresponding value of $\phi(u)$.

The following example shows how a problem may require the solution of many equations and that there may be several alternative ways of choosing these equations.

Example 2.1.3: The Shortest Interval with a Given Probability: Consider a probability distribution $F(x)$ whose density $\phi(x)$ has the following properties:

a. $\phi(x)$ is positive on an interval (A, B) (which may be infinite) and zero outside that interval.
b. The density is continuous on (A, B) and has a unique maximum at some point ξ, *interior* to the interval (A, B).
c. To the left of ξ, the density $\phi(x)$ is strictly increasing; to the right of ξ, it is strictly decreasing.

We want to determine the *shortest* interval (a, b) with $A \leq a < b \leq B$, for which

$$\int_a^b \phi(u)\, du = \beta. \tag{1}$$

That is equivalent to minimizing $L(a, b) = b - a$, subject to the constraints (1) and $A \leq a < b \leq B$. This problem can be solved by using Lagrange multipliers. We consider the function

$$K(a, b, \lambda) = b - a + \lambda\left[\int_a^b \phi(u)\, du - \beta\right].$$

Setting the partial derivatives of K with respect to λ, a, and b, equal to zero, we obtain the equality (1) and the equations

$$1 + \lambda\phi(a) = 1 + \lambda\phi(b) = 0.$$

These do not yield a and b explicitly, but they provide the important information that, at the points a and b, the density must have equal values, viz. $\phi(a) = \phi(b)$. With the stated assumptions on the density $\phi(\cdot)$, it is clear that intervals with that property exist. It remains to find such a and b for which also the equality (1) is satisfied.

To do this, we try to find intervals (a', a'') and (b'', b'), for which $\phi(a'') = \phi(a')$ and $\phi(b'') = \phi(b')$, and such that β lies between the probability contents of these intervals. Such intervals can be constructed as follows: Let a' and b' be the unique solutions to the equations

$$F(a') = \frac{1}{2}(1 - \beta), \quad \text{and} \quad F(b') = \frac{1}{2}(1 + \beta).$$

The solution of such equations is discussed in Example 2.1.1. Next, we solve the equations

$$\phi(a'') = \phi(a'), \quad \text{and} \quad \phi(b'') = \phi(b'),$$

respectively for a'' and b''. The assumptions on $\phi(\cdot)$ again imply that these have unique solutions.

Barring very special cases, such as a symmetric density $\phi(\cdot)$, the probability content of (a', a'') will be smaller than β and that of (b'', b') larger. In addition, the desired value of a must lie between a' and b'', and that of b lies between b' and a''. (Why?)

There are now several alternative ways of finding a and b. We can apply bisection to the interval with endpoints a' and b''. For each current value a° of a, we compute b° satisfying $\phi(b^\circ) = \phi(a^\circ)$, and next, the probability $F(b^\circ) - F(a^\circ)$. a° is adjusted depending on whether that probability is smaller or larger than β. We halt when differs from β by a sufficiently small amount. The same argument and procedure apply verbatim to the interval with endpoints b' and a''.

In a different approach, let $F(a'') - F(a') = c_1$, and $F(b') - F(b'') = c_2$. Recalling that $\phi(a') = \phi(a'')$, and $\phi(b') = \phi(b'')$, we set

$$v^\circ = \frac{1}{2}[\phi(a') + \phi(b')],$$

and we solve the equations $\phi(a^\circ) = v^\circ$, and $\phi(b^\circ) = v^\circ$, and we compare $F(b^\circ) - F(a^\circ)$ to β. It is now obvious how the subsequent values of v° should be chosen. (For purely esthetic reasons, I prefer that approach to the one-sided ones).

The implementation of this algorithm requires subroutines to solve several different equations and each routine is called repeatedly. The practical difficulty of the implementation depends on how easily the density $\phi(\cdot)$ and the distribution $F(\cdot)$ can be computed. In principle, it should be verified that the solution obtained by the Lagrange multiplier argument is indeed the (global) minimum. That is, in general, a daunting task. For the problem at hand, we can (safely) accept that assertion as intuitively clear.

2.2. IMPLICIT EQUATIONS

In probability problems, the quantities of interest are often solutions to equations that are difficult, if not impossible, to write in an explicit analytic form. One of the most challenging tasks in algorithmic probability is to try to get at the values of important quantities by embedding their computation in a feasible numerical scheme. There is no general theory of how that can be done. The following example, and many of the problems at the end of this chapter, serve to illustrate the art.

Example 2.2.1: The Game of Billiards: In some games, of which Billiards is a familiar example, two players compete by successively trying to score points. To win, Player I must score n_1 points before Player II scores n_2 points. It clearly matters who starts the game, so a toss-up is performed to decide who goes first. The players now alternatingly get turns that last as long as they successfully score points. In billiards, hitting the red and the second white ball with the white ball assigned to each player, is called a *carom*. There are variations on how a carom is to be made, as in three-cushion billiards. Some of the problems at the end of the chapter deal with such variants.

In a probability model for Billiards, the successive attempts are treated as independent Bernoulli trials with success probabilities p_1 for Player I and p_2 for Player II. We assume that $0 < p_2 \le p_1 < 1$. γ is the probability that Player I gets the first turn.

Our objective is to set the *handicap* for the game. That is, for given values of p_1, p_2, and n_1, we wish to compute the value of n_2 for which the game is balanced. The advantage in skill of the better player I is to be offset by requiring Player II to score fewer points.

We denote by $P_1(n_1, n_2)$ the probability that Player I wins a game with target scores n_1 and n_2, given that he goes first. Similarly, $P_2(n_1, n_2)$ is the probability that Player II wins given that he gets the first turn. The unconditional probability that Player I wins is then given by

$$P(n_1, n_2) = \gamma P_1(n_1, n_2) + (1 - \gamma)[1 - P_2(n_1, n_2)].$$

The handicap problem therefore amounts to finding the integer n_2 for which $P(n_1, n_2)$ is as close to 0.5 as possible. A common approach to such questions is to derive a formula for $P(n_1, n_2)$ and to apply root-finding techniques to find n_2. Analytic expressions for $P_1(n_1, n_2)$ and $P_2(n_1, n_2)$ are known. They were obtained in two very nice papers, published in 1943 in Dutch. The formulas are, however, quite complicated and not very convenient for a numerical solution of the handicap problem.

Our solution consists in embedding the search for n_2 in a recursive scheme. To visualize it, we think of the progress of the two scores as a random walk on the lattice points (j_1, j_2) with nonnegative integer coordinates. The walk describes a path with steps of one unit, horizontally to the right whenever I scores a point; vertically upward for every point scored by II. Player I wins if the line $x = n_1$ is reached before the line $y = n_2$, and vice versa. The number of points scored by each player during each turn has a geometric density.

In terms of generic variables j_1 and j_2, the conditional probabilities $P_1(j_1, j_2)$ and $P_2(j_1, j_2)$ satisfy the equations

$$P_1(j_1, j_2) = p_1 P_1(j_1 - 1, j_2) + q_1[1 - P_2(j_1, j_2)], \tag{1}$$

$$P_2(j_1, j_2) = p_2 P_2(j_1, j_2 - 1) + q_2[1 - P_1(j_1, j_2)], \tag{2}$$

for all $j_1 \geq 1$ and $j_2 \geq 1$, where $q_1 = 1 - p_1$, and $q_2 = 1 - p_2$. The equations (1) and (2) are obtained from the law of total probability, by considering the various outcomes of the first trial. In addition, we have the boundary conditions

$$P_1(0, j_2) = P_2(j_1, 0) = 1, \quad \text{and} \quad P_1(j_1, 0) = P_2(0, j_2) = 0,$$

for $j_1 \geq 1$ and $j_2 \geq 1$. From (1) and (2), we readily obtain the recurrence relations

$$P_1(j_1, j_2) = \frac{p_1}{1 - q_1 q_2} P_1(j_1 - 1, j_2) + \frac{q_1 p_2}{1 - q_1 q_2}[1 - P_2(j_1, j_2 - 1)], \tag{3}$$

and

$$P_2(j_1, j_2) = \frac{p_2}{1 - q_1 q_2} P_2(j_1, j_2 - 1) + \frac{q_2 p_1}{1 - q_1 q_2}[1 - P_1(j_1 - 1, j_2)]. \tag{4}$$

The quantities $P_1(j_1, j_2)$ and $P_2(j_1, j_2)$ for $j_1 + j_2 = k + 1$ are simply related to those for which $j_1 + j_2 = k$. These quantities may therefore be computed by an iterative scheme similar to Pascal's triangle for the binomial coefficients. By cautious programming, the quantities for $j_1 + j_2 = k + 1$ may also be written in the same array as those for k, so that we do not need large storage arrays. We also note that, for fixed $n_1 > 1$, $P(n_1, n_2)$ decreases in n_2 from the value 1 at $n_2 = 0$ to zero at infinity.

We now evaluate $P_1(j_1, j_2)$ and $P_2(j_1, j_2)$ for successive values of $j_1 + j_2 = k$. When k exceeds n_1, we start checking whether $P(n_1, k - n_1)$ lies in a small interval (α, β) about 0.5. If so, we print the values of $n_2 = k - n_1$, $P_1(n_1, n_2)$, $P_2(n_1, n_2)$ and $P(n_1, n_2)$ for later examination. The interval (α, β) can conveniently be chosen as (0.48, 0.52). That will typically result in few candidate values for n_2 to be examined from the printout. Choosing such an interval is necessary as there is usually no integer n_2 for which $P(n_1, n_2)$ is exactly 1/2. (α, β) should not be chosen too small as there is then the possibility of jumping over it.

The largest value of k that needs to be examined is $2n_1$. It is clear that the desired n_2 is never larger than n_1. An even better stopping criterion on k is to halt as soon as for some m_2, the smaller of the quantities $P_1(n_1, m_2)$ and $1 - P_2(n_1, m_2)$ exceeds β. (Why?) That stopping criterion can easily be encoded into the program.

References: Bottema, O. and Van Veen, S. C., "Kansberekeningen bij het biljartspel I, II" (Probabilities for the game of billiards, in Dutch), *Nieuw Archief voor Wiskunde,* 22, 16 - 33 and 123 - 158, 1943, and Neuts, M. F., "Setting the handicap in billiards: A numerical investigation," *Math. Magazine,* 46, 119 - 127, 1973.

2.3. MAIN PROBLEM SET FOR CHAPTER 2

The skills to be developed by the problems in this section are: the precise formulation of equations to be solved, the use of appropriate judgment in the choice of the solution method, the selection of good starting solutions, and the verification through accuracy checks, of the correctness of the computed numerical values.

EASIER PROBLEMS

Problem 2.3.1: Solve the equation

$$F(x) = [\Phi(x)]^3 \left[\frac{1}{2} + \frac{1}{\pi} \arctan x \right] = \alpha,$$

for $\alpha = 0.9, 0.95$, and 0.99, using both Newton's method and bisection. Obtain the values of the desired percentiles $x(\alpha)$ to within 10^{-6}. Print the given values of α and the corresponding $x(\alpha)$ and $F[x(\alpha)]$ computed by both methods. $F(\cdot)$ is the distribution of $\max[X_0, X_1, X_2, X_3]$, where X_i are independent, X_0 has an standard Cauchy distribution, and the other three random variables have the standard normal distribution.

Problem 2.3.2: $\Phi(x)$ is the standard normal distribution. Show that

$$\Phi(x - 1) + 3\Phi(2x) = 1,$$

has a unique solution x_0. By Newton's method, compute x_0 to within 10^{-8}.

Problem 2.3.3: The purpose of this problem is to make some comparisons of the normal and the Cauchy distributions. That can be done in many ways; the following are some of them. $\Phi(x)$ is the standard normal distribution and $G(x; a)$, the Cauchy distribution

$$G(x; a) = \frac{1}{2} + \frac{1}{\pi}\arctan(\frac{x}{a}),$$

with median at 0 and scale parameter $a > 0$.

a. Determine the value of a for which the *quartiles* of both distributions agree. First find the value of x_0 for which $\Phi(x_0) = 0.75$, and next the value a_0 of a for which $G_{a_0}(x_0) = 0.75$. After a_0 has been found, compute and interpret the values of the ratios

$$A_k = \frac{\Phi(k + 1) - \Phi(k)}{G(k + 1; a_0) - G(k; a_0)}, \quad \text{for } 0 \le k \le 5.$$

b. Find the value a_1 of a for which the probability $G(1; a) - G(-1; a)$ of the interval $(-1, +1)$ equals $\Phi(1) - \Phi(-1)$. Show that

$$a_1 = \cot\{\pi[\Phi(1) - \frac{1}{2}]\}.$$

Next, compute the ratios A_k corresponding to a_1.

c. The densities of $\Phi(x)$ and $G(x; a)$ both have inflection points located symmetrically about 0. Determine the value a_2 for which the inflection points of both densities coincide. Evaluate the corresponding ratios A_k.

Problem 2.3.4: The beta density $\phi(u; a, b)$ is defined for $0 < u < 1$, by

$$\phi(u; a, b) = \frac{1}{B(a, b)} u^{a-1}(1-u)^{b-1}.$$

The parameters a and b are positive and $B(a, b)$ is Euler's beta function evaluated at a and b. Write a program to compute and plot the density and to evaluate the $5k$–percentiles of the beta distribution for $1 \leq k \leq 19$.

Remark: This problem requires library routines for the *beta function* and for the *incomplete beta function ratio.* Some libraries also have a routine for the *reciprocal beta function,* $[B(a, b)]^{-1}$. Look up these routines in a software library and learn to use them.

Problem 2.3.5: The gamma density $\gamma(u; \lambda, \alpha)$ is defined for $u > 0$, by

$$\gamma(u; \lambda, \alpha) = \frac{1}{\Gamma(\alpha)} e^{-\lambda u} \lambda^{\alpha} u^{\alpha-1}.$$

The parameters λ and α are positive and $\Gamma(\alpha)$ is Euler's gamma function at α. Write a program to compute and plot the density and to evaluate the $5k$–percentiles of the gamma distribution for $1 \leq k \leq 19$.

Remark: This problem requires library routines for the *gamma function* and for the *incomplete gamma function ratio.* Some libraries also have a routine for the *reciprocal gamma function,* $[\Gamma(\alpha)]^{-1}$. Look up these routines in a software library and learn to use them.

Problem 2.3.6: If you perform Bernoulli trials with success probability p, the time T until the first success has the geometric distribution

$$P\{T \leq k\} = 1 - (1-p)^k, \quad \text{for } k \geq 1.$$

You are offered an alternative whereby, at the n–th trial, you perform n Bernoulli trials with probability of success $n^{-1}p_1$. If any of these trials is a success, you have scored a success at time n. Clearly, that amounts to replacing the original Bernoulli trials by a new sequence in which the success probability at the nth trial is

$$p(n) = 1 - \left[1 - \frac{p_1}{n}\right]^n.$$

To find out if that alternative is to your advantage, compare the mean $E(T_1)$ of its time T_1 of the first success to that for ordinary Bernoulli trials. Show that $p(n)$ is decreasing in n and that

$$p_1 > p(n) > 1 - e^{-p_1}.$$

For $p = 0.05k$, with $1 \leq k \leq 19$, compute the value p_1 for which $E(T_1) = p^{-1}$. Based on the numerical results, what you think of the proposed alternative?

Problem 2.3.7: X has a gamma density

$$\phi(u) = e^{-\lambda u} \frac{(\lambda u)^{\alpha - 1}}{\Gamma(\alpha)} \lambda, \quad \text{for } u > 0.$$

Find the value a for which $f(a) = P\{a \leq X \leq a + 1\}$ is maximum and compute the maximum value. Execute your program for $\alpha = 5$, and $\lambda = 0.75$. Determine a to within 10^{-5}.

Hint: Solve the equation $f'(a) = 0$, by Newton's method. A good initial guess is $a_0 = \max(0, c - 0.5)$, where c is the point where $\phi(u)$ attains its maximum. c can be found explicitly.

Problem 2.3.8: Prove that the equation

$$u = \frac{1}{2}\log\frac{\pi}{2} + \log(1 + 2u),$$

has a unique positive root u^*. Show that the sequence $\{u_n\}$, defined by

$$u_0 = \frac{1}{2}\log\frac{\pi}{2}, \quad \text{and} \quad u_{n+1} = u_0 + \log(1 + 2u_n),$$

converges monotonically to u^*. Compute u^* to within 10^{-7} and prove that the standard normal density $\phi(v)$ is larger than the standard Cauchy density $[\pi(1 + v^2)]^{-1}$, if and only if $|v| < \sqrt{2u^*}$.

Problem 2.3.9: For $x > 0$, the function $f(x)$ is differentiable and increasing. $f(0) = 0$ and $f(x)$ tends to infinity as $x \to \infty$. Discuss the solution set of the equation

$$\Phi[a + f(x)] - \Phi[a - f(x)] = b,$$

where a is a real number and $0 < b < 1$. Write a program to solve that equation, allowing the function $f(x)$ to be specified by the user. Build plausible safeguards into your program to ensure the user of the correctness of the program and of the numerical accuracy of the solution. Use the program to solve

$$\Phi[1 + 0.20\sqrt{x}\,] - \Phi[1 - 0.20\sqrt{x}\,] = 0.90.$$

Problem 2.3.10: The given constants a and b are positive and $\Phi(x)$ is the standard normal distribution. Use the bisection method to compute the unique solution $x = \theta(a, b)$ of the equation

$$\Phi(ax) - \Phi(-bx) = 0.999,$$

on a regular grid of points (a, b) in the rectangle $0.25 \le a \le 4.00$, $0.25 \le b \le 1.00$. The purpose of this problem is the efficient organization of the solution of an equation for many values of its parameters. Select an adequate number of points (a, b) to obtain a clear three-dimensional plot of the graph of the function $\theta(a, b)$.

Problem 2.3.11: X has a normal distribution with mean $\mu = 0$ and variance σ^2. Constants a and b satisfying $0 < a < b$ are given. Compute the value of σ for which the probability $P\{a \le X \le b\}$ is largest. For $a = 1$, $b = 2$, find σ to within 10^{-4}.

Problem 2.3.12: N points are placed independently in the interval (0, 1), according to the probability density $\phi(u) = 12u^2(1 - u)$. The probability that a point lands in the subinterval $(a, a + 1/3)$ is $p(a)$.

a. Show that $p(a)$ is maximum when a is the unique root in (0, 1) of the equation $a^2 - a/3 - 2/27 = 0$. Find that root and the corresponding value of $p(a)$ explicitly. How large should N be so that the probability that at least three points fall in the interval will be at least 0.6?

b. With the insight gained from part *a*, solve the same problem for the probability density $\phi(u) = 42u^5(1 - u)$. Appropriate equations must now be solved numerically.

Hint: Study the general discussion in Example 2.1.3.

Problem 2.3.13: Determine to within 10^{-9}, the value of λ for which the probability

$$\phi(\lambda; m, n) = \sum_{r=m}^{n} e^{-\lambda} \frac{\lambda^r}{r!},$$

is maximum. m and n are integers satisfying $0 \leq m \leq n$. Your code should be able to handle values of n up to 200. Specifically find the maximum of $\phi(\lambda; 75, 200)$.

Problem 2.3.14: X_1 and X_2 are independent Poisson random variables with mean λ. Write a program to find the value of λ for which $P\{X_1 > k_1, X_2 \leq k_2\}$ is largest. Implement your code for $\lambda = 5$, $k_1 = 2$, and $k_2 = 3$.

AVERAGE PROBLEMS

Problem 2.3.15: The purpose of this problem is to compare the negative binomial density

$$p_k = \binom{m+k-1}{m-1} p^m (1-p)^k, \quad \text{for } k \geq 0,$$

with parameters p, $0 < p < 1$, and m, a positive integer, to a Poisson density matched to it according to various approximation strategies.

For each approximation procedure, print the density $\{p_k\}$ and the matched Poisson density, as well as the first two moments of the two densities. Comment on how the various approximations differ. Implement each procedure for the data: $p = 0.5, 0.6, 0.75$ and $1 \leq m \leq 5$.

a. Match the parameter λ to the mean $mp^{-1}(1-p)$ of $\{p_k\}$.
b. Match the variances of both densities.
c. Choose λ such that the 90–th percentiles agree.
d. Compare $\{p_k\}$ to the Poisson density with λ chosen to minimize

$$\sum_{k=0}^{\infty} \left[e^{-\lambda} \frac{\lambda^k}{k!} - p_k \right]^2.$$

Problem 2.3.16: Consider a sequence of independent Bernoulli trials with probability p of success. Integers k, n and r, satisfying $1 \le k < r < n$ are given. The random variable T_j, $j \ge 1$, is the time of the j–th success. Use the bisection method to compute to within 10^{-7}, the value p^* of p for which the probability $P\{T_k \le n < T_r\}$ is maximum. Execute your program for $n = 15$, $k = 4$ and $r = 7$.

Problem 2.3.17: In this problem, you are asked to examine two distortions of the exponential distribution in which either the behavior near 0 or that for large x is quite different from that of the exponential. For each, discuss how to compute the density and distribution, the mode, mean, and variance, as well as the $5k$–percentiles for $1 \le k \le 19$.

a. The distribution $F_1(x)$ is given by

$$F_1(x) = \frac{1 - e^{-ax}}{1 + be^{-ax}}, \quad \text{for } x \ge 0.$$

Show that it is sufficient to study its behavior for $a = 1$. In that case, the mode is at $x_0 = \log b$. Obtain an explicit formula for the mean.

b. The density of the distribution $F_2(x)$ is given by

$$\phi_2(x) = Ke^{-[ax + \gamma x^2]}, \quad \text{for } x \ge 0,$$

where K is a constant guaranteeing that $\phi_2(\cdot)$ is a proper density. Show that it is again sufficient to study the behavior for $a = 1$. Verify that, for $a = 1$, $\phi_2(x)$ may be rewritten as

$$\phi_2(x) = \left[1 - \Phi\left(\frac{1}{\sqrt{2\gamma}} \right) \right]^{-1} \sqrt{\frac{\gamma}{\pi}}\, e^{-\gamma\left[x + \frac{1}{2\gamma}\right]^2}, \quad \text{for } x \ge 0,$$

and the distribution $F_2(x)$ as

$$F_2(x) = 1 - \frac{1 - \Phi\left(\frac{1}{\sqrt{2\gamma}} + x\sqrt{2\gamma}\right)}{1 - \Phi\left(\frac{1}{\sqrt{2\gamma}}\right)}, \quad \text{for } x \geq 0.$$

The last two formulas offer examples of explicit expressions that are ill-suited for numerical computation. Discuss what happens when γ is small.

c. To handle the potential numerical difficulties for small γ, use the following classical approximation to the normal distribution. It is particularly appropriate for large values of x.

The ratio of $1 - \Phi(x)$ and the asymptotic expansion

$$\frac{1}{\sqrt{2\pi}} e^{-\frac{1}{2}x^2} \left\{ \frac{1}{x} - \frac{1}{x^3} + \frac{1 \cdot 3}{x^5} - \cdots + (-1)^k \frac{1 \cdot 3 \cdots (2k-1)}{x^{2k+1}} \right\},$$

tends to 1 as $x \to \infty$. For every $x > 0$, the expansion overestimates or underestimates $1 - \Phi(x)$, depending on whether k is even or odd.

Use this to obtain an upper and a lower bound on $F_2(x)$ and use these bounds to compute the $5k$-percentiles of $F_2(x)$ for $1 \leq k \leq 19$. Do this for two distinct values of k and see what accuracy you so get in the numerical values of the percentiles.

Problem 2.3.18: The random variables X_k, $k \geq 1$, are independent and have a common (continuous) distribution $F(\cdot)$. Two real numbers a and b, satisfying $0 < a < b$, are specified. In order for the event A to occur, the value of some X_n must fall in the interval $[a, b]$, *all* X_k with $k < n$ take values outside of $[a, b]$ and there must be at least one such X_k that is less than a and another one that is greater than b. In other words, when the interval $[a, b]$ is visited for the first time, we must have seen at least one observation smaller than a and at least one larger than b.

a. Show that the probability $P(a, b)$ of the event A is given by

$$P(a, b) = [F(b) - F(a)] \left[\frac{[1 - F(b) + F(a)]^2}{F(b) - F(a)} - \frac{[F(a)]^2}{1 - F(a)} - \frac{[1 - F(b)]^2}{F(b)} \right].$$

b. For $F(x) = 1 - e^{-\lambda}$, find the value of λ for which $P(1, 2)$ is largest.

c. If the X_k are normal with mean 0 and variance σ^2, determine the value of σ for which $P(1, 2)$ is maximum.

Problem 2.3.19: In a game, players A and B, respectively, do $m + n$ and n independent Bernoulli trials with probability p. If A has k successes, his score X is defined by $X = \min(k, n)$. B's score is his number Y of successes. Player A wins if and only if $X > Y$. The advantage of extra trials for A is offset by the fact that B also wins when $X = Y$. $P(p; m + n, n)$ is the probability that A wins.

a. Show that

$$P(1/2; n + 1, n) = \frac{1}{2} - \frac{1}{2^{2n+1}}.$$

If A is allowed only one more trial, the game is to the advantage of B. (The proof can be given by a quick, insightful argument or by some manipulation of binomial coefficients.)

b. Show that, in general,

$$P(p; m + n, n) = (1 - p^n) \sum_{k=n}^{m+n} \binom{m + n}{k} p^k (1 - p)^{m + n - k}$$

$$+ \sum_{k=1}^{n-1} \binom{m + n}{k} p^k (1 - p)^{m + n - k} \sum_{\nu=0}^{k-1} \binom{n}{\nu} p^{\nu} (1 - p)^{n - \nu}.$$

c. Prove that, for fixed $n \geq 1$, there exists a value of m for which $P(p; m + n, n)$ exceeds 1/2 if and only if $p < 2^{-1/n} = p^*(n)$. Under that condition, construct an algorithm to find the smallest such value of m.

d. For given values of m and n, determine p° and $p_{\max}$ in (0, 1), for which respectively $P(p^\circ; m + n, n) = 1/2$ and $P(p_{\max}; m + n, n)$ is maximum.

e. For part *c*, implement your code for $p = 0.05r$, $1 \leq r \leq 15$, and for $n = 10$. For part *d*, determine p° for $n = 10$ and $m = 4$.

Reference: This problem was suggested by Problem 1098, *Math.*

Magazine, 53, p. 180, 1980.

Problem 2.3.20: The random variables X and Y are independent, normally distributed, respectively with means 0 and $\mu > 0$, and variances 1 and σ^2. Show that the distribution function of $Z = \max\{X, Y\}$ is given by

$$F(x) = \Phi(x)\,\Phi\left\{\frac{x-\mu}{\sigma}\right\}, \quad \text{for all } x.$$

For given values of μ and σ, find the modes (local maxima) of the probability density $\psi(\cdot)$ of $F(\cdot)$. Examine the derivative of $\psi(\cdot)$. Try to find as small an initial interval $[a, b]$ as possible that contains all the zeros of that function. Next, search the interval $[a, b]$ for the modes by a method of your choice.

Problem 2.3.21: At time $t = 0$, n items with independent, identically distributed lifetimes are simultaneously activated. Their common lifetime distribution is exponential with mean one. The successive failures of the items are recorded by a monitoring device that records (only) the numbers of items still active at the time points $a, 2a, \cdots, ma$. An interval $(ja, ja+a)$ is called a slot. Each slot contains a random number of failures. $E(a; m, n)$ is the expected number of slots among the first m that contain exactly one failure. Show that for $m \geq 1$ and $n \geq 1$, $E(a; m, n)$ is given by

$$E(a; m, n) = n\sum_{j=1}^{m} [e^{-a(j-1)} - e^{-aj}][1 - e^{-a(j-1)} + e^{-aj}]^{n-1}.$$

Write a program to find the value a^* of a for which $E(a; m, n)$ is maximum. Having found a^*, compute the probabilities p_k, $0 \leq k \leq n$, where p_k is the probability that k failures occur after time ma^*. Execute your program for $n = 10$ and $10 \leq m \leq 20$.

Remark: This problem deals with a highly simplified *monitoring* model in which a is to be chosen such that the expected number of *isolated* failures recorded, is as large as possible. In practice, the restriction that the system be observed at equally spaced times is severe. Over time, the rate at which failures occur is decreasing. It would seem more appropriate to choose progressively longer intervals between observations.

Problem 2.3.22: By integration by parts or by a direct probability argument, show that the Erlang distribution of order k with scale parameter λ may be written for $x \geq 0$ as

$$E_k(x;\lambda) = \int_0^x e^{-\lambda u} \frac{(\lambda u)^{k-1}}{(k-1)!} \lambda \, du = 1 - \sum_{i=0}^{k-1} e^{-\lambda x} \frac{(\lambda x)^i}{i!}.$$

If the independent random variables $X_1, \cdots, X_m$ have Erlang distributions of orders k_j with scale parameters λ_j, show that for $Y = \max(X_1, \cdots, X_m)$,

$$P\{Y \leq x\} = \prod_{j=1}^{m} \left[1 - \sum_{r=0}^{k_j-1} e^{-\lambda_j x} \frac{(\lambda_j x)^r}{r!} \right], \quad \text{for } x \geq 0.$$

For a given probability $0 < \beta < 1$, write a subroutine to determine the value of x for which $P\{Y \leq x\} = \beta$. Recall that the mean of the Erlang distribution $E_k(x;\lambda)$ is $k(\lambda)^{-1}$. Use multiples of the largest mean of the m given Erlang distributions to find an interval (x', x''), that includes the desired value of x. When such an interval has been found, solve for x by the bisection method. Test your code for a variety of parameters of your choice and for $\beta = 0.9, 0.95, 0.99$.

Remark: The solution to this problem is a step in the rather complex algorithmic task in Example 3.5.1.

HARDER PROBLEMS

Problem 2.3.23: Two units, I and II, have m and n components respectively. All components have independent, exponentially distributed lifetimes and operate in parallel. Components of Unit I have a mean lifetime of 1; those in group II of λ^{-1}. Failed components cannot be replaced.

The design objective is the following: It is important that all components of Unit I fail before those of Unit II. In addition, when Unit I fails at least n_1 components of II should still be functioning. That could trivially be accomplished by making n very large, but that is not cost-effective. Instead, we want to guarantee a certain probability, say, 0.75 that Unit I fails first and when that occurs, there remain between n_1 and $n_2 > n_1$ working components in Unit II.

For given values of λ, n_1 and n_2, it may not be possible to find a value of n for which that is possible. In that case, we want to find the maximum value (obviously < 0.75) of the probability of interest.

To solve this problem, proceed as follows: First, find the smallest n_0, for which the probability that I fails before II is at least 0.75. n_0 is a lower bound on n. The probability $P(n_0)$ of the event of interest is less than 0.75. Next, increase n, keeping track of $P(n)$, the probability of interest. Stop when $P(n)$ reaches 0.75 or starts decreasing. Print the numerical answers to the design problem. Write a general program, but execute your code for $m = 3$, $n_1 = 1$, $n_2 = 4$, and λ successively equal to 0.75, 1.00, and 1.25.

Problem 2.3.24: An experimenter observes the maximum X_n^* of n independent normal random variables $X_1, \cdots, X_n$, with common mean 0 and variance 1. To find an interval (a, b) for which

$$P\{a \le X_n^* \le b\} = 0.95,$$

he first *incorrectly* reasons as follows: "The observation X_n^* is 'after all' one of the given normal variables, so its distribution is standard normal." He looks up a table and concludes that (a, b) should be (–1.96, 1.96). After his error is pointed out, he realizes that a and b should satisfy

$$\Phi^n(b) - \Phi^n(a) = 0.95,$$

but for obscure reasons he decides to choose a symmetric interval $(-c, +c)$. c is found by solving the equation

$$\Phi^n(c) - [1 - \Phi(c)]^n = 0.95.$$

One more learned in probability points out that the distribution $\Phi^n(\cdot)$ is, particularly for large n, highly skewed to the right and that a shorter interval (a, b) is obtained by setting a and b equal to the unique solutions of

$$\Phi^n(a) = 0.025, \quad \text{and} \quad \Phi^n(b) = 0.975.$$

In support of that suggestion the person points out that a similar procedure is used by statisticians in calculating confidence intervals for skewed distributions, such as the χ^2 and the F-distributions. Though that procedure does not yield the *shortest* interval (a, b), it is computationally simple and satisfactory at least for small n. Write a program to compute for $n = 1, \ldots, 25$:

1. The true probability of the interval $(-1.96, +1.96)$
2. The length $2c$ of the symmetric interval
3. The length $b - a$ of the interval found by the third procedure.

The lengths should be computed to within 10^{-4} by the bisection method. The computation of the shortest interval is the subject of Problem 2.3.25. See also Example 2.3.1.

Problem 2.3.25: For the standard normal distribution $\Phi(\cdot)$ with density $\phi(\cdot)$ and a given positive integer n, find the shortest interval $[a, b]$ for which $\Phi^n(b) - \Phi^n(a) = \beta$, where $\beta = 0.90, 0.95, 0.99$. Show that a and b are the unique real numbers $b > a$, for which also

$$\phi(b)\Phi^{n-1}(b) - \phi(a)\Phi^{n-1}(a) = 0.$$

Write an efficient code to compute a and b for $n = 1, \cdots, 25$ and for the given values of β. See Example 2.3.1. How does one construct the corresponding intervals for the normal distribution with mean μ and variance σ^2?

Problem 2.3.26: Compute the locations of the mode and of the two inflection points of the density of $\Phi^n(x)$, where $\Phi(x)$ is the standard normal distribution. Do this for n ranging from 1 to 50. If for a given n, the mode is at c_n and the inflection points at $a_n < c_n < b_n$, compute also the probabilities $\Phi^n(b_n) - \Phi^n(c_n)$, and $\Phi^n(c_n) - \Phi^n(a_n)$. How are your results related to the maximum of n independent $N(0, 1)$ random variables?

Problem 2.3.27: Write a code to determine the mode of the probability density

$$\phi(u) = pe^{-\lambda_1 u}\frac{(\lambda_1 u)^{k-1}}{(k-1)!}\lambda_1 + (1-p)e^{-\lambda_2 u}\frac{(\lambda_2 u)^{k-1}}{(k-1)!}\lambda_2,$$

on $[0, \infty)$, a mixture of two Erlang densities of the same order. Use the results in Example 2.1.1. Allow the user to supply general values of the parameters. Then rearrange and/or redefine them, so that the results of the example may be efficiently implemented. Report your numerical answers in terms of the original data.

a. Use Newton's method starting at the mode of the first component.
b. Use the secant method with 0 and the mode of the first component as starting values.
c. Use the bisection method with the same starting values as in part *b*.
d. Implement your code for the following parameter values:

$$\lambda_1 = 2.5, \quad \lambda_2 = 5, \quad k = 10, \quad p = 0.25,$$

$$\lambda_1 = 1.75, \quad \lambda_2 = 2.5, \quad k = 20, \quad p = 0.1,$$

$$\lambda_1 = 4, \quad \lambda_2 = 2.5, \quad k = 1, \quad p = 0.75,$$

$$\lambda_1 = 1.25, \quad \lambda_2 = 1.26, \quad k = 6, \quad p = 0.01.$$

Discuss the relative performance qualities of these methods. What numerical difficulties, if any, can you expect with each of them?

Problem 2.3.28: For the *gamma density*

$$\phi(x) = \frac{\lambda^{\alpha+1}}{\Gamma(\alpha+1)} e^{-\lambda u} u^{\alpha}, \quad \text{for } x > 0,$$

with parameters λ and α satisfying $\lambda > 0$, $\alpha > -1$, write a program to find, for any given values of λ, α, and β, the *shortest* interval of probability content β.

Show that, without loss of generality, we may set $\lambda = 1$. Note that for some α, the conditions on $\phi(x)$ in Example 2.1.3 are not satisfied. Handle such cases separately. For the other cases, use the method in Example 2.1.3. Explain which of the alternative procedures you choose to implement. Execute your program for $\alpha = -0.75$, 0, 1.25, 2.50, 7.50, 12.00, and for $\beta = 0.75$, 0.90, 0.95, 0.99.

Problem 2.3.29: Consider the set of ordered n-tuples $\{a_1, a_2, \cdots, a_n\}$ of nonnegative integers whose sum is m.

a. If such an n-tuple is chosen at random, show that for $0 \leq j \leq m$ and $1 \leq r \leq n$,

$$P(j, r; m, n) = P\{a_1 + \cdots + a_r = j\}$$

$$= \frac{\binom{j+r-1}{j}\binom{m+n-j-r-1}{m-j}}{\binom{m+n-1}{m}}.$$

b. If m and n tend to infinity in such a manner that $m/n \to \alpha$, a positive constant, show that for all fixed j and r,

$$P(j, r; m, n) \to \binom{j+r-1}{j}\alpha^j(1+\alpha)^{-j-r}.$$

c. A common way of describing the model in this problem is as the placement of m identical items in n boxes, such that the resulting n-tuples of counts are equally likely (Bose - Einstein statistics). Suppose you are asked to find the smallest r for which the probability that the first r boxes contain at least j items is at least 0.95. How would you compute r?

d. Write a code to answer the question in c for $m = 100$, $n = 20$, and $j = 15$. Find r exactly and also using the (negative binomial) limit probabilities as an approximation.

Problem 2.3.30: The Game of Billiards: Write a computer program to implement the procedure in Example 2.2.1 to set the handicap in the game of billiards. Implement your program for various choices of p_1 and p_2 and for $n_1 = 100, 200, 300$.

Problem 2.3.31: Show that for the billiards model of Example 2.2.1, the quantities $P_1(n_1, n_2)$ also satisfy the recurrence relation

$$P_1(n_1, n_2) = p_1^{n_1} + \sum_{\nu_1=0}^{n_1-1}\sum_{\nu_2=0}^{n_2-1} p_1^{\nu_1}q_1p_2^{\nu_2}q_2P_1(n_1-\nu_1, n_2-\nu_2),$$

with obvious initial conditions. Write a computer program to compute $P_1(n_1, n_2)$ for given values of p_1, p_2, n_1 and n_2. Execute your program for $p_1 = 0.45$, $p_2 = 0.40$, $n_1 = 15$ and $n_2 = 10$. Compare the merits of that recurrence relation to that in Example 2.2.1. In particular, compare the numbers of multiplications needed in each methods.

Problem 2.3.32: The parameters p_1, p_2, n_1 and n_2 have the same significance as for the billiards model in Example 2.2.1. A different recursive scheme is obtained by considering the random path traced out by the successive pairs $\{(X_\nu, Y_\nu)\}$ where X_ν and Y_ν are the numbers of points scored by Players I and II after the νth attempt. For Player I to win, that path must eventually reach one of the points with coordinates (n_1, m) with $0 \le m < n_2$. Also, the ordinates of the points where the path meets the lines $x = r$ for $r = 1, \cdots, n_1$ are nondecreasing.

By $c_k^{(r)}$, denote the probability that the score path $\{(X_\nu, Y_\nu)\}$ meets the line $x = r$ at (r, k), given that Player I starts. Show that

$$c_0^1 = p_1 + q_1 q_2 c_0^1, \qquad c_k^1 = q_1 \sum_{\nu=0}^{k} p_2^\nu q_2 c_{k-\nu}^1, \quad \text{for } 1 \le k < n_2,$$

and for $1 \le k < n_2$, $r \ge 1$,

$$c_0^{r+1} = c_0^1 c_0^r = [c_0^1]^{r+1}, \qquad c_k^{r+1} = \sum_{\nu=0}^{k} c_\nu^{(r)} c_{k-\nu}^1.$$

Finally, $P_1(n_1, n_2) = \sum_{k=0}^{n_2-1} c_k^{(n_1)}$. Find the corresponding recurrence relations for the probability $P_2(n_1, n_2)$ that Player II wins given that he gets the first turn. These recurrence relations are slightly different from the preceding ones.

Use the recurrence relations you have established to compute $P_1(n_1, m)$ for $1 \le m \le n_2$ and $P_2(m, n_2)$ for $1 \le m \le n_1$. The values of p_1, p_2, n_1, and n_2 are given.

Problem 2.3.33: Three - Cushion Billiards: This problem deals with a variant of the billiards model in Example 2.2.1. The parameters p_1, p_2, n_1, and n_2 have the same significance, except that p_1 and p_2 are now the probabilities of making ordinary caroms. If within a turn, a player makes m points, he must successfully make three-cushion shots to score additional points within that turn. The probabilities that the players make successful three-cushion shots are δp_1 and δp_2, respectively, where δ satisfies $0 < \delta < 1$. The integer m is much smaller than n_1 and n_2.

Write a program to compute the probabilities $P_1(n_1, n_2)$ and $P_2(n_1, n_2)$ for given values of p_1, p_2, n_1, n_2, m, and δ. Execute your program for $p_1 = 0.80$, $p_2 = 0.75$, $n_1 = n_2 = 25$, $m = 4$, and $\delta = 0.20$.

Problem 2.3.34: In this variant of the billiards model, Player II is allowed one fault per turn, that is, he loses his turn only upon missing a second time. The symbols p_1, p_2, n_1, and n_2 have the same meaning as in Example 2.2.1. Player I gets the first turn with probability 1/2.

a. Write an efficient program to compute the probability $P(n_1, n_2)$ that Player I wins.

b. A good Player I is asked to play a game under these terms against a weaker Player II, who also insists that n_1 be equal to n_2. Player I will agree to play if and only if his probability of winning is at least 1/2. Develop and code an algorithm to find the smallest value p^* of p_1 for which Player I agrees to play. The values of p_2 and of $n_1 = n_2$ are given. Execute your program for $p_2 = 0.40$, $n_1 = n_2 = 25$.

Hint: $P(n_1, n_1)$ clearly increases with p_1 and $P(n_1, n_1) < 1/2$, for $p_1 = p_2$. The unique value p^* in the interval $(p_2, 1)$ for which $P(n_1, n_1) = 1/2$ may be found by an appropriate bisection method. Determine p^* to within 0.01. Note that Newton's method would not be practical here as the equation to be solved is given only implicitly. Each step of the algorithm requires an implementation of one of the algorithms to compute $P(n_1, n_1)$. For large n_1 a substantial amount of computation is therefore required.

Problem 2.3.35: With the parameters p_1, p_2, n_1, and n_2 having the same significance as in Example 2.2.1, consider the following variant with the possibility of a draw. Player I gets the first turn. If he accumulates n_1 points before Player II reaches n_2, he stops playing. Player II then gets one more turn. If Player II manages to reach n_2 during that turn, the game is a draw. If he does not reach n_2, he loses. In all other cases, Player II wins. Under these rules, both players always get the same number of turns.

Write an efficient program to compute the probabilities $P(n_1, n_2)$ that Player I wins, $Q(n_1, n_2)$ that Player II wins, and $R(n_1, n_2)$ that the game is a draw. Execute your program for the sets of data:

$$p_1 = p_2 = 0.70, \quad n_1 = n_2 = 25.$$

$$p_1 = p_2 = 0.30, \quad n_1 = n_2 = 25.$$

$$p_1 = 0.30, p_2 = 0.40, \quad n_1 = 40, n_2 = 30.$$

Problem 2.3.36: The Probability of a Draw: For the model in Problem 2.3.35, show that the game ends in a draw with probability

$$R(n_1, n_2) = p_1^{n_1} p_2^{n_2} \sum_{r=0}^{\infty} \binom{n_1 + r - 1}{r} \binom{n_2 + r - 1}{r} (q_1 q_2)^r .$$

Verify that, in terms of the hypergeometric function $F(a, b; c; x)$ defined by

$$F(a, b; c; x) = 1 + \sum_{\nu=1}^{\infty} \frac{a(a - 1) \cdots (a - \nu + 1) \cdot b(b - 1) \cdots (b - \nu + 1)}{c(c - 1) \cdots (c - \nu + 1)\nu!} x^{\nu},$$

the probability $R(n_1, n_2)$ may be written as

$$R(n_1, n_2) = p_1^{n_1} p_2^{n_2} F(n_1, n_2; 1; q_1 q_2).$$

Also, by using the classical equality

$$F(a, b; c; x) = (1 - x)^{c - a - b} F(c - a, c - b; c; x),$$

rewrite $R(n_1, n_2)$ as

$$R(n_1, n_2) = p_1^{n_1} p_2^{n_2} (1 - q_1 q_2)^{1 - n_1 - n_2} F(1 - n_1, 1 - n_2; 1; q_1 q_2).$$

a. Use a library subroutine for the hypergeometric function to evaluate $R(n_1, n_2)$ by means of the two preceding expressions.

b. In this part of the problem, we ask to carry out a careful analysis of the error involved in truncating the explicit formula for $R(n_1, n_2)$ at the term with $r = m$. We denote the sum of the remaining terms by $T(m)$ and rewrite $T(m)$ as

$$T(m) = p_1^{n_1} p_2^{n_2} \sum_{r=m+1}^{\infty} \binom{n_1 + r - 1}{r} \binom{n_2 + r - 1}{r} (q_1 q_2)^r$$

$$= p_1^{n_1} p_2^{n_2} \binom{n_1 + m}{m+1} \binom{n_2 + m}{m+1} (q_1 q_2)^{m+1} \sum_{r=m+1}^{\infty} \theta_r,$$

where for $r \geq m + 1$, θ_r is defined by

$$\theta_r = \frac{\binom{n_1 + r - 1}{r} \binom{n_2 + r - 1}{r}}{\binom{n_1 + m}{m + 1} \binom{n_2 + m}{m + 1}} (q_1 q_2)^{r-m-1}.$$

It is readily seen that $\theta_{m+1} = 1$, and that for $r \geq m+1$,

$$\frac{\theta_{r+1}}{\theta_r} = \frac{(n_1 + r)(n_2 + r)}{(r + 1)^2} q_1 q_2.$$

We now show that for every η satisfying $q_1 q_2 < \eta < 1$, it is possible to find an integer m such that for all $r > m$,

$$\frac{\theta_{r+1}}{\theta_r} < \eta.$$

If m is so chosen, we see that

$$T(m) < (1 - \eta)^{-1} p_1^{n_1} p_2^{n_2} \binom{n_1 + m}{m+1} \binom{n_2 + m}{m+1} (q_1 q_2)^{m+1}.$$

To prove the existence of m with the stated property, note that

$$\frac{(n_1 + r)(n_2 + r)}{(r + 1)^2} q_1 q_2 < \eta,$$

for all r that exceed the largest zero of the quadratic function

$$(\eta - q_1 q_2) r^2 + [2\eta - (n_1 + n_2) q_1 q_2] r - n_1 n_2 q_1 q_2,$$

or explicitly, for

$$r > \frac{(n_1 + n_2)q_1q_2 - 2\eta + [4q_1q_2\eta(n_1 - 1)(n_2 - 1) + (q_1q_2)^2(n_1 - n_2)^2]^{1/2}}{2(\eta - q_1q_2)}.$$

It suffices to choose m as the integer part of the quantity on the right in the preceding inequality. A useful choice of η for which the formulas simplify considerably is

$$\eta = q_1q_2 + \frac{1}{2}(1 - q_1q_2) = \frac{1}{2}(1 + q_1q_2).$$

For that η, compute the value of m and of the upper bound for $T(m)$. If the upper bound exceeds 10^{-4}, reduce η and compute the corresponding values of m and $T(m)$ until the upper bound is less than 10^{-4}. Print the successive upper bounds and finally compute $R(n_1, n_2)$ by summing the first $m + 1$ terms of the series in the first formula.

Problem 2.3.37: This model is an idealization of a decision problem common in clinical trials or quality engineering. N independent Bernoulli trials in all will be performed with one of two devices, say, coins. Coin i has probability $p(i)$ of success, $i = 1, 2$. In general, the probabilities $p(1)$ and $p(2)$ are unknown, but we would like to use the coin with the higher probability as often as we can.

One experimental protocol is as follows: An integer m, with $0 \leq m \leq N/2$, is specified and each of the coins is tossed m times. The resulting numbers of successes are $M(1)$ and $M(2)$. If $M(1) > M(2)$, we use Coin 1 for the remaining $N - 2m$ tosses. If $M(1) = M(2)$, one of the two coins is chosen at random for use in the remaining tosses. In the remaining case, Coin 2 is used.

a. Our purpose is to construct an algorithm to find the value of m for which the *expected number* $S[m; N, p(1), p(2)]$ of successes in the N trials is maximized. In the application, values of N between 100 and 300 are of primary interest. Your code should therefore be able to handle values of N up to 300. With N, $p(1)$, and $p(2)$ as data, the code should compute and print the maximum of $S[m; N, p(1), p(2)]$ and the corresponding m.

b. After testing that code, fix N at 250. Explore how the maximum of $S[m; N, p(1), p(2)]$ and the corresponding m depend on $p(1)$ and $p(2)$. Do that by setting $p(1)$ successively equal to $0.10k$, where $5 \leq k \leq 9$ and $p(2) = p(1) + \delta$, where $\delta = 0.01, 0.05, 0.10$. Discuss the qualitative implications of your numerical results.

Problem 2.3.38: For the experimental protocol in Problem 2.3.37, for each m, we would like to announce an integer $L(m)$ such that, with probability at least 0.75, the total number K of successes in the N trials is $L(m)$ or better. $L(m)$ is one less than the 25th percentile of the distribution of K. We also want to find the smallest m for which the maximum of $L(m)$ is attained. Develop an efficient algorithm to compute $L(m)$ and the desired optimal value of m. Implement your algorithm for the data in Problem 2.3.37. For the m that maximizes $L(m)$ also print the expected value $S[m; N, p(1), p(2)]$ of K.

The criteria for selecting m here and in Problem 2.3.37 represent different attitudes towards risk taking. Do they result in significantly different values of m? If you were faced with such a problem, say in testing a medical procedure, which criterion would you use and why?

Remark: The density of K can be written in a closed form as a weighted sum of binomial densities. That expression is computationally harder to use than a nice, elementary recursive scheme. For large N, the density of K must be computed many times. The algorithm then requires an appreciable processing time. We suggest using small values of N in your initial tests.

CHALLENGING PROBLEMS

Problem 2.3.39: From a box containing m red and n green balls, we randomly draw balls, one at a time and without replacement. $P_G(m, n; k)$ is the probability that, when the last red ball is drawn, $k \geq 1$ green ones remain. Show that

$$P_G(m, n; k) = \frac{\binom{m+n-k-1}{m-1}}{\binom{m+n}{n}}.$$

Clearly, $n \geq k$. Prove that, for fixed m and k, $P_G(m, n; k)$ increases in n for $k \leq n \leq km - 1$. If p_1 and p_2 are the smallest and largest values of $P_G(m, n; k)$ in that range, find the smallest n for which $P_G(m, n; k) \geq$

$0.25p_1 + 0.75p_2$. Obtain the numerical results for $m = 50$ and $k = 5$.

Problem 2.3.40: Pitching Pennies: Players I and II initially have m and n coins. Player I starts and attempts to throw all his coins, one after another, into a box placed a distance away. Any coins that miss the box are picked up for his next round. Player II then tries to do the same with his n coins. Any coins outside the box are again gathered up. Both players so continue until one of them gets all his coins into the box. If Player I is the first to do so, Player II gets one more turn. If in that final trial, he gets all coins in the box, the game is a draw; otherwise he loses. If Player II lands all coins in the box first, he wins.

Successive throws of individual coins are modelled as independent Bernoulli trials with probabilities p_1 and p_2 for Players I and II. $P(m, n)$, $Q(m, n)$, and $R(m, n)$ are the probabilities that Player I wins, that Player II wins, and the game is a draw.

a. Give analytic expressions for $P(m, n)$, $Q(m, n)$, and $R(m, n)$, as functions of p_1 and p_2.

b. Derive a formula for the probability $T(k)$ that the game lasts k turns. Distinguish between k even and odd.

c. By considering all possible situations after the first turn, derive recurrence relations for the probabilities of winning and of a draw. As in billiards, you will need to define the probabilities $P_1(m, n)$ and $P_2(m, n)$, (and similarly for the others) according to whether Player I or Player II starts.

d. Derive recurrence relations for $P(m, n)$, $Q(m, n)$, and $R(m, n)$ by considering all possible situations after both players have had one turn each.

e. Select one of the three procedures to compute the probabilities $P(m, n)$, $Q(m, n)$, and $R(m, n)$, (the explicit formulas or the two recursive schemes). Give reasons for your choice. Also compute the probabilities $\{T(k)\}$. Write a program to compute these quantities for given m, n, p_1, and p_2. Execute your code for $m = 20$, $n = 15$, $p_1 = 0.60$, and $p_2 = 0.40$.

f. If $m > n$ and p_2 is given, compute the value of p_1 for which $P(m, n) = Q(m, n)$. Compute the corresponding probability $R(m, n)$. Execute your code for $m = 20$, $n = 15$, and $p_2 = 0.40$.

Hint: In deriving the explicit formulas note that the number of throws required to get any given coin into the box has a geometric density. This model is also related to the failure times of m or n items in parallel with geometric lifetime distributions.

Reference: Schuster, E. F., "An integer programming handicap system in a write ring tossing game," *Mathematics Magazine,* 48, 134 - 142, 1975.

Problem 2.3.41: Pitching Pennies: Suppose that in the model in Problem 2.3.40, Player II is not allowed the final turn. Whoever gets all coins in the box first wins. There is no possibility of a draw. Consider only the case with $m = n$ and $p_1 = p_2 = p$. The advantage to Player I of getting to go first is measured by $\phi(p) = P(m, m) - Q(m, m)$. For $p \geq 0.10$, compute and plot $\phi(p)$ and also the expected number of turns in that game.

Problem 2.3.42: A particle moves counterclockwise and at a uniform speed on the circle with equation $u^2 + v^2 = 1$. At a randomly chosen point on the circle, the particle escapes along the tangent. $G(y; b)$ is the probability that the particle meets the line $u = b$, where $b > 1$, *below* the point with coordinates (b, y).

a. Show that for all real y,

$$G(y; b) = \frac{1}{2\pi}\left[\pi - \arccos\left(\frac{1}{\sqrt{y^2 + b^2}}\right) + \arctan\left(\frac{y}{b}\right)\right],$$

and verify that $G(-\infty; b) = 0$ and $G(+\infty; b) = 1/2$.

b. We wish to place a screen of length $2d$ along the line $u = b$, so as to maximize the probability that the particle will hit the screen. If the midpoint of the screen is placed at the point (b, y), the probability $P(y)$ of a hit is $G(y + d; b) - G(y - d; b)$. Show that $P(y)$ is largest at the unique solution y^* of

$$y = \frac{1}{4bd}\left[\frac{(y - d)[(y + d)^2 + b^2]}{[(y - d)^2 + b^2 - 1]^{1/2}} - \frac{(y + d)[(y - d)^2 + b^2]}{[(y + d)^2 + b^2 - 1]^{1/2}}\right].$$

c. Write a program to find y^* for given values of $b \geq 1$ and $d > 0$. Also compute $P(y^*)$ and the conditional probability density of the point where the particle will hit the screen, given that it hits. Execute your program for the data:

1. $b = 2,\ \ d = 2,$ 2. $b = 5,\ \ d = 2,$ 3. $b = 3,\ \ d = 4.$

d. $R(b) = G(2d;\, b) - G(0;\, b)$ is the hitting probability when the screen stands upright on the u-axis at $u = b$. Compute the value b^* for which $R(b)$ is largest. Tabulate b^* and $R(b^*)$ for $d = 0.25k$, and $1 \leq k \leq 10$.

Hint: The conditional distribution of the point of impact, given a hit, is

$$H(y) = [P(y^*)]^{-1}[G(y;\, b) - G(y^* - d;\, b)],$$

for $y^* - d \leq y \leq y^* + d$. Use Newton's method to solve the equation for y^*. The case $b = 1$ is somewhat special. Show analytically that then

$$\begin{aligned} G(y;\, 1) &= \frac{1}{2} + \frac{1}{\pi}\arctan y, && \text{for } y \leq 0, \\ &= \frac{1}{2}, && \text{for } y \geq 0, \end{aligned}$$

so that $y^* = -d$. For $b > 1$, one can show that $y^* > -d$. We suggest choosing $y = -d$ as the starting solution in the iterative computation of y^*.

Problem 2.3.43: A Selection Model for Binomial Trials: This problem deals with the sample size in a selection model for binomial trials. There are $m + 1$ coins of which one has probability p_1 of showing heads. All the others have probability $p_2 = \delta p_1$, $0 < \delta < 1$. Each coin is tossed n times and the numbers of heads $X_1, \cdots, X_{m+1}$ are recorded for each coin. The coins yielding the maximum number of heads are retained. If there are several, one of them is chosen at random.

a. Show that the probability $P^*(n)$ that the best coin (with probability p_1) is selected by that procedure is given by

$$P^*(n) = \sum_{r=0}^{n} \binom{n}{r} p_1^r (1 - p_1)^{n-r}$$

$$\times \sum_{j=0}^{m} \frac{1}{j+1} \binom{m}{j} \left[\binom{n}{r} p_2^r (1 - p_2)^{n-r}\right]^j \left[F(r - 1, p_2)\right]^{m-j},$$

where $F(k, p_2)$ is defined by

$$F(k, p_2) = \sum_{i=0}^{k} \binom{n}{i} p_2^i (1 - p_2)^{n-i}, \quad \text{for } 0 \leq k \leq n.$$

b. Write an efficient code to evaluate the probabilities $P^*(n)$. Recurrence relations for the binomial probabilities should be fully exploited. Specify a value θ between 0.80 and 0.95. Stop at the first n for which $P^*(n)$ exceeds θ. Implement your code for various choices of p_1, δ, m, and θ. What do the numerical results teach about the preformance of that procedure?

Hint: We suggest using two arrays containing the binomial densities with parameters p_1 and p_2 for successive values of n. In evaluating the first sum in the formula for $P^*(n)$, terms for which the leftmost binomial probability is small, say, less than 10^{-9}, may be neglected. That avoids many computations of the inner sums. These can be computed by using a convenient recursive scheme. The computation involved increases significantly for larger θ. In testing the code, use a small value of θ such as 0.5.

Problem 2.3.44: A Maximum-Likelihood Estimator: $X_1, \cdots, X_n$ are independent, identically distributed with the *gamma density*

$$\phi(x) = \frac{1}{\Gamma(\alpha)} \beta^{-\alpha} e^{-x/\beta} x^{\alpha - 1} \quad \text{for } x > 0.$$

The parameters α and β are positive. The *maximum likelihood estimator* of the pair (α, β), for given observations $x_1, \cdots, x_n$, is the pair of values for which the *likelihood function*

$$L(\alpha, \beta; x_1, \cdots, x_n) = \prod_{i=1}^{n} p(x_i),$$

attains its global maximum.

a. By taking the logarithm of L and differentiating, show that the maximum likelihood estimator is obtained by solving the equations

$$\alpha\beta = \bar{x}, \quad \text{and} \quad n\frac{\Gamma'(\alpha)}{\Gamma(\alpha)} + n \log \beta - \sum_{i=1}^{n} \log x_i = 0,$$

where $\bar{x}$ is the sample mean of the n observations. The first equation yields that $\beta = \alpha^{-1}\bar{x}$. Upon substitution into the other equation, we see that α satisfies

$$\log \alpha = \frac{\Gamma'(\alpha)}{\Gamma(\alpha)} - \log \left[\frac{1}{\bar{x}} \left(\prod_{i=1}^{n} x_i \right)^{1/n} \right],$$

an equation that is to be solved numerically. The term

$$\frac{\Gamma'(\alpha)}{\Gamma(\alpha)} = [\log \Gamma(\alpha)]'$$

is a transcendental function for which library routines are available. In solving for α, note that only the ratio

$$\frac{1}{\bar{x}} \left(\prod_{i=1}^{n} x_i \right)^{1/n}$$

of the geometric and arithmetic means of the data is passed to the subroutine that computes α.

b. Call an appropriate library routine, and write and test a subroutine to compute the maximum likelihood estimator of the pair (α, β).

c. Look up and code a procedure to generate random variates with the gamma density $\phi(\cdot)$. For given α and β, generate 500 sets of $n = 20$ observations. Compute the corresponding estimates $\hat{\alpha}$ and $\hat{\beta}$. Use these pairs to calculate summary statistics on the empirical joint distribution of the estimators. Calculate the sample means and standard deviations of the estimates. Form a scatter plot of the 500 pairs to see how they cluster around the values of α and β you have chosen. Execute your program for $\alpha = 10$, $\beta = 5$, and also for $\alpha = 0.50$ and $\beta = 4$.

References: For a discussion of this statistical problem, see: Dudewicz, E. J. and Mishra, S. N. *"Modern Mathematical Statistics,"* New York: J. Wiley & Sons, 1988 (p. 359.) Procedures to generate random variates with a gamma distribution are given, among others references, in: Fishman, G. S. *"Principles of Discrete Event Simulation,"* New York: J. Wiley and Sons, 1978, and Devroye, L. *"Non-Uniform Random Variate Generation,"* New

York: Springer-Verlag, 1986.

Problem 2.3.45: Consider the same set-up as for Problem 2.3.44, but suppose that α is a positive *integer*. Then $\phi(\cdot)$ is an Erlang density. To find the maximum likelihood estimator, we locate the global maximum of L, which is now a function of the integer α and a positive real variable β.

a. Discuss the numerical procedure to find the maximum likelihood estimator of (α, β). The equation for α is simpler than in Problem 2.3.44.

b. Erlang variates are easily generated as sums of α independent exponential variates. For a given β, generate 500 sets of $n = 20$ Erlang variates of order α for $\alpha = 1, \cdots, 10, 15, 20$. Compute the corresponding estimates $\hat{\alpha}$ and $\hat{\beta}$. Calculate summary statistics of the empirical distribution of these estimates. In particular, print a table of the empirical frequencies of the values $\hat{\alpha}$. We are mainly interested in how often the order α of the Erlang distribution is correctly estimated. Implement your program for $\beta = 5$.

c. Suppose that β is fixed (rather than a parameter in the simulation experiment in part *b*). L is now a function of α only. Adapt the estimator in part *a* to handle that case. Do the same experiment as in part *b*. One would expect that, with only one parameter to estimate, the maximum likelihood estimator for α would perform better than in the preceding case. Is that borne out by your empirical results?

Problem 2.3.46: This problem deals with the small sample properties of the *maximum likelihood estimator* of the parameter λ of a left-conditioned Poisson distribution. Exact algorithmic and experimental approaches are suggested. Consider n independent observations $X_1, \cdots, X_n$ with the common probability density

$$p(j) = e^{-\lambda}\frac{\lambda^j}{j!} \cdot [1 - \sum_{i=0}^{K-1} e^{-\lambda}\frac{\lambda^i}{i!}]^{-1}, \quad \text{for } j \geq K.$$

K is a known positive integer. The maximum likelihood estimator for λ is the value for which the *likelihood function* L, defined as

$$L(x_1, \cdots, x_n; \lambda) = \prod_{k=1}^{n} p(x_k),$$

is maximum. The data $x_1, \cdots, x_n$ are the observed numerical values of $X_1, \cdots, X_n$. We set $\bar{x}_n = n^{-1}\sum_{j=1}^{n} x_i$, and

$$P(j;\lambda) = e^{-\lambda}\frac{\lambda^j}{j!}, \quad \text{for } j \geq 0, \qquad F(r;\lambda) = \sum_{i=0}^{r} P(i;\lambda), \quad \text{for } r \geq 0.$$

a. Show that the maximum likelihood estimator $\hat{\lambda}(n)$ of λ is the unique positive solution

$$\lambda = \bar{x}_n \frac{1 - F(K-1;\lambda)}{1 - F(K-2;\lambda)}.$$

Note that $\hat{\lambda}(n) \geq \bar{x}_n \geq K$.

b. Write a first subroutine to generate n variates $x_1, \cdots, x_n$ with the density $\{p(j)\}$ and a second subroutine to compute the corresponding $\hat{\lambda}(n)$.

c. Next, fix n at 10 and K at 8. (Other values may be chosen.) To examine the qualities of the estimator $\hat{\lambda}(n)$, *by simulation,* proceed as follows: Let λ successively take the values $4 + 0.5\nu$, $0 \leq \nu \leq 12$. For each such value of λ, generate 1,000 samples of size $n = 10$. Use the subroutines in part b to compute 1,000 sample values of $\hat{\lambda}(n)$ and plot a histogram of these estimates. Plot the sample mean of these 1,000 observations as a function of λ. That conveys statistical information on the *bias* of the maximum likelihood estimator. Also, plot the sample standard deviation of these values as a function of λ. That yields insight into the variability of the maximum likelihood estimator. Give detailed interpretations of your numerical results.

d. The following is an exact computation of the distribution of the maximum likelihood estimator $\hat{\lambda}(n)$. It is feasible for small sample sizes such as $n = 10$. The proposed computation is to be performed for the values of λ in part c. First, compute the probabilities $p(j)$ until the remaining tail probability is smaller than 10^{-9}. Suppose that the computed density concentrates on the $M(\lambda)$ integers $K, \cdots, K + M(\lambda) - 1$. Next, compute the n-fold convolution $\{p^{(n)}(j)\}$ of the density $\{p(j)\}$. It is stored in an array of length $nM(\lambda)$. Except for terms with a negligible probability, that array contains density of the sum $X_1 + \cdots + X_n$. The entry corresponding to the value ν of the sum is also the probability that the sample mean $\bar{X}_n$ takes equals νn^{-1}. Now solve the equations

$$\xi = \frac{\nu}{n} \frac{1 - F(K - 1; \xi)}{1 - F(K - 2; \xi)},$$

as efficiently as possible for $K \leq \nu \leq K + M(\lambda) - 1$. Except for the far tail, you now have the exact discrete probability density of the estimator $\hat{\lambda}(n)$. Numerically compute its mean and standard deviation for the given values of λ. Compare your results to those obtained in part *c*.

Remark: Similar investigations of the small sample properties of other estimators for parameters of special probability distributions are feasible. Analogous problems can be formulated for truncated or conditional densities derived from the binomial, negative binomial, or other classical densities.

Problem 2.3.47: Optimal Redundancy: This problem deals with optimal redundancy, a subject of considerable interest in reliability theory. A system consists of m units in series. Unit i, $1 \leq i \leq m$, consists of a number n_i of identical components, placed in parallel. The system is to operate during a fixed time period. The components of Unit i have a probability p_i of failing during that period. All component failures occur according to independent Bernoulli trials with probabilities depending on the unit to which they belong.

a. Show that the probability $P(n_1, \cdots, n_m)$ that the system *does not fail* is

$$P(n_1, \cdots, n_m) = \prod_{i=1}^{m} (1 - p_i^{n_i}).$$

b. The components of Unit i each cost A_i dollars. Develop an algorithm to find the values of $n_1, \cdots, n_m$, for which $P(n_1, \cdots, n_m) \geq \alpha$, where $0 < \alpha < 1$, and for which the cost $C(n_1, \cdots, n_m) = \sum_{i=1}^{m} n_i A_i$ is minimum. Find the least cost design for $\alpha = 0.95$ and

i	1	2	3	4	5
p_i	0.50	0.40	0.50	0.30	0.20
A_i	5	5	4	3	10

CHAPTER 3

FUNCTIONS OF RANDOM VARIABLES

3.1. INTRODUCTION

Many practical situations call for the derivation of the probability distributions of functions of given random variables. Examples are: "Given two or more (independent) random variables, what is the distribution of their sum? Of their maximum?" The general conceptual approach to answering such questions is deceptively simple. We recall the derivation of the distribution $G(\cdot)$ of a function $f(X)$ of X, a random variable with distribution $F(\cdot)$. The general argument, which sometimes gets lost in the calculus of special cases, notes that the event $\{f(X) \leq x\}$ is simply the image of the interval $(-\infty, x]$ under the inverse function f^{-1} of f. The problem therefore amounts to evaluating the probability $G(x) = P\{f(X) \leq x\} = P[X \,\varepsilon\, f^{-1}\{(-\infty, x]\}]$, for all real x. Exactly the same argument applies for discrete or integer valued random variables. It also extends to the multivariate case. For the bivariate case, $G(\cdot)$, the distribution of $f(X_1, X_2)$ is found by identifying the events $(X_1, X_2) \,\varepsilon\, f^{-1}\{(-\infty, x]\}$ and computing their probabilities.

Although the general idea is almost obvious, its implementation can be difficult. Identifying the necessary set of events can be a delicate and complex task. The explicit calculation of their probabilities, by any available means, can be done only for reasonably nice cases. We encourage the reader to review the derivations of the classical χ^2-, t-, or F-distributions. These are carried out in steps that involve functions of one or two (independent) random variables only. For these, the events are readily described and the calculation amounts to evaluating especially tractable single or double integrals. To illustrate the general argument, we consider a nontraditional example:

Example 3.1.1: A Function of Poisson Random Variables: Let X and Y be independent Poisson-distributed random variables with parameters λ and μ. For a given positive integer m, we wish to compute the (discrete) density of $Z = X(Y + m)^{-1}$.

The possible values (the support) of Z are all fractions of the form

$i(j+m)^{-1}$, where i and j are nonnegative. If x is such a fraction, the event $\{Z = x\}$ is the infinite set of pairs (i, j) for which $i = x(j+m)$. To give a precise interpretation of the problem, we shall determine a subset E of the support of Z for which $P(Z \varepsilon E) > 1 - 10^{-10}$. Eventually, we shall print a list of the ranked fractions in E with their corresponding probabilities. For $x \varepsilon E$, the probability $P\{Z = x\}$ is found by summing the quantities

$$e^{-\lambda}\frac{\lambda^i}{i!} \times e^{-\mu}\frac{\mu^j}{j!}$$

over all pairs (i, j) for which $x(j+m) = i$. Practically, this could be done as follows: Run through the indices i and j in a systematic way, filling an array U with the values of $i(j+m)^{-1}$ and an array V with the corresponding products of Poisson probabilities. The upper values of i and j should be chosen so that a probability of at most 10^{-10} is neglected in each Poisson density.

The array U is now sorted in order of increasing magnitude and V is rearranged accordingly. The array U contains several runs of successive fractions with the same value. Due to limited machine accuracy, these may have slightly different stored values. We scan the array U, collapse all terms differing by a negligible amount, say, $\varepsilon = 10^{-10}$, and sum the corresponding terms in V.

A more cautious way of dealing with fractions of equal values is to carry along two arrays of the indices (i, j) for each computed fraction. These arrays are rearranged in the same order as U. For any x, the Poisson products for the pairs for which $i(j_1+m) = i_1(j+m)$ are accumulated.

In the next example, the general argument is illustrated by the computation of the distribution of a polynomial of a random variable X. The solution is straightforward, but computation-intensive.

Example 3.1.2: A Polynomial of a Random Variable: X is a random variable with distribution $F(\cdot)$. In some structural engineering applications, the distribution $G(\cdot)$ of the random variable $Y = P(X) = a_0 + a_1X + \cdots + a_nX^n$ arises. We assume that X takes its values in an interval $[c, d]$, where c or d may be infinite.

The first step consists in determining the smallest and the largest values, $y_{\min}$ and $y_{\max}$ of $P(x)$ in $[c, d]$, as well as all the local minima and maxima of $P(x)$ in that interval. That requires computing the roots of the equation $P'(x) = 0$ in $[c, d]$, possibly with their multiplicities, and the evaluation of $P(c)$ and $P(d)$. The second derivative $P''(x)$ is used to classify the critical points as local maxima or minima and to eliminate possible inflection points.

In order to tabulate $G(\cdot)$, the range $[y_{\min}, y_{\max}]$ is subdivided by a set of equidistant points y_j, $0 \le j \le N$. N depends on the desired accuracy. If $y_{\min}$ is $-\infty$, or $y_{\max}$ is ∞, we choose y_1 so that $P\{Y \le y_1\}$ is small, or y_N so that $P\{Y > y_N\}$ is small.

For each y_j, we find all roots of the equation $P(x) = y_j$ in $[c, d]$. The set where $P(x) \le y_j$ is a union of disjoint subintervals of $[c, d]$. For each y_j, the number of such intervals depends on the graph of $P(x)$. $G(y_j)$ is found by summing the probability contents $G(z'') - G(z')$ of all such intervals $(z', z'']$. Unless $F(\cdot)$ is a simple or classical distribution, each of these last steps may require a numerical integration. In the problem section, we list a few such problems of a rather simple type. The development of a general code to evaluate $G(\cdot)$ is clearly a major and potentially delicate task.

Remark *a.* Given the practical difficulty of finding distributions of functions of random variables, this chapter will deal with some special methods of wide applicability, such as the *convolution product,* and some techniques based on order statistics. The related problems call for more complex recursive schemes than those in the earlier chapters. Many require recursions of sequences. Later subjects will involve matrix recurrence equations, so that more abstract insights needed to recognize similarities in the algorithms will be nurtured here.

Remark *b.* Building on the idea in Chapter 2 that implicitly defined equations are common and important, we shall consider several questions of optimal design of decision experiments and other applications.

Remark *c.* We do not list many problems requiring *numerical integration.* Such problems require an excellent acquaintance with classical numerical analysis to handle the additional approximations that are needed. The structural reasoning underlying algorithm development can also be learned from discrete problems. Readers familiar with numerical integration techniques can find a special section of problems that call for such knowledge at the end of this chapter.

3.2. SUMS OF RANDOM VARIABLES

The distribution $G(\cdot)$ of $X_1 + X_2$, where (X_1, X_2) has the joint distribution $F(x_1, x_2)$ is found by evaluating, for every x, the probability assigned by the distribution F to the half-plane $\{(x_1, x_2): x_1 + x_2 \leq x\}$. If F has a density $\phi(u_1, u_2)$, then for all real x,

$$G(x) = \int_{-\infty}^{x} dv \int_{-\infty}^{\infty} \phi(u_1, v - u_1)\, du_1,$$

The corresponding density $g(x)$ is obtained by differentiation and is therefore

$$g(x) = \int_{-\infty}^{\infty} \phi(u_1, x - u_1)\, du_1.$$

Similarly, for integer-valued random variables X_1 and X_2 with joint density $\{p_{i_1 i_2}\}$, the probability density $\{g_n\}$ of their sum $X_1 + X_2$ is given by

$$g_n = \sum_{k=-\infty}^{\infty} p_{k,n-k}, \qquad \text{for all integers } n.$$

The Convolution Product: In the important case where X_1 and X_2 are *independent* these formulas lead to

$$g(x) = \int_{-\infty}^{\infty} \phi_1(u_1)\phi_2(x - u_1)\, du_1,$$

where ϕ_1 and ϕ_2 are the densities of X_1 and X_2, and

$$g_n = \sum_{k=-\infty}^{\infty} p_{1,k} p_{2,n-k},$$

in terms of the densities $\{p_{1,k}\}$ and $\{p_{2,k}\}$ of X_1 and X_2.

The operation that associates the function $g(\cdot)$ with $\phi_1(\cdot)$ and $\phi_2(\cdot)$, – or the sequence $\{g_n\}$ with the sequences $\{p_{1,k}\}$ and $\{p_{2,k}\}$, – is the *convolution product*. It is one of the most important mathematical operations of probability theory. Except for rare tractable cases, familiar from introductory textbooks, the convolution product of continuous densities requires numerical integration techniques. The general procedure of choice is the *Fast Fourier Transform,* which has excellent algorithmic properties. It is discussed in most modern texts on numerical analysis and well-written library routines are available for its execution. For the reasons stated earlier, we shall not discuss this method here. The points of algorithmic care, to be discussed here for the discrete case, should obviously also be taken into account in the continuous case.

The Discrete Case: A common source of errors in analytical and numerical computations is a lack of care in the definition of the range of the indices. We illustrate this by discussing the care needed in carrying out the convolution of two discrete densities.

In numerical computation, we are mostly working with truncated densities. We may think of the random variables X_i, $i = 1, 2$, as concentrating on the integers between M_i and N_i, $i = 1, 2$. The sum $X_1 + X_2$ therefore takes only values between $M_1 + M_2$ and $N_1 + N_2$. Let n be an index in that range. Thus g_n is now given by

$$g_n = \sum_{k = L(k)}^{K(k)} p_{1,k} p_{2,n-k},$$

where $L(k)$ and $K(k)$ are chosen to avoid computing products that are zero and, most importantly, to avoid indices that may be out of range in the storage arrays. To find the values of these summation limits, we note that $p_{1,k} p_{2,n-k}$ cannot differ from zero unless $M_1 \le k \le N_1$, and $M_2 \le n - k \le N_2$. Combining both inequalities leads to

$$L(k) = \max\,[M_1, n - N_2], \quad \text{and} \quad K(k) = \min\,[N_1, n - M_2].$$

We should also remember that the indices in this discussion are not necessarily the same as those of the memory arrays in which the densities are stored. When they differ, the indices used in the code should be appropriately shifted.

Repeated Convolutions: Usually, sums of independent random variables have more than two summands. The practical need for information on sums with many summands has led to the beautiful mathematical developments collectively known as the *Central Limit Problem.* When numerical, nonapproximate computation of repeated convolution products is feasible and desirable, the following comments are helpful. In order to discuss these, we introduce some concise notation. By $\mathbf{p}(i)$, we denote the density $\{p_{i,k}\}$, and by $\mathbf{p}(i) * \mathbf{p}(j)$, the convolution of the densities $\mathbf{p}(i)$ and $\mathbf{p}(j)$. Since $*$ is a commutative and associative binary operation, the convolution product $\mathbf{p}(1) * \mathbf{p}(2) * \cdots * \mathbf{p}(m)$ can be evaluated by pairwise convolution products taken in any order. The *trimming* operation, described later in this section, is useful in some cases.

Convolution Powers: When all densities in the product are the same, the m-fold product, $\mathbf{p}^{(m)}$, is called the *m-fold convolution* or the *mth convolution power* of the density $\mathbf{p}$. For consistency, the 0th power $\mathbf{p}^{(0)}$ is defined as the sequence $\{1, 0, 0, \cdots\}$.

If an application calls for the computation of one, or a few, convolution powers of a density $\mathbf{p}$, the following is a useful idea: To compute the mth power, we determine the binary digits $i(1), \cdots, i(r)$ of m, where $i(1)$ is the rightmost digit. The array $\mathbf{c}$ that will eventually hold $p^{(m)}$ is initialized by $\mathbf{p}^{(0)}$ if $i(1) = 0$ and by $\mathbf{p}^{(1)}$ if $i(1) = 1$.

Next, by taking single convolution products, we successively form the powers $\mathbf{p}^{(2^j)}$. If $i(j) = 1$, we form the convolution of the current $\mathbf{c}$ with $\mathbf{p}^{(2^j)}$ to obtain the next $\mathbf{c}$; otherwise we form the next even power of $\mathbf{p}$. While that is not the optimal procedure of this type, it greatly reduces the number of convolution products needed to compute $\mathbf{p}^{(m)}$. For example, $\mathbf{p}^{(25)}$ is computed by six convolution products using the following steps:

$$\mathbf{c} \leftarrow \mathbf{p},$$

$$\mathbf{p} * \mathbf{p} = \mathbf{p}^{(2)},$$

$$\mathbf{p}^{(2)} * \mathbf{p}^{(2)} = \mathbf{p}^{(4)},$$

$$\mathbf{p}^{(4)} * \mathbf{p}^{(4)} = \mathbf{p}^{(8)},$$

$$\mathbf{c} \leftarrow \mathbf{c} * \mathbf{p}^{(8)} = \mathbf{p}^{(9)},$$

$$\mathbf{p}^{(8)} * \mathbf{p}^{(8)} = \mathbf{p}^{(16)},$$

$$\mathbf{c} \leftarrow \mathbf{c} * \mathbf{p}^{(16)} = \mathbf{p}^{(25)}.$$

Convolution Polynomials: Another computation that can be efficiently organized is that of a density with either of the forms:

$$a_0\mathbf{p}^{(0)} + a_1\mathbf{p}^{(1)} + \cdots + a_m\mathbf{p}^{(m)},$$

where the coefficients a_j, $1 \le j \le m$, are constants, or

$$\mathbf{b}_0 * \mathbf{p}^{(0)} + \mathbf{b}_1 * \mathbf{p}^{(1)} + \cdots + \mathbf{b}_m * \mathbf{p}^{(m)},$$

where the coefficients $\mathbf{b}_j$, $1 \le j \le m$, are sequences themselves. The first arise in studying the distribution of a random sum of independent, identically distributed random variables; among other applications, the second arise in the study of waiting time distributions in discrete queues.

For both cases, we adapt the familiar *Horner algorithm* for polynomials. We show the details for the first form; those for the second are entirely analogous. We initialize an array by setting $\mathbf{c}(1) = a_{m-1}p^{(0)} + a_m p^{(1)}$. Next, we form $\mathbf{c}(1) * \mathbf{p}$ and add $a_{m-2}\mathbf{p}^{(0)}$ to it to obtain $\mathbf{c}(2)$. We continue in this manner until the desired density is obtained in $\mathbf{c}(m)$.

For example, the sequence $\mathbf{c}(3) = 0.25\mathbf{p}^{(0)} + 0.40\mathbf{p}^{(2)} + 0.35\mathbf{p}^{(3)}$ is computed by the following steps:

$$\mathbf{c}(1) \leftarrow 0.40\mathbf{p}^{(0)} + 0.35\mathbf{p}$$

$$\mathbf{c}(2) \leftarrow \mathbf{c}(1) * \mathbf{p}$$

$$\mathbf{c}(3) \leftarrow 0.25\mathbf{p}^{(0)} + \mathbf{c}(2) * \mathbf{p}$$

Adding $a\,\mathbf{p}^{(0)}$ to an array, as in the last step, is trivially easy. We just add a to the contents of the location containing the term with true index zero.

Remark: The techniques for evaluating convolution powers and polynomials also apply to matrices. Entire similar steps are carried out to compute the mth power A^m of a square matrix A. Horner's algorithm can be fruitfully applied to matrix polynomials of the form $\sum_{k=1}^{m} C_k A^k$ or $\sum_{k=1}^{m} A^k C_k$, where the coefficients C_j are either scalars or matrices. Such computations occur in important applications.

Trimming: Computations involving repeated convolutions can rapidly lead to very long arrays. The following is a heuristic procedure, useful in computations for probability models, that frequently can save on storage and computer processing time. We describe it for the convolution product

$$\mathbf{p}(1) * \mathbf{p}(2) * \cdots * \mathbf{p}(m),$$

but it also applies to the other constructions we have discussed. The $\mathbf{p}(i)$ are sequences of probabilities whose sum is s_i. Often, the initial and/or the final terms of such sequences are very small. The sum of the terms in $\mathbf{p}(i) * \mathbf{p}(j)$ is $s_i + s_j$ and, in many cases, the initial and/or final terms of that convolution are significantly smaller than those of the factor sequences.

In trimming, we choose an ε, say, equal to 10^{-12}, and in the successive convolutions, we delete all initial terms whose sum is at most ε and all final terms whose sum is at most ε. The sum of the neglected terms is computed; it is at most 2ε. For the subsequent convolutions, we need to keep a careful account of the new starting and ending indices of the trimmed sequences. Better still, the trimmed sequence can be shifted so that its starting index is 1. We must then keep track of the true initial and final indices of the current sequence.

The total amount of probability neglected is simply the sum of all trimmed terms. The induced error affects primarily the outer tails of the final convolution product. In cases where it is preferable to "spread the error around," we can, after forming, say, the convolution $\mathbf{p}(1) * \mathbf{p}(2)$, normalize that sequence so that the sum of its terms is again $s_1 + s_2$. We do the same for the subsequent convolution products.

Whether to apply trimming or not is clearly a matter of judgment. We have applied it to many problems for which we have examined the exact numerical results to those with trimming, both with and without normalization. Unless very high accuracy of the computed probabilities was required, trimming had no significant effect, yet it greatly reduced the computer processing time.

3.3. MAXIMA AND MINIMA OF RANDOM VARIABLES

Using again the bivariate case as an illustration, the distribution of $Z = \max(X_1, X_2)$ amounts to finding for all x, the probability $G(x)$ of the event $\{X_1 \le x,\ X_2 \le x\}$. In terms of the joint distribution $F(x_1, x_2)$, $G(x) = F(x, x)$. The same argument holds for the multivariate case. For several *independent* random variables, the distribution $G(x)$ of $Z =$

$\max(X_1, \cdots, X_m)$ is just the product

$$G(x) = \prod_{i=1}^{m} F_i(x).$$

For the minimum, a little extra caution is needed. $Y = \min(X_1, X_2)$ is at most x, if $X_1 \le x$ *or* $X_2 \le x$. The distribution $H(x)$ of Y is now given by

$$H(x) = 1 - F(x, \infty) - F(\infty, x) + F(x, x).$$

The corresponding formula for the multivariate case has many more terms. For *independent* random variables, the probability distribution $H(x)$ of $Y = \min(X_1, \cdots, X_m)$ simplifies to

$$H(x) = 1 - \prod_{i=1}^{m} [1 - F_i(x)].$$

Probability distributions of other *order statistics* of independent *continuous* random variables can be written as integrals of functions of their distributions. The computation of these distributions is challenging. Even for normally distributed random variables, the tabulation of order statistics distributions was a major task. It could be undertaken only after the advent of the modern computer.

Order statistics of discrete, and specifically integer-valued, random variables $X_1, \cdots, X_m$ offer special challenges. Their maximum, minimum, median, and other order statistics are no longer necessarily attained by one and only one of the given random variables. Therefore, a careful consideration of cases is necessary. Several problems related to that issue are listed at the end of the chapter.

A few words need to be said about the *exponential* and *geometric* distributions. For these, some calculations of order statistics are particularly simple. Some problems at the end of the chapter are tractable only under these distributional assumptions. The source of all these simplifications is the *memoryless* property shared by these distributions. If X has an exponential distribution with parameter λ, then for all positive x and y,

$$P\{X \le x + y \mid X > x\} = 1 - e^{-\lambda y}.$$

Similarly, if $P\{X = k\} = q^k p$, for $k \geq 0$, with $q = 1 - p$, then

$$P\{X = n + k \mid X > n\} = q^k p, \quad \text{for all } k \geq 0,\ n \geq 0.$$

This implies that for independent exponential random variables $X_1, \cdots, X_m$ with parameters $\lambda_1, \cdots, \lambda_m$, the minimum is again exponential and with the parameter $\lambda_1 + \cdots + \lambda_m$. Similarly, the minimum of geometrically distributed random variables with parameters $p_i = 1 - q_i$, $1 \leq i \leq m$ (on the nonnegative integers) has the geometric density

$$r_k = (q_1 \cdots q_m)^k [1 - q_1 \cdots q_m], \quad \text{for } k \geq 0.$$

The probability that the minimum is k and that of the given random variables exactly ν are equal to k, is given by

$$r_k(\nu) = (q_1 \cdots q_m)^k \sum_{E_\nu} c_1 \cdots c_m,$$

where for $1 \leq \nu \leq m$, the summation extends over all terms obtained by choosing ν factors c_j from among the p_j and $m - \nu$ factors c_j from among the q_j. If all the parameters p_j are equal, $r_k(\nu)$ simplifies to

$$r_k(\nu) = q^{mk} \binom{m}{\nu} p^\nu q^{m-\nu}.$$

We next illustrate the use of the memoryless property of the exponential distribution by an example from reliability theory.

Example 3.3.1: Consider a system with Units I and II. Unit I consists of m identical components in parallel. This means that Unit I is operative as long as at least one component functions. Similarly, Unit II consists of n parallel components. All components have independent, exponentially distributed lifetimes; those of Unit I, with parameter λ; those of II, with parameter μ. We wish to compute the probability that Unit I outlives Unit II.

The probability distributions $G_1(x)$ and $G_2(x)$ of the lifetimes of the two units are $G_1(x) = (1 - e^{-\lambda x})^m$, and $G_2(x) = (1 - e^{-\mu x})^n$, respectively. Each unit fails when its component with the longest failure time expires. Unit I outlives Unit II if it has the longer of the two lifetimes. The desired probability is therefore

$$p = \int_0^\infty m(1 - e^{-\lambda x})^{m-1}\lambda e^{-\lambda x}\,(1 - e^{-\mu x})^n\,dx.$$

We shall give a probabilistic argument to show how p can be computed *without numerical integration.* Let $p(i, j)$ be the probability that I outlives II if the units have respectively i and j functioning components of each type left. Then clearly

$$p(i, 0) = 1, \quad p(0, j) = 0, \quad \text{for all } i > 0,\ j > 0.$$

Next, for positive i and j, the time to the first component failure is the minimum of the first failure times (of components) in I and II. Each of the latter have exponential distributions with parameters λi and μj, respectively. The probability that the first component failure occurs in Unit I is therefore $\lambda i(\lambda i + \mu j)^{-1}$. The corresponding probability for Unit II is the complementary probability $\mu j(\lambda i + \mu j)^{-1}$.

After the first component failure there are either $(i - 1, j)$ or $(i, j - 1)$ components left and, *by the memoryless property,* the remaining lifetimes are independent, exponential random variables with the same parameters as before. It therefore follows that for all positive i and j,

$$p(i, j) = \frac{\lambda i}{\lambda i + \mu j} p(i - 1, j) + \frac{\mu j}{\lambda i + \mu j} p(i, j - 1).$$

That recurrence is easily implemented to compute $p = p(m, n)$.

3.4. THE INCLUSION - EXCLUSION FORMULA

The classical combinatorial formula, to be discussed next, can be used to obtain a variety of derived probabilities. In most cases, its direct implementation also presents serious numerical difficulties. Because of these, various approximations and bounds have been investigated since the early days of modern probability theory. What follows is a summary of the most salient results.

Let $A_1, A_2, \cdots, A_n$ be n events belonging to a common probability space. For each outcome ω, $X(\omega)$ is the number of these events to which ω belongs. Informally, the random variable X is the number of the n events that occur. The *inclusion - exclusion formula* gives an expression for $p_m = P\{X = m\}$, for $0 \le m \le n$. In order to state the formula concisely, we define some notation.

$P(A_{i_1}A_{i_2} \cdots A_{i_r})$ is the probability of the event $A_{i_1}A_{i_2} \cdots A_{i_r}$, where $i_1 < i_2 < \cdots < i_r$ are indices chosen from among $1, \cdots, n$. The quantities S_r are defined by $S_0 = 1$, and

$$S_r = \sum P(A_{i_1}A_{i_2} \cdots A_{i_r}), \quad \text{for } 1 \le r \le n.$$

The summation extends over all possible choices of the indices $1 \le i_1 < i_2 < \cdots < i_r \le n$. The inclusion - exclusion formula states that

$$p_m = P\{X = m\} = \sum_{r=m}^{n} (-1)^{r-m} \binom{r}{m} S_r, \quad \text{for } 0 \le m \le n.$$

In addition, for $1 \le m \le n$,

$$P\{X \ge m\} = \sum_{r=m}^{n} p_m = \sum_{r=m}^{n} (-1)^{r-m} \binom{r-1}{m-1} S_r.$$

The numerical difficulties of these formulas are obvious; the quantities S_r often take on very large values and the probabilities of interest are obtained by taking sums and differences of such quantities. Overflow in the computation of the S_r and inaccuracies due to loss of significance are common and very real difficulties. To illustrate these, as well as the very nice approximations that have been found for some classical combinatorial applications, we consider some examples.

Example 3.4.1: Suppose we have an ordered string with k locations. Each location is marked with one of n different labels in such a manner that all n^k random labelings are equally likely. The random variable X is the number of labels that do *not* occur in the string. We wish to derive expressions for the probabilities $p(m; k; n) = P\{X = m\}$ for $0 \le m < n$.

Let A_j be the event that label j is not used. $A_{j_1}A_{j_2} \cdots A_{j_r}$ is then the event that the labels $j_1 < j_2 < \cdots < j_r$ are not used. The number of ways the locations can be labeled using the remaining $n - r$ labels is $(n-r)^k$. Therefore,

$$P(A_{j_1} A_{j_2} \cdots A_{j_r}) = \frac{(n-r)^k}{n^k} = \left[1 - \frac{r}{n}\right]^k .$$

That probability depends only on r and not on the labels $j_1, \cdots, j_r$ we have selected. All terms in the sum S_r are therefore equal. The number of terms in S_r is the number of ways that r out of n distinct labels can be chosen. It follows that

$$S_r = \binom{n}{r}\left[1 - \frac{r}{n}\right]^k, \quad \text{for } 0 \le r \le n.$$

Using the inclusion - exclusion formula, we now obtain that

$$p(m;k;n) = \sum_{r=m}^{n} (-1)^{r-m} \binom{r}{m}\binom{n}{r}\left[1 - \frac{r}{n}\right]^k,$$

which may be rewritten as

$$p(m;k;n) = \binom{n}{m}\sum_{\nu=0}^{n-m} (-1)^{\nu}\binom{n-m}{\nu}\left[1 - \frac{m+\nu}{m}\right]^k, \quad \text{for } 0 \le m < n.$$

To illustrate a use of that formula, consider the number X of dates of the year that do not occur as birth dates in a random group of k persons. To see the difficulties of a direct implementation, think through the delicate numerical steps required, say, for $k = 1{,}000$ and $n = 365$.

For parameter values of that size, an excellent numerical approximation is obtained by use of a limit theorem of von Mises, published in 1939. For moderate values of m, the $p_m(0)$ are numerically close to the Poisson probabilities with

$$\lambda = ne^{-k/n}.$$

The precise statement is the following limit theorem: If k and n tend to infinity in such a manner that $\lambda = ne^{-k/n}$ remains bounded, then

$$p(m;k;n) - \frac{\lambda^m}{m!}e^{-\lambda} \to 0.$$

An outline of the proof may be found in Feller, W. *"An Introduction to Probability Theory and Its Applications, Vol. I."* Third Edition; New York: J. Wiley & Sons, 1967, pp. 103 - 104, which also contains other classical applications of the inclusion - exclusion formula and approximation results.

Bonferroni's Inequalities: In some applications, suitable approximations to the probabilities $p_m = P\{X = m\}$ and $P\{X \geq m\}$ can be obtained by evaluating only a few terms in the inclusion - exclusion formula. Bonferroni's inequalities state that the successive partial sums in the expressions for $p_m = P\{X = m\}$ and $P\{X \geq m\}$ alternatingly over- and underestimate the desired quantities. In cases where the difference between successive partial sums becomes satisfactorily small, the midpoint between them can serve as an approximation to the desired probabilities.

3.5. A WAITING TIME IN MULTINOMIAL TRIALS

In this section, we discuss a substantial computational problem that involves many ideas about recursive schemes discussed in Chapters 1 and 2.

Example 3.5.1: A Waiting Time in Multinomial Trials: Successive independent multinomial trials with m alternatives of positive probabilities $p_1, \cdots, p_m$ are performed. Positive integers $i_1, \cdots, i_m$ are specified. The random variable N is the number of trials required until for $1 \leq j \leq m$, the outcomes j have occurred at least i_j times. A formula for the distribution of N is easily written, but somewhat difficult to implement computationally.

The event $\{N \leq n\}$ occurs if in the first n multinomial trials, each of the outcomes j has occurred at least i_j times. Its probability is therefore

$$P\{N \leq n\} = \sum \frac{n!}{k_1! \cdots k_m!} p_1^{k_1} \cdots p_m^{k_m}, \tag{1}$$

where the summation extends over all m-tuples $k_1, \cdots, k_m$ satisfying $k_1 + \cdots + k_m = n$ and $k_1 \geq i_1, \cdots, k_m \geq i_m$. Clearly, that probability is positive only if $n \geq i_1 + \cdots + i_m = i_m^*$. A direct evaluation by summation of the multinomial probabilities is extremely belabored. In cases of interest, the distribution of N may have a long tail. Typically $P\{N \leq n\}$ must be computed for several hundreds of indices n.

A more efficient and elegant solution is possible. To implement it, we need an estimate of the largest index n^* for which $P\{N \leq n\}$ is to be computed. To obtain such an estimate, we first consider another problem.

Multinomial Trials at Poisson Events: Suppose that the multinomial trials are performed at the events of a Poisson process of rate 1 instead of at integer time points. The random variable T is the first time we have seen each of the outcomes j, i_j times, $1 \leq j \leq m$. By a classical property of the Poisson process the occurrences of the outcomes of types j, $1 \leq j \leq m$, form m *independent* Poisson processes of rates p_j. The waiting time T_j until outcome j has occurred i_j times therefore has the Erlang distribution

$$P\{T_j \leq x\} = 1 - \sum_{k_j=0}^{i_j-1} e^{-p_j x} \frac{(p_j x)^{k_j}}{k_j!}, \quad \text{for } x \geq 0.$$

Clearly, T is the maximum of the random variables T_j, $1 \leq j \leq m$, so that

$$P\{T \leq x\} = \prod_{j=1}^{m} \left[1 - \sum_{k_j=0}^{i_j-1} e^{-p_j x} \frac{(p_j x)^{k_j}}{k_j!} \right], \quad \text{for } x \geq 0.$$

Moreover,

$$P\{T \leq x\} = \sum_{n=i_m^*}^{\infty} e^{-x} \frac{x^n}{n!} P\{N = n\}.$$

If we now specify a probability α, say, $\alpha = 0.01$, it is a fairly easy matter (see Problem 2.3.22) to find the value x^* of x for which $P\{T \leq x^*\} = 1 - \alpha$. x^* is computed by the solution techniques of Chapter 2. While it is difficult to relate x^* precisely to the smallest index n^* for which $P\{N \leq n^*\} \geq 1 - \alpha$, it is intuitively clear that x^* and n^* are not very different. A conservative bound on n^* was derived in the reference at the end of this example. We do not need n^* exactly. For practical purposes, we compute x^*, increase it, say, by 10%, and set n^* equal to the integer closest to the value so obtained.

A Recurrence Between Sequences: We are now ready for the more substantial task of computing $P\{N \leq n\}$ for $i_m^* \leq n \leq n^*$. That is done by a recurrence on j between successive *sequences*. These are related by what are essentially convolution formulas. For $j = m$, the desired array of probabilities is obtained. Let us write $F(n; m; p_1, \cdots, p_m)$ for the sum in formula (1). If we let k_m range over the integers between i_m and $n - i_m^* + i_m$, and introduce factors

$$\frac{n!p_m^{k_m}}{k_m!},$$

to complete the binomial probability $\binom{n}{k_m} p_m^{k_m}(1 - p_m)^{n-k_m}$, we obtain that

$$F(n; m; p_1, \cdots, p_m)$$

$$= \sum_{k_m=i_m}^{n-i_m^*+i_m} \binom{n}{k_m} p_m^{k_m}(1 - p_m)^{n-k_m} F(n - k_m; m - 1; p'_1, \cdots, p'_{m-1}).$$

The last factor has the same form as in (1), but with only $m - 1$ alternatives and the $m - 1$ parameters

$$p'_j = \frac{p_j}{1 - p_m}, \quad \text{for } 1 \leq j \leq m - 1.$$

We see that to compute $F(n; m; p_1, \cdots, p_m)$ for $i_m^* \leq n \leq n^*$, we need the quantities $F(\nu; m - 1; p'_1, \cdots, p'_{m-1})$ for $i_m^* - i_m \leq \nu \leq n^* - i_m$. That is the basis for a recursive scheme to solve the problem with m alternatives in terms of a similar problem with $m - 1$ alternatives.

Let us write $i_r^* = i_1 + \cdots + i_r$. For $3 \leq r \leq m$, the algorithm consists in recursively computing the arrays $\{F(\nu; r; \pi_1, \cdots, \pi_r)\}$ with indices $i_r^* \leq \nu \leq n^* - i_m^* + i_r^*$ by the convolution-type formula

$$F(\nu; r; \pi_1, \cdots, \pi_r) = \sum_{j=i_r}^{\nu-i_{r-1}^*} \binom{\nu}{j} \pi_r^j(1 - \pi_r)^{\nu-j} F(\nu - j; r - 1; \pi'_1, \cdots, \pi'_{r-1}),$$

where $\pi_j = p_j(p_1 + \cdots + p_r)^{-1}$, for $1 \le j \le r$, and $\pi_j' = p_j(p_1 + \cdots + p_{r-1})^{-1}$, for $1 \le j \le r-1$. The array for $r = 2$, which initializes the recurrence, is easily computed by the formula

$$F(\nu;\, 2;\, \pi_1,\pi_2) = \sum_{j=i_1}^{\nu-i_2} \binom{\nu}{j} \pi_1^j (1-\pi_1)^{\nu-j}, \quad \text{for } i_2^* \le \nu \le n^* - i_m^* + i_2^*.$$

At each iteration, the necessary binomial coefficients are efficiently computed by the recurrence relations discussed in Chapter 1. These coefficients are computed and stored at the beginning of each step in the recursion.

Trimming: Without significant loss of accuracy, we can save much processing time by the trimming discussed in Section 3.2 for repeated convolutions. The elements in the initial arrays $\{F(\nu;\, r;\, \pi_1, \cdots, \pi_r)\}$ very rapidly tend to 1. As soon as the terms exceed, say, $1 - 10^{-9}$, we fill the rest of the array with ones. Computational runs, with and without trimming, have shown that this has an insignificant effect on the computed probabilities, while reducing the processing time by as much as 50%.

Reference: Carson, C. C. and Neuts, M. F., "Some computational problems related to multinomial trials," *Canadian Journal of Statistics,* 3, 235 - 248, 1975.

Remark *a.* It is clear that the proposed algorithm will not be efficient if some of the parameters p_j are very small, or some of the i_j large. The easily computed quantity x^* gives an idea of the required computational effort. The number m of alternatives has a moderate effect, since the computational work grows linearly in m. We have implemented the algorithm for m as large as 30 and for cases where n^* was up to 500, with only a modest computation time.

Remark *b.* The recursive scheme is easily modified to handle the case where some parameters i_j are allowed to be zero. That modification is the subject of Problem 3.6.48. The number of iterative steps is then one more than the number of non-zero i_js.

3.6. MAIN PROBLEM SET FOR CHAPTER 3

The following problems deal with derived distributions and other probabilities related to a variety of models. Each problem requires prior analysis to set up appropriate recurrence relations, to obtain expressions for the desired quantities, and, occasionally, equations to be solved numerically. The skills developed in Chapters 1 and 2 are now to be used in settings of greater complexity.

EASIER PROBLEMS

Problem 3.6.1: X and Y are independent Poisson-distributed random variables with means λ and μ. Compute the (discrete) density, the mean, and the variance of the random variable $Z = X^2(X^2 + Y^2 + 1)^{-1}$. Execute your code for $\lambda = 10$ and $\mu = 20$.

Problem 3.6.2: X and Y are independent, Poisson-distributed random variables with parameters λ and μ, respectively. As efficiently as possible, evaluate the probability that

$$\frac{X^2}{a^2} + \frac{Y^2}{b^2} \leq 1.$$

Implement your code for

$$\lambda = 4, \quad \mu = 8, \quad a = 10, \quad b = 8,$$

and

$$\lambda = 12, \quad \mu = 10, \quad a = 8, \quad b = 2.$$

Problem 3.6.3: The random variables $X_1, \cdots, X_n$ are independent, Poisson-distributed with means $\lambda_1, \cdots, \lambda_n$, and $Y = \max\{X_1, \cdots, X_n\}$.

a. Show that

$$P\{Y \leq k\} = \prod_{j=1}^{n} \left[\sum_{i=0}^{k} e^{-\lambda_j} \frac{\lambda_j^i}{i!}\right], \quad \text{for} \quad k \geq 0,$$

$$E(Y) = \sum_{k=0}^{\infty} \left\{1 - \prod_{j=1}^{n} \left[\sum_{i=0}^{k} e^{-\lambda_j} \frac{\lambda_j^i}{i!}\right]\right\}, \quad \text{for} \quad k \geq 0.$$

b. For given values of $\lambda_1, \cdots, \lambda_n$, develop as efficient an algorithm as possible to compute the probability density of Y and the mean $E(Y)$. The values of the λ_j can be as large as 75, so use special care in computing the Poisson probabilities.

c. Use your code to compute the density of Y for $n = 50$ and $\lambda_j = 24 + j$. Discuss and implement tests to validate your numerical results.

Problem 3.6.4: Under the same assumptions as in Problem 3.6.3, develop an algorithm to compute the probability density and the mean of the random variable $Z = \min\{X_1, \cdots, X_n\}$. Implement your program for the same parameter values.

Problem 3.6.5: Write an efficient computer program to tabulate the probability distribution $G(x)$ of $Y = AX^2 + BX + C$, where X has a normal distribution with mean μ and variance σ^2. The parameters A, B, C, μ, and σ^2 are data. Your program should be able to handle all parameter values in the range $0 < A < 10^3$, $|B| < 10^4$, $|C| < 10^4$, $\mu < 10^3$, $\sigma^2 < 10^4$. The table should list the value of $G(x)$ at 201 equidistant points $x_0 < x_1 < \cdots < x_{200}$, where x_0 is the minimum of the quadratic polynomial and x_{200} is chosen such that $P\{Y \le x_{200}\} = 0.9999$. Test your code against the explicitly computable distributions of $Y = X^2$ and $Y = (X - 2)^2 + 5$, with $\mu = 0$ and $\sigma^2 = 1$.

Hint: The minimum of $Au^2 + Bu + C$ is attained for $u = -B(2A)^{-1}$ and is equal to $x_0 = (4AC - B^2)(4A)^{-1}$. For $x > x_0$, the equation $Au^2 + Bu + C = x$ has two real roots $u_1(x) < u_2(x)$. The distribution $G(x)$ is given by

$$G(x) = P\{Y \le x\} = P\{u_1(x) \le x \le u_2(x)\} = \Phi\left[\frac{u_2(x) - \mu}{\sigma}\right] - \Phi\left[\frac{u_1(x) - \mu}{\sigma}\right].$$

To find the value of x_{200} we solve the equation

$$\Phi\left[\frac{u_2(x) - \mu}{\sigma}\right] - \Phi\left[\frac{u_1(x) - \mu}{\sigma}\right] = 0.9999.$$

With x_0 and x_{200} known, the remaining steps are routine.

Problem 3.6.6: X has a uniform distribution on [0, 1]. Express the probability distribution of $Y = 6X^3 - 9X^2 + 4X$ in terms of the roots of a polynomial equation. If $G(x) = P\{Y \le x\}$, then evaluate $G(x_j)$ for $u_2 = x_0 <$

$x_1 < \cdots x_{100} = u_1$. The x_js are 101 equally spaced points. The values u_1 and u_2 are respectively the maximum and minimum of $6x^3 - 9x^2 + 4x$ in [0, 1].

Hint: Draw a sketch of the graph of $f(x) = 6x^3 - 9x^2 + 4x$ on [0, 1]. For each x_j, $G(x_j)$ is readily expressed in terms of the roots of $f(u) = x_j$. Efficiently compute those roots.

Problem 3.6.7: The random variable X is uniformly distributed on $[0, 2\pi)$. Compute a table of the distribution $F(x)$ of $Y = f(X) = 2\sin^2 X + \sin^2(2X)$. Y takes values in an interval of the form $[0, A]$, where A is to be computed. Print the entries in the table at $x_k = k(A/100)$, for $k = 0, \cdots, 100$.

Hint: Carefully determine all critical points of $f(x)$ in $[0, 2\pi)$. Their locations can be found explicitly.

Problem 3.6.8: The probability densities $\{p_k\}$ and $\{r_k\}$ concentrate on the integers $M_1, \cdots, N_1$ and $M_2, \cdots, N_1$, respectively. The $N_1 - M_1 + 1$ terms of the first density are stored in the consecutive locations, starting at the index J_1 of an array. The second density starts at the location J_2 of a different array. Write and test a general purpose subroutine to compute the convolution of the two densities. A good test is obtained by choosing both densities to be uniform. For that case, it is easy to figure out, by a simple count, what the convolution should be.

Problem 3.6.9: In forming the convolution of a density $\{p_k\}$ on the integers M to N with itself, one can reduce the number of multiplications to roughly one half of that required by a straightforward coding of the defining formula. However, some extra care is needed to handle the indices. Once you see how to do that, write and test a general purpose subroutine to perform that convolution. Use that subroutine in problems calling for the m-fold convolution of a density on a set of integers.

Problem 3.6.10: The random variables X_i, $1 \le i \le m$, are independent, identically distributed, and take values in the set of integers $-k \le \nu \le +k$. Their probability density $\{p_\nu\}$ is given by

$$p_\nu = p_{-\nu} = \frac{|\nu|}{k(k+1)}, \quad \text{for } 1 \le \nu \le k,$$

and $p_\nu = 0$ elsewhere. S_m is the sum $X_1 + \cdots + X_m$.

a. Write an efficient code to compute the probabilities $P\{S_m = r\}$, $0 \le r \le mk$, for a given number m of summands.

b. Verify that the variance σ^2 of the density $\{p_n\}$ is given by

$$\sigma^2 = \frac{2}{k(k+1)} \sum_{\nu=1}^{k} \nu^3 = \frac{1}{2} k(k+1).$$

c For $x_j = 0.25j$, $0 \le j \le 20$, compute the exact probabilities

$$\theta_j = P\{x_{j-1} < \frac{S_m}{\sigma\sqrt{m}} \le x_j\},$$

and compare the θ_j to the approximate values obtained by applying the central limit theorem with continuity correction. Execute your program for $k = 6$, $m = 10$ and also for $k = 6$, $m = 50$.

Problem 3.6.11: In n independent Bernoulli trials the success probability at the jth trial is p_j. For a given n and an array $\{p_j\}$ compute a table of the probabilities r_k, $0 \le k \le n$, that exactly k successes occur.

a. Use your program to compute the probabilities r_k, $0 \le k \le 40$, for the case where $n = 1{,}000$ and $p_j = 1/j$. For that particular case, the number of successes Y_n is asymptotically normal with mean and variance $\log n$.

b. Compare the exact probabilities $P\{Y_n \ge k\}$ to the values obtained from the normal approximation with continuity correction.

c. Compare the exact probabilities $P\{Y_n \ge k\}$ to the corresponding Poisson probabilities with parameter $\lambda = \log n$. See Problem 3.6.12.

Remark: If the random variables $X_1, \cdots, X_n$ are independent with a common *continuous* distribution $F(\cdot)$, then Y_n has the same distribution as the number of records among $X_1, \cdots, X_n$. X_j is a record if it exceeds the preceding $j - 1$ observations.

Problem 3.6.12: The General Binomial Density: Consider n independent Bernoulli trials with success probability p_i at the ith trial. For a given array $\{p_i\}$ of length n, develop a code to compute the probability $r_k(n)$ of k successes in the n trials. Set up a recursion for the arrays $\{r_k(\nu)\}$, for $1 \le$

$\nu \le n$. Your code should be able to handle values of n up to 1,000. Implement your code for $p_i = i^{-1}$, for $i \ge 1$. For that case,

$$P\left\{\frac{Y_n - \log n}{(\log n)^{1/2}} \le x\right\} \to \Phi(x), \qquad \text{as } n \to \infty.$$

$\Phi(\cdot)$ is the standard normal distribution. Compare the quantities r_k to their normal approximation with continuity correction. What do you think of the quality of that approximation?

Problem 3.6.13: In a sequence of independent Bernoulli trials the success probability at the jth trial is p_j. For a given array $\{p_j\}$, develop and code an algorithm to compute a table of the probabilities, $P_k(\nu)$, that the kth success occurs at the νth trial. For $p_j = 1/j$, compute the first 10,000 positive terms of the sequence $P_3(\nu)$.

Problem 3.6.14: X and Y are independent, integer-valued and

$$P\{X = k\} = p_k, \quad \text{for } 1 \le k \le M, \qquad \text{and} \qquad P\{Y = k\} = r_k, \quad \text{for } 1 \le k \le N,$$

a. Show that for $1 - N \le k \le M - 1$, $s_k = P\{X - Y = k\}$ is given by

$$s_k = \sum_{i = L(k)}^{K(k)} p_{k+i} r_i,$$

where $L(k) = \max(1, 1 - k)$ and $K(k) = \min(N, M - k)$.

b. Write a code to compute the probability density $\{s_k\}$. Implement it for the binomial densities with parameters $p(1) = 0.75$, $n(1) = 10$, and $p(2) = 0.80$, $n(2) = 15$. Be careful in handling the parameters M and N.

Problem 3.6.15: Geometric mixtures of the successive convolutions of a probability density $\{r_\nu\}$ arise naturally in many applications. If $R(z)$ is the generating function of $\{r_\nu\}$, the generating function of such a mixture is given by $\Phi(z) = p[1 - qR(z)]^{-1}$, where $0 < p = 1 - q \le 1$. Show that by redefining p we may, without loss of generality, assume that $r_0 = 0$. In what follows, consider that done.

a. By expanding both sides of the equation

$$\Phi(z) = p + qR(z)\Phi(z),$$

in power series, show that the density $\{\phi_n\}$ with generating function $\Phi(z)$ can be recursively evaluated by

$$\phi_0 = p, \qquad \phi_n = q\sum_{k=1}^{n} r_k \phi_{n-k}, \quad \text{for } n \geq 1.$$

b. For $\lambda = 2$ and $p = 1/3$, compute the probabilities

$$\phi_n = \sum_{\nu=0}^{\infty} pq^{\nu} e^{-\lambda\nu} \frac{(\lambda\nu)^n}{n!}, \quad \text{for } n \geq 0.$$

as efficiently as possible.

Problem 3.6.16: The side AB of a rectangular triangle has unit length. The length of the hypotenuse BC is $1 + X$. X has a gamma distribution:

$$P\{X \leq x\} = \frac{1}{\Gamma(\alpha+1)} \int_0^{\lambda x} e^{-u} u^{\alpha+1}\, du, \quad \text{for } x \geq 0.$$

Tabulate the probability distribution $G(\cdot)$ of the area Y of the triangle ABC, for $\lambda = 0.5$ and $\alpha = 4$. Do this by evaluating $G(0.01k)$, for $0 \leq k \leq k^*$, where k^* is the smallest integer for which $P\{Y > 0.01k^*\} < 10^{-6}$.

Problem 3.6.17: For a standard deck of 52 cards we assign a value of 4 points to an Ace, 3 to a King, 2 to a Queen, and 1 to a Jack. A hand consists of five randomly drawn cards. The most valuable hand therefore contains 19 points (four Aces and one King).

A set of n, $5 \leq n \leq 52$, cards is drawn at random and without replacement. By P_n, we denote the probability that this set of n cards contains four Aces and at least one King. Q_{nj} is the probability that the set contains the four Aces and j Kings, $1 \leq j \leq 4$, and R_{nm} is the conditional probability that the set of n cards contains a total of m points, given that it contains at least one most valuable hand of five cards.

Set up explicit or recursion formulas to evaluate the quantities P_n, Q_{nj}, and R_{nm} for all indices for which they are positive. Write a computer program to implement these formulas. The derivations for the first two quantities are elementary, but the R_{nm}s require a careful consideration of cases.

Problem 3.6.18: Consider a game played with a standard deck of 52 cards. Assume that m persons, numbered $1, \cdots, m$, successively turn over one card at a time. The first person to turn over an ace wins. $P_1(j, m)$ is the probability that person j wins.

$P_1(j, m)$ decreases in j for every m. Persons with a higher number are therefore at a disadvantage. The game can be modified so that the person who turns over the ith ace wins, where $i =$ 1, 2, 3, or 4. Denote the corresponding probabilities that player j wins by $P_i(j, m)$. Compute a table of the probabilities $P_i(j, m)$ for $i = 1, 2, 3, 4$, $1 \leq j \leq m$ and all appropriate values of m.

For each case, which player is the most likely winner? Print the number of that player and his corresponding probability of winning.

It is suggested that all players have the same probability of winning if one of the four possible values of i is picked at random and the m players play the game in which the one to draw the ith ace wins. Check numerically whether that assertion is valid. Do so by examining whether for all appropriate j, the probabilities $(1/4)\sum_{r=1_4} P_r(j, m)$ are within 10^{-9} of each other.

AVERAGE PROBLEMS

Problem 3.6.19: If the random variable X has a uniform distribution on [0, 1], the probability distribution $F(x)$ of $Y = 256(X - 1/2)^4 - 32(X - 1/2)^2 + 1$ is

$$
\begin{aligned}
F(x) &= 0, && \text{for } x \leq 0, \\
&= \frac{1}{2}[(1 + x^{1/2})^{1/2} - (1 - x^{1/2})^{1/2}], && \text{for } 0 \leq x \leq 1, \\
&= \frac{1}{2}(1 + x^{1/2})^{1/2}, && \text{for } 1 \leq x \leq 9, \\
&= 1, && \text{for } x \geq 9.
\end{aligned}
$$

Verify that formula and also show directly from the definition of Y that $E(Y) = 23/15$. Next, show that

$$E(Y) = \int_0^9 [1 - F(x)]\, dx,$$

and use Simpson's rule to evaluate that integral numerically.

Problem 3.6.20: The random variables $X_1, \cdots, X_n$ are independent, Poisson-distributed with mean λ. The random variable J_n is defined by

$$J_n = \min_{1 \le \nu \le n} \left\{ \nu : X_\nu = \max\{X_1, \cdots, X_n\} \right\}.$$

J_n is the index for which the largest observation first occurs.

a. Show that for $1 \le j \le n$,

$$P\{J_n = j\} = \delta_{1j} e^{-n\lambda} + \sum_{k=1}^{\infty} e^{-\lambda} \frac{\lambda^k}{k!} \left[\sum_{r=0}^{k-1} e^{-\lambda} \frac{\lambda^r}{r!} \right]^{j-1} \left[\sum_{\nu=0}^{k} e^{-\lambda} \frac{\lambda^\nu}{\nu!} \right]^{n-j}.$$

b. Write a program to evaluate the probabilities $P\{J_n = j\}$ for $1 \le j \le n$ to within 10^{-8}, for given values of n and λ, where λ may be up to 50 and n up to 200.

c. Execute your program for $\lambda = 10$ and for $n = 20k$, $1 \le k \le 10$. Do your numerical results suggest something interesting about the growth of the successive maxima for independent, identically distributed Poisson random variables? If you offer qualitative conclusions, examine them for at least five more data sets.

Remark: When the X_r have a common *continuous* distribution, the density of the corresponding random variable J_n is *uniform* on $1, \cdots, n$. (Prove this.) For discrete random variables, that is, in general, not the case. The reader is encouraged to examine the density of J_n for distributions other than Poisson.

Problem 3.6.21: This problem deals with a procedure to compute the distribution of the *range* of m independent, identically distributed, *discrete* random variables $X_1, \cdots, X_m$. These take values $x_1 < x_2 < \cdots < x_n$, with the corresponding probabilities $p_1 < p_2 < \cdots < p_n$. Without loss of generality, assume that $x_1 \geq 0$. The p_j are positive and sum to one. The range R of $X_1, \cdots, X_m$ is defined by

$$R = \max\{X_1, \cdots, X_m\} - \min\{X_1, \cdots, X_m\}.$$

It takes values in the set $\{0, x_j - x_i, \ i > j\}$. The distribution $F(x)$ of the X_j is given by

$$F(x) = \sum_{\{i : x_i \leq x\}} p_i .$$

Show that $P\{R = 0\} = \sum_{i=1}^{n} p_i^m$.

For $i < j$, show that the probability $w(x_i, x_j)$ that the minimum X_1^* and the maximum X_m^* of the m random variables take on the values x_i and x_j is given by

$$w(x_i, x_j) = [F(x_j) - F(x_{i-1})]^m - [F(x_j) - F(x_i)]^m$$

$$- [F(x_{j-1}) - F(x_{i-1})]^m + [F(x_{j-1}) - F(x_i)]^m .$$

For $i = 1$, the terms with subscript $i - 1$ are omitted. The numbers in the set $\{0, x_j - x_i, \ i > j\}$ are in general not all distinct. If a is any number in that set, the probability $P\{R = a\}$ is obtained by summing the quantities $w(x_i, x_j)$ over all pairs (i, j) for which $x_j - x_i = a$.

Write an efficient routine to form two lists, one of the numbers in the set $\{x_j - x_i, i > j\}$, and the other of the corresponding probabilities $w(x_i, x_j)$. Do not recompute repeated terms in the formulas for $w(x_i, x_j)$. Sort both lists so that the first is increasing and the second contains the corresponding probabilities. Because of the finite precision of the computer, some of the values of $x_j - x_i$ may be mathematically equal yet represented by different numbers in the computer. Appropriately defining machine equality, find terms that should be lumped to obtain a nice printout of the computed discrete density and distribution of R. For $m = 50$, generate a random data set by choosing the numbers x_j as independent exponential variates with an exponential distribution of mean 10. Partition the interval (0, 1) into 50 random portions by choosing 49 uniform random numbers. Use the interval lengths as the parameters p_j.

Remark: In testing an algorithm, it is often easier to generate data randomly than to enter long lists of numbers.

Problem 3.6.22: For the random variables defined in Problem 3.6.21, compute the distribution of the *mid-range*

$$Y = \frac{1}{2}\left[\max\{X_1, \cdots, X_m\} + \min\{X_1, \cdots, X_m\}\right].$$

Write the appropriate code as a subroutine that can be appended to the program for Problem 3.6.21.

Problem 3.6.23: Bold Play: A person desperately needs N dollars, but currently has only $k < N$ dollars. The person decides to play a game of chance that offers a probability p of doubling the ante and a probability $q = 1 - p$, of losing it. He decides to resort to *bold play,* that is, at every play he either bets his entire current capital or the sum needed to reach N dollars should he win that play. Only integer-valued bets are allowed. Successive games are independent trials. The probability p remains the same throughout and does not depend on the amount wagered. Betting continues until either the gambler is ruined or the amount N is reached. $P(k; N)$ is the probability that the amount N is reached.

For $p = 0.46 + 0.01i$, $1 \le i \le 7$, and $51 \le N \le 100$, compute $P(k; N)$ for $1 \le k \le N - 1$. Draw informative graphs of the computed $P(k; N)$. Discuss their implications.

Problem 3.6.24: A Two-Stage Experiment: In the first stage of an experiment, Bernoulli trials with probability p of success are performed until m successes have occurred. X is the number of trials required in that stage. In the second stage, X more Bernoulli trials are performed, also with probability p. Y is the number of successes in these.

a. Show that the joint probability density $P(n, k)$ of X and Y is given by

$$P(n, k) = P\{X = n, Y = k\} = \binom{n}{k}\binom{n-1}{m-1} p^{m+k}(1-p)^{2n-m-k},$$

for $0 \le k \le n$, for every $n \ge m$. $P(n, k) = 0$ for all other pairs of integers.

b. Compute a table of the probabilities $\{P(n, k)\}$, for all $m \le n \le n_0$, where n_0 is such that the sum of the terms not computed is at most 10^{-7}. Print the probabilities $P(n, k)$, $0 \le k \le n$, in separate tables for successive values of n. Print five entries per line, along with the appropriate indices n and k.

c. For $m = 4$, $p = 0.65$, and $j = 7$, compute and print a table of the conditional probabilities $P\{X = n \mid Y = j\}$, for $n \ge \max(j, m)$. Compute enough terms of that conditional density to ensure that their sum exceeds $1 - 10^{-5}$. Also print the corresponding conditional distribution $P\{X \le n \mid Y = j\}$.

d. Compute and print tables of the density and distribution of $X + Y$. The sum of the computed terms of that density should exceed $1 - 10^{-5}$.

e. Compute the density and distribution of $Z = \max(X, Y)$. The sum of the computed terms of that density should exceed $1 - 10^{-5}$.

f. Compute the means $E(X)$ and $E(Y)$, the variances of X and Y, as well as the covariance and the correlation coefficient of X and Y.

Hint: The joint density concentrates on the lattice points of an infinite wedge-shaped region. It is helpful to sketch that region, as well as the lattice points corresponding to the events related to $X + Y$ and $\max(X, Y)$. By exploiting recurrence relations, the program can be made efficient.

Problem 3.6.25: X has an exponential distribution with parameter λ. Show that the probability distribution $F(x)$ of $Y = \cos X$ is given by:

$$F(x) = 0, \quad \text{for } x < -1,$$

$$= \frac{\exp(-\lambda \arccos x) - \exp(\lambda \arccos x - 2\lambda\pi)}{1 - e^{-2\lambda\pi}}, \quad \text{for } -1 \le x \le 1,$$

$$= 1, \quad \text{for } x \ge 1.$$

a. Compute a table of $F(x)$ for $x = 0.01k$, for $-100 \le k \le 100$.

b. Show that

$$E(\cos X) = 1 - \int_{-1}^{1} F(x)\, dx.$$

Using that formula, compute $E(\cos X)$ by numerical integration.

c. Use the characteristic function of the exponential distribution to show that

$$E(\cos X) = \mathrm{Re}\left[\frac{\lambda}{\lambda - i}\right] = \frac{\lambda^2}{\lambda^2 + 1}.$$

Compare the exact value of $E(\cos X)$ to that obtained from the numerical integration. Do so for $\lambda = 1, 2, 3, 4, 5$.

Problem 3.6.26: Let $P(z)$ be the probability generating function of the density $p(j)$ on the nonnegative integers. In a number of applications, we need to compute the probability density $\{\theta(j)\}$ with generating function

$$\Theta(z) = \exp\{\lambda[1 - P(z)]\}.$$

a. Show that $\{\theta(j)\}$ is the convolution of an infinite sequence of Poisson densities $\{\pi_r(\nu)\}$, where the rth density places Poisson probability masses at the points kr, $k \ge 0$. Calculate the means of the densities $\{\pi_r(\nu)\}$.

b. The mean ψ of $\{\theta(j)\}$ is $\lambda\eta$, where $\eta = P'(1)$. We can compute $\{\theta(j)\}$ by successive convolution products of the $\{\pi_r(\nu)\}$. We stop at the index ν^*, chosen so that the sum of the means of all neglected sequences is less than 0.001ψ. Find ν^* numerically for given λ and $\{p(j)\}$. Write and test a code to compute $\{\theta(j)\}$ by successive convolutions.

c. Suppose that the number of offspring of one individual of a species has a Poisson distribution with parameter λ. Each of the descendants again produces independent numbers of offspring according to Poisson distributions with the same λ. K is the number of second-generation offspring of the original individual. For a given value of λ, determine integers $n_1(\lambda)$ and

$n_2(\lambda)$ such that

$$P\{n_1(\lambda) \leq K \leq n_2(\lambda)\} \geq 0.90,$$

and the difference $n_2(\lambda) - n_1(\lambda)$ is as small as possible. Plot graphs of the integers $n_1(\lambda)$ and $n_2(\lambda)$ for $\lambda = 2 + 0.25k$, for $k = 0, \cdots, 10$.

Problem 3.6.27: A recurrence relation for the terms of the probability density $\{\theta(j)\}$ in Problem 3.6.26 is derived as follows: By differentiation in the definition of $\Theta(z)$, we get that $\Theta'(z) = \lambda P'(z)\Theta(z)$.

Equate the coefficients of z^j in that equality, to obtain the recursion

$$\theta_0 = e^{-\lambda(1 - p_0)}, \qquad \theta_{j+1} = \lambda \sum_{r=0}^{j} \left[1 - \frac{r}{j+1}\right] p_{j+1-r}\,\theta_r, \quad \text{for } j \geq 0.$$

Use that recurrence as another approach to solving Problem 3.6.26.

Reference: M. Pourahmadi "Taylor expansion of $\exp(\sum_{k=0}^{\infty} a_k z^k)$ and some applications." *American Mathematical Monthly,* 91, 303 - 308, 1984.

Problem 3.6.28: A person does up to $n \leq 15$ independent Bernoulli trials with probability p. A success yields a reward of $a > 0$ units; a failure entails a loss of $b > 0$. Let c be a number satisfying $0 < c \leq na$. For convenience, a, b, and c are integers. After the kth trial, the person's total gain is $V_k = \nu a - (k - \nu)b$, where ν is the number of successes in the first k trials. The person decides to stop at the first index k for which $V_k \geq c$. If there is such an index $k \leq n$, we set the *stopping time* T equal to k.

a. Identify the sample space with the set of strings of n symbols, with 1 or 0, standing for success or failure. For each such string compute the successive values of V_r, $1 \leq r \leq n$. Label the string by k, if and only if $V_r \geq c$ for the first time when $r = k$. Next, compute and print tables of the probabilities $P\{T = k\}$, for $1 \leq k \leq n$, and of the conditional probabilities $P\{V_k = j \mid T = k\}$.

b. The procedure in part a is clearly inefficient. It requires more storage than is needed and some unnecessary computation. Modify the procedure by systematically generating the strings of zeros and ones without storing them. As each string is generated, compute and store only the essential information to compute the required probabilities. Implement both procedures for $n = 15$, $p = 0.2$, $a = 20$, $b = 5$, and $c = 10j$, $1 \le j \le 80$.

Remark: Solving probability problems by complete enumeration of the sample space is rarely of practical interest. The limitations inherent in this problem make that clear. Yet, the problem has some interesting algorithmic aspects related to sample path behavior and to efficient enumeration.

Problem 3.6.29: Suppose the person in Problem 3.6.28 also places a lower bound on the loss. He will also stop if for some integer d, $0 > d \ge -nb$, there is a first integer k' such that $V_{k'} \le d$. Adapt the procedure in part b of Problem 3.6.28 to that case. Starting from the data given there, select some interesting values of d for numerical implementation. Offer qualitative interpretations for your numerical results by imagining a gambler using such a game plan in a casino. Is the gambler's objective of winning c dollars likely to be met?

HARDER PROBLEMS

Problem 3.6.30: Consider independent sequences $\{X_n\}$ and $\{Y_n\}$ of independent Poisson random variables. All X_js have the common mean λ; the Y_js all have mean μ. We say that the sequence $\{X_n\}$ *leads at time* k if and only if $\max\{X_1, \cdots, X_k\} > \max\{Y_1, \cdots, Y_k\}$. If the reversed inequality holds, we say that $\{Y_n\}$ leads. If both maxima are equal, the sequences are tied at time k. The probabilities of these three events are $P(k)$, $Q(k)$, and $R(k)$. Obviously, their sum is one.

a. Show that

$$P(k) = \sum_{\nu=1}^{\infty} [G^k(\nu) - G^k(\nu - 1)]H^k(\nu - 1),$$

$$Q(k) = \sum_{\nu=1}^{\infty} [H^k(\nu) - H^k(\nu - 1)]G^k(\nu - 1),$$

$$R(k) = G^k(0)H^k(0) + \sum_{\nu=1}^{\infty} [G^k(\nu) - G^k(\nu - 1)][H^k(\nu) - H^k(\nu - 1)],$$

for $k \ge 1$, where for $\nu \ge 0$, $G(\cdot)$, and $H(\cdot)$ are defined by

$$G(\nu) = \sum_{r=0}^{\nu} e^{-\lambda} \frac{\lambda^r}{r!}, \quad \text{and} \quad H(\nu) = \sum_{r=0}^{\nu} e^{-\mu} \frac{\mu^r}{r!}.$$

b. With full attention to numerical accuracy, write a program to evaluate the probabilities $P(k)$, $Q(k)$, and $R(k)$ for k up to 5,000. Execute your program for $\lambda = 5$, $\mu = 6$ and also for $\lambda = \mu = 10$. Draw informative plots of the probabilities $P(k)$ and $R(k)$.

Problem 3.6.31: X has a uniform distribution of [0, 1]. Write a computer program to evaluate the distribution $F(\cdot)$ of $Y = \sum_{j=0}^{n} a_j X^j$, where the coefficients a_j, $1 \le j \le n$, are given nonnegative numbers. The coefficient a_0 is real. Evaluate $F(x)$ at 101 equidistant points x_k, such that $a_0 = x_0 < x_1 < \cdots < x_{100} = \sum_{j=0}^{n} a_j$.

For each of the polynomials

$$Y = 1 + 3X + 3X^2 + X^3,$$

$$Y = 1 + 2X^2 + X^4,$$

$$Y = 4 + 4X + X^2,$$

derive an explicit formula for $F(\cdot)$. For each of these test cases, compare your numerical results to the exact distributions.

Hint: The given polynomial is nondecreasing in X. Use Newton's method efficiently to solve any equations that may arise in your numerical procedure. The explicit distributions for the test cases are easily derived. For the first example,

$$F(x) = P\{(1 + X)^3 \le x\} = P\{X \le x^{1/3} - 1\} = x^{1/3} - 1, \quad \text{for } 1 \le x \le 8.$$

Problem 3.6.32: The positive-integer-valued random variables $X_1, \cdots, X_n$ are independent and identically distributed. Their probability density is $\{p_k\}$. $\alpha > 1$ is a given constant. The minimum of $X_1, \cdots, X_n$ is denoted by X^*. N is the number of indices j among $1, \cdots, n$ for which $X_j \le \alpha X^*$. Obviously, $1 \le N \le n$. Set $r_j = P\{N = j\}$. Show that for $1 \le j \le n$,

$$r_j = \sum_{m=1}^{\infty} \binom{n}{j} [(p_m + \cdots p_{m^*})^j - (p_{m+1} + \cdots p_{m^*})^j][1 - \sum_{\nu=1}^{m^*} p_\nu]^{n-j},$$

where $m^* = [\alpha m]$, the integer part of αm. Write an efficient program to evaluate the probabilities $\{r_j\}$ for an arbitrary given density $\{p_k\}$. Execute your program for the following $\{p_k\}$ and with the specified data.

a. $\{p_k\}$ is the Poisson density with $\lambda = 10$, but shifted to the positive integers, i.e.,

$$p_k = e^{-\lambda}\frac{\lambda^{k-1}}{(k-1)!}, \quad \text{for} \quad k \geq 1.$$

In addition, $n = 20$ and $\alpha = 2.5$.

b. $\{p_k\}$ is the binomial density on $1, \cdots, M$, i.e.,

$$p_k = \binom{M}{k-1} p^{k-1}(1-p)^{M-k+1}, \quad \text{for} \quad 1 \leq k \leq M+1,$$

with $M = 30$ and $p = 0.75$; $n = 20$ and $\alpha = 2$.

Problem 3.6.33: The integer-valued random variables $X_1, \cdots, X_{2n+1}$ are independent and identically distributed with common density $\{p_k\}$. For every outcome ω, the integers $X_1(\omega), \cdots, X_{2n+1}(\omega)$ are ranked in increasing order of magnitude. The $(n+1)$st integer $Y(\omega)$ in that ranking is the *median.* Note that the median may be attained by more than one of the $X_1, \cdots, X_{2n+1}$. The random variable K is the number of $X_1, \cdots, X_{2n+1}$ that are equal to the median Y.

a. Develop an efficient algorithm to compute the probabilities $r(j;k) = P\{K = j, Y = k\}$, and the mean and variance of K and Y. Also compute the marginal densities of K and Y.

b. Examine the probability densities of K and Y numerically for 15 Poisson random variables X_i with $\lambda = 12$.

c. Execute your program for 17 binomial random variables X_i with

$$p_k = \binom{m}{k} p^k (1-p)^{m-k}, \quad \text{for} \quad 0 \leq k \leq m,$$

and $p = 0.75$, $m = 20$.

The next problem illustrates the importance of exploring alternative approaches to a solution. We first exhibit a recursion formula that is somewhat delicate to program. Next, we present a different solution. The reader is asked to consider the merits of each solution.

Problem 3.6.34: A box initially contains n identical, unmarked items. We successively draw out random samples of size $m \leq n$. The items drawn are marked by Bernoulli trials of probability p. All m items are returned to the box. Once an item is marked, it remains marked. We are interested in the probabilities $P_j(k) = P\{X_j = k\}$ that, after the jth set of Bernoulli trials, k items are marked.

a. If T is the time until all items are marked, show that $P\{T \leq j\} = P_j(n)$.

b. Show that for $j \geq 0$ and $0 \leq k \leq n$, the recurrence relation

$$P_{j+1}(k) = \sum_{\nu=0}^{k} P_j(\nu) \sum_{r=L_\nu}^{n-\nu} \frac{\binom{n-\nu}{r}\binom{\nu}{m-r}}{\binom{n}{m}} \binom{r}{k-\nu} p^{k-\nu}(1-p)^{r-k+\nu}$$

holds. The lower summation index $L(k) = \max(k - \nu, m - \nu)$.

c. For a different solution, consider an arbitrary item. For that item to become marked at a given stage, the item must belong to the sample and be marked. Show that therefore the number of trials until that item becomes marked has the geometric density

$$\phi_j = \left(1 - \frac{mp}{n}\right)^{j-1} \frac{mp}{n}, \qquad \text{for } j \geq 1.$$

Deduce further that

$$P_j(k) = \binom{n}{k}\left[1 - \left(1 - \frac{mp}{n}\right)^j\right]^k \left(1 - \frac{mp}{n}\right)^{j(n-k)},$$

which are clearly binomial probabilities. Discuss the care needed in implementing both algorithms for the $P_j(k)$. Do not write a code for the recursion in part *b*.

d. Code the formulas in *c*. Exercise due care in evaluating the binomial probabilities. Start from the modal term and use logarithms where needed.

e. For $j \geq 1$, $F(j) = P\{T \leq j\}$ is given by

$$F(j) = \left[1 - \left(1 - \frac{mp}{n}\right)^j\right]^n .$$

Clearly for large n or small mp, those probabilities increase very slowly in j. For such cases, it is useful to have an estimate of the 95th (or another high) percentile of $F(\cdot)$. Use the logarithmic series to show that, for α close to 1, the α-quantile ψ_α is approximately

$$\psi_\alpha = \frac{\log\,[-\log \alpha] - \log n}{\log\,(1 - \frac{mp}{n})} .$$

f. For the data $n = 100$, $m = 10$, $p = 0.75$, first find the approximate value of the 95th percentile. Compute up to 5,000 terms of the probability distribution $F(\cdot)$ or until the 95th percentile is reached. For 10 large values of j, compute also the probabilities $P_j(k)$. Select these js so that the numbers obtained are informative.

Problem 3.6.35: A Selection Problem for Poisson Random Variables: $X_1, \cdots, X_{n+1}$ are independent and Poisson-distributed. It is known that n of these random variables have the same mean λ. The remaining one has a smaller mean $\lambda_1 = \delta\lambda$, $0 < \delta < 1$. We are interested in identifying the random variable with the mean λ_1. A procedure for this is as follows: When observing the values of $X_1, \cdots, X_{n+1}$, record all the indices for which the minimum of the $n + 1$ observations is attained. Choose one of these indices at random and declare that the corresponding random variable has mean λ_1. There is a *correct selection* (CS) if that procedure indeed selects the population with the smallest mean. The probability of a correct selection is denoted by $P(CS;\lambda,n,\delta)$.

a. Show that

$$P(CS;\lambda,n,\delta) = \sum_{\nu=0}^{\infty} e^{-\lambda_1}\frac{\lambda_1^\nu}{\nu!}\sum_{j=0}^{n}\frac{1}{j+1}\binom{n}{j}[e^{-\lambda}\frac{\lambda^\nu}{\nu!}]^j\,[1 - \sum_{r=0}^{\nu} e^{-\lambda}\frac{\lambda^r}{r!}]^{n-j} .$$

b. Write a program to evaluate $P(CS; \lambda, n, \delta)$ for a fixed value of λ and for $1 \le n \le 50$. Set $\delta = 0.9$.

c. Write a program to evaluate $P(CS; \lambda, n, \delta)$ for a fixed value of n and for $\lambda = 0.25k$, with $1 \le k \le 20$. Set $\delta = 0.9$.

d. Select particular values for λ and n and examine the sensitivity of the procedure by varying the value of δ.

Hint: In evaluating the series for $P(CS; \lambda, n, \delta)$ note that each term is bounded above by

$$e^{-\lambda_1}\frac{\lambda_1^{\nu}}{\nu!}\,[1 - \sum_{r=0}^{\nu-1} e^{-\lambda}\frac{\lambda^r}{r!}]^n.$$

The quantity in square brackets decreases as ν increases. If we compute the sum in the formula for $P(CS; \lambda, n, \delta)$ for ν up to ν^*, then the remainder is at most

$$\sum_{\nu=\nu^*+1}^{\infty} e^{-\lambda_1}\frac{\lambda_1^{\nu}}{\nu!}\,[1 - \sum_{r=0}^{\nu-1} e^{-\lambda}\frac{\lambda^r}{r!}]^n,$$

and that quantity is less than

$$[1 - \sum_{\nu=0}^{\nu^*} e^{-\lambda_1}\frac{\lambda_1^{\nu}}{\nu!}][1 - \sum_{r=0}^{\nu^*} e^{-\lambda}\frac{\lambda^r}{r!}]^n = R_{\nu^*}(\lambda).$$

This is not the best possible bound on the remainder of the series, but it is easily computed from intermediate quantities used in evaluating $P(CS; \lambda, n, \delta)$. It should be used to define a stopping rule in the DO-loop to compute $P(CS; \lambda, n, \delta)$.

Problem 3.6.36: There are two groups, I and II, respectively, of m and n items. All items have independent, exponential lifetimes and all are activated at time 0. Items in I have a mean lifetime 1; those in Group II last on average for λ^{-1}. Failed items are not replaced. We denote by T_1 the time of the mth failure in Group I and by T_2, the time of the nth failure in Group II. The probability $P\{T_1 \le T_2\}$ is denoted by $p(m, n)$. It is clearly a function also of λ.

Show by a direct probability argument that for all $m \geq 1$ and $n \geq 1$,

$$p(m, n) = \frac{\lambda m}{\lambda m + n} p(m-1, n) + \frac{n}{\lambda m + n} p(m, n-1),$$

where $p(m, 0) = 0$, for $m \geq 1$ and $p(0, n) = 1$, for $n \geq 1$.

a. Suppose that λ and n are given. Develop an efficient algorithm to compute the largest value of m for which the probability that Group II outlives Group I is at least 0.9. Execute your program for $\lambda = 4$ and $n = 20$.

b. With m and n given, develop an algorithm to compute the value λ^* of λ for which $p(m, n) = 0.5$. By using the bisection method, evaluate λ^* to within 10^{-4}. Execute your program for $m = 10$ and $n = 20$. Test your program for $m = n$ when the answer is obvious.

Problem 3.6.37: Testing a Hypothesis: Our objective is to study a test of hypothesis formalized as follows: We have m coins of which it is claimed that at least one has a higher probability p_1 of coming up heads than the accepted value p_0. For example, think of a group of m putative psychics who claim they can predict the outcome of a roll of a balanced die with higher probability p_1 than $p_0 = 1/6$ achieved by ordinary individuals. We state the null-hypothesis H_0 as "all m coins have probability p_0" and the (simplest) alternate hypothesis H_1 as "one coin has probability $p_1 > p_0$ and all others have probability p_0." The experiment consists in tossing each coin n times. $X_1, \cdots, X_m$ are the numbers of heads scored with the various coins.

We shall *reject* the null-hypothesis H_0, if and only if $\max(X_1, \cdots, X_m) > r$. r is the smallest positive integer for which, if H_0 is valid,

$$P_{H_0}\{\max(X_1, \cdots, X_m) > r\} \leq \alpha.$$

The *size*, α, of the test, is usually chosen to be 0.1, 0.05, or 0.01. Rejecting a valid null-hypothesis is called a *Type I error*. We are choosing r such that the probability of a Type I error is at most α.

If the hypothesis H_1 is valid, it may still occur (due to chance) that $\max(X_1, \cdots, X_m) \leq r$. We then accept the null-hypothesis H_0, even though it is invalid. This is a *Type II error*. $\beta = P_{H_1}\{\max(X_1, \cdots, X_m) \leq r\}$ is the probability of a Type II error.

a. For given m, n, p_0, $p_1 > p_0$, and α, determine r most efficiently. For that value of r, evaluate β. Plot β as a function of p_1 on $[p_0, 1]$ and explain the significance of that graph.

b. Whereas α can be chosen by the experimenter, β can be reduced only by increasing the number n of tosses performed with each coin. Supposing that both α and β are specified, discuss an algorithm to find r first and then the smallest n for which the probabilities of Type I and Type II errors are at most α and β.

c. Implement your code for $m = 10$, $n = 25$, $p_0 = 1/6$, and $\alpha = 0.05$ to carry out the tasks in part *a*.

d. For $\alpha = \beta = 0.05$, $m = 10$, $p_0 = 1/6$, $p_1 = 1/4$, what is the value of r and the required sample size n?

e. For fixed α, m, n, and p_0, the probability $\beta(p_1)$ decreases from $\beta(p_0) \geq 1 - \alpha$, to zero at $p_1 = 1$. Suppose that proponents of H_1 (believers in ESP?) insist on a sample size of $n = 10{,}000$, whereas proponents of H_0 (skeptics?) insist that α be equal to 0.01. For $m = 10$ and $p_0 = 1/6$, find the values of p_1 for which $\beta(p_1)$ equals 0.1, 0.05, and 0.01, respectively. Use the normal approximation to the binomial distribution. Discuss the significance of your numerical results.

f. Are you satisfied with the way H_0 and H_1 have been formulated? What can we say if possibly more than one coin has the higher probability p_1?

Problem 3.6.38: Successive, independent trinomial trials with outcomes 1, 2, 3, having positive probabilities p_1, p_2, and p_3, $p_1 + p_2 + p_3 = 1$, are performed. For given positive integers, m_1, m_2, and m_3, we say that an exit of Type 1 occurs if the outcome 1 occurs m_1 times before either outcome 2 is seen m_2 or outcome 3 m_3 times. Exits of Types 2 and 3 are similarly defined.

Furthermore, upon an exit of Type 1, we continue *binomial* trials with the probabilities $p_2(1 - p_1)^{-1}$ and $p_3(1 - p_1)^{-1}$, until a total of m_2 outcomes 2 or a total of m_3 outcomes 3 have occurred. We say that the exits occur in the order (1, 2, 3) if the quotas are met in that order. Other orders are similarly defined.

a. For given p_1, p_2, p_3 and m_1, m_2, m_3, compute the probabilities of exits of Types 1, 2, and 3 as well as the probabilities of the exits occurring in any of the six possible orders.

b. Units 1, 2, and 3 have independent, Erlang-distributed lifetimes, respectively with parameters λ_i and m_i, for $i = 1, 2, 3$. Show how the algorithm of part *a* can be used to compute the probabilities $P(j)$, $j = 1, 2, 3$, that Unit j fails first. Also, the probabilities $P(j_1, j_2, j_3)$ that the units fail in the order (j_1, j_2, j_3), for any permutation of 1, 2, and 3.

c. Compute the probabilities defined in part *b* for the Erlang distributions with parameters $m_1 = \lambda_1 = 2, \quad m_2 = \lambda_2 = 7, \quad m_3 = \lambda_3 = 12$.

Problem 3.6.39: You are one of m players who each perform independent Bernoulli trials with a positive probability p of success. The successive trials are performed simultaneously, one trial per player per time unit. Those players who first accumulate $k \geq 1$ successes win. The random variable $T(m)$ is the number of trials until at least one player wins. Write a program to compute tables of

a. The probability $P_1(r, n)$ that $T(m) = n$ and there are r winners exactly
b. The probability $P_2(n)$ that $T(m) = n$
c. The probability $P_3(n)$ that $T(m) = n$ and you are among the winners
d. The probability $P_4(n)$ that $T(m) = n$ and you are the only winner.

Compute all these probabilities for $k \leq n \leq N$, where N is the smallest integer for which $\sum_{n=N}^{\infty} P_2(n) < 10^{-4}$. Write an efficient program by selecting the most appropriate quantities for intermediate computation. Implement your program for $m = 5$, $k = 4$, and $p = 0.25$.

Remark: This problem is related to the minimum of m independent negative binomial random variables with probability density

$$p_\nu = \binom{\nu - 1}{k - 1} p^k (1 - p)^{\nu - k}, \quad \text{for } \nu \geq k.$$

By setting $\theta_\nu = \sum_{j=k}^{\nu} p_j$, for $\nu \geq k$, we obtain for instance that

$$P_1(r, n) = \binom{m}{r} p_n^r (1 - \theta_n)^{m - r}, \quad \text{for } n \geq k,$$

since at time n, r of the m players must score their kth success and the other $m - r$ players must have fewer than k successes. Formulas for the other probabilities can be derived in the same manner. Careful organization is essential to their successful implementation.

Problem 3.6.40: Consider m individuals numbered $1, \cdots, m$, who perform independent Bernoulli trials in cyclic order, i.e., for all $n \geq 0$, the $(nm + j)$th trial is performed by Player j. At each trial performed by Player j, he has a probability p_j, $0 < p_j < 1$, of success and a probability $q_j = 1 - p_j$ of failure. In addition to the probabilities $p_1, \cdots, p_m$, m positive integers $k_1, \cdots, k_m$ are specified. In order to win the game, the individual j must accumulate k_j successes before any other Player $i \neq j$, scores k_i successes.

a. Show that the probability $P(nm + j)$ that Player j wins the game at the $(nm + j)$th trial is

$$A(n + 1, j) \prod_{i<j} B(n + 1, i) \prod_{i>j} B(n, i), \quad \text{for all } n \geq 0,$$

where for $1 \leq i \leq m$ and $1 \leq j \leq m$, $A(n, j)$ and $B(n, j)$ are defined by

$$A(n, j) = \binom{n - 1}{k_j - 1} p_j^{k_j} q_j^{n - k_j}, \quad \text{if and only if } n \geq k_j.$$

$$B(n, i) = 1, \quad \text{for } n < k_i,$$

$$= \sum_{\nu=0}^{k_i - 1} \binom{n}{\nu} p_i^{\nu} q_i^{n - \nu}, \quad \text{for } n \geq k_i.$$

b. Write a program to compute the probabilities $P(nm + j)$ for Nm trials, where N is chosen such that the probability that the game lasts more than Nm trials is less than 10^{-5}. Print the probabilities $P(nm + j)$ for $0 \leq n \leq N$ and for $1 \leq j \leq m$ in m separate tables for different js.

c. Also print the probabilities $P^*(j)$, $1 \leq j \leq m$, that Player j wins and the probabilities $R(\nu)$ that the game requires a total of ν Bernoulli trials. Execute your code using the data:

j	1	2	3	4
p_j	0.75	0.60	0.50	0.30
k_j	5	5	4	3

Hint: Compute the quantities $A(n, j)$ and $B(n, j)$ recursively on n, using the recurrence relations for the binomial probabilities.

The following three problems deal with a systematic exploration of a two-parameter family of probability densities. Such explorations are useful and common in many applied models. The background of the family of densities is unimportant here, but they are obtained by setting $\theta = \rho^{-1}$ in the third explicit solution stated in Problem 4.4.76.

Problem 3.6.41: Let p, with $0 < p = 1 - q < 1$ and θ, with $0 < \theta \le 1$, be parameters of the probability generating function, $\Psi(z)$, given by:

$$\Psi(z) = p(p\theta)^{q(1-p\theta)^{-1}}[1 - (1-\theta)z][1 - (1-p\theta)z]^{-1-q(1-p\theta)^{-1}}.$$

a. Expand $\Psi(z)$ in a power series to verify that it is the generating function of the density $\{\psi_i\}$ with

$$\psi_0 = p(p\theta)^{q(1-p\theta)^{-1}},$$

$$\psi_i = \psi_0 q\left[\theta + \frac{1}{i}\right]\prod_{j=1}^{i-1}\left(1 - p\theta + \frac{q}{j}\right), \quad \text{for } i \ge 1.$$

Identify the density obtained for $\theta = 1$.

b. Show that the mean ψ_1^* and the variance ψ_2^* of the density $\{\psi_i\}$ are given by

$$\psi_1^* = 2q(p\theta)^{-1}, \quad \text{and} \quad \psi_2^* = (2 - p - p^2)(p\theta)^{-2} - q(p\theta)^{-1}.$$

Show that its coefficient of variation γ^* can take on all values in the interval $(1/\sqrt{2}, \infty)$ and no others. If we specify θ and an allowable value for γ^* show that the corresponding density is obtained by setting

$$p = \frac{4\gamma^{*\,2} - 2}{1 - \theta + 4\gamma^{*\,2}}.$$

c. Write an efficient code to compute the probabilities $\{\psi_i\}$ for given parameters p and θ. By using the formulas for the mean and variance, test your code over a wide range of parameter values.

d. If $\psi_0 < 0.01$, compute the density $\{\psi_i\}$ by determining the modal index i^* and the corresponding probability. Next, evaluate terms on either side of that index until a total probability mass of 0.999 is attained. Does that resolve any difficulties you may have encountered in part *c*?

Problem 3.6.42: Successively setting $\theta = k/10$ for $1 \le k \le 10$, calculate p for which the mean $\psi_1^* = 10$. Next, compute and plot the corresponding densities $\{\psi_i\}$ and the cumulative distributions.

Problem 3.6.43: Fix an allowable value for the coefficient of variation γ^*. Compute and plot the corresponding densities $\{\psi_i\}$ and cumulative distributions for $\theta = k/10$ for $1 \le k \le 10$. Implement your code for $\gamma^* = 1, 5, 10$.

Problem 3.6.44: Consider K groups of experimenters. In the jth group, there are M_j individuals, each of whom performs N_j independent Bernoulli trials with probability p_j of success. For $1 \le j \le K$, $1 \le i \le M_j$, X_{ij} is the number of successes of the ith experimenter in Group j.

a. For given K, M_j, N_j, and p_j, compute the marginal probability densities of the random variables

$$U = \max_{1 \le j \le K} \min_{1 \le i \le M_j} X_{ij}, \quad \text{and} \quad V = \min_{1 \le j \le K} \max_{1 \le i \le M_j} X_{ij}.$$

Note that $0 \le U \le \max_{1 \le j \le K} N_j$, and $0 \le V \le \min_{1 \le j \le K} N_j$.

b. Execute your program for the data:

$$K = 4, \quad M_1 = M_2 = M_3 = M_4 = 5,$$

$$N_1 = 10, \; p_1 = 0.10, \; N_2 = 20, \; p_1 = 0.05,$$

$$N_3 = 5, \; p_1 = 0.20, \; N_4 = 50, \; p_1 = 0.02.$$

Remark: This problem requires careful handling of the indices of the positive elements in various binomial distributions to obtain the correct summations and products. That is its challenge and also the point to be learned from it.

CHALLENGING PROBLEMS

Problem 3.6.45: The integer-valued random variables $X_1, \cdots, X_n$ are independent, identically distributed with common density $\{p_j\}$ on $0, \cdots, K$. p_K is positive. Their partial sums are $S_1 = X_1$, $S_2 = X_1 + X_2$, $\cdots$, $S_n = X_1 + \cdots + X_n$. Independently of the $X_1, \cdots, X_n$, an index N is selected from among $0, \cdots, n$ and the random variable Y is defined by $Y = S_N$. The probabilities $\alpha_j = P\{N = j\}$ are given for $1 \le j \le n$.

Show that the probability density of Y is given by

$$P\{Y = \nu\} = \alpha_1 p_\nu + \alpha_2 p_\nu^{(2)} + \cdots + \alpha_n p_\nu^{(n)},$$

where $p_\nu^{(j)} = P\{S_j = \nu\}$. That probability density may symbolically be written as

$$\alpha_1 \mathbf{p} + \alpha_2 \mathbf{p}^{(2)} + \cdots + \alpha_n \mathbf{p}^{(n)},$$

where $\mathbf{p}^{(j)}$ denotes the j-fold convolution of the density $\{p_j\}$ with itself. The density of Y can be efficiently computed by the analogue of Horner's algorithm discussed in Section 3.2. The recurrence for the density of Y is as follows:

Step 1: Multiply the density $\{p_\nu\}$ by α_n and add the quantity α_{n-1} to the term with index zero. Call the resulting sequence $\{c_\nu^1\}$.

Step 2: Form the convolution $\{w_\nu^1\} * \{p_\nu\}$ and add the quantity α_{n-2} to the term with index zero. Call the resulting sequence $\{c_\nu^2\}$.

Repeat Step 2 successively to form the sequences

$$\{w_\nu^3\} = \{w_\nu^2\} * \{p_\nu\} + \alpha_{n-3}\varepsilon,$$

up to

$$\{w_\nu^{n-1}\} = \{w_\nu^{n-2}\} * \{p_\nu\} + \alpha_1 \varepsilon,$$

where ε denotes the sequence with $\varepsilon_0 = 1$, and $\varepsilon_j = 0$, for $j \geq 1$. In the final step compute the convolution $\{w_\nu^n\} = \{w_\nu^{n-1}\} * \{p_\nu\}$.

Show that $\{w_\nu^n\}$, which concentrates on the integers $0, \cdots, Kn$, is the density of Y. Use the associative property of the convolution product. For given integers K and n calculate the number of multiplications needed to compute $\{w_n^n\}$. Write an efficient computer program to evaluate the density $\{w_\nu^n\}$. Use only one array W to hold the successive densities $\{w_\nu^1\}, \cdots,$ $\{w_\nu^n\}$. That can be accomplished by computing the successive convolutions downward from the highest index. Implement your program for

$$n = 10, \quad \alpha_1 = \cdots \alpha_5 = 0.05, \quad \alpha_6 = \cdots = \alpha_{10} = 0.15,$$

$$K = 5, \; p_0 = 0.05, \; p_1 = 0.15, \; p_2 = 0.15, \; p_3 = 0.20, \; p_4 = 0.25, \; p_5 = 0.25.$$

Check the validity of your program by verifying that the mean of Y is $\alpha\mu$. α and μ are the means of the densities $\{\alpha_j\}$ and $\{p_j\}$ respectively.

Problem 3.6.46: For the model in Problem 3.6.40, develop and code an algorithm to adjust the values of the parameters k_j so that the probabilities $P^*(j)$, $1 \leq j \leq m$, that Player j wins are as close to equal as possible. The requirement "as close to equal as possible" can be formalized in several ways. Use the *least square deviation* and the *maximum entropy* criteria. In the least square deviation criterion, minimize

$$\sum_{j=1}^{m} [P^*(j) - \bar{P}^*]^2,$$

where $\bar{P}^*$ is the arithmetic mean of the $P^*(j)$. For the maximum entropy criterion, the value of $H = -\sum_{j=1}^{m} P^*(j) \log P^*(j)$ is maximized. Start from given values of the k_j. Systematically increase k_j for j corresponding to the largest $P^*(j)$ or decrease k_j for the smallest $P^*(j)$. Execute your code for the data:

j	1	2	3	4
p_j	0.75	0.60	0.50	0.30
k_j	7	5	2	3

Problem 3.6.47: For the model in Problem 3.6.40, develop and code an algorithm to determine the values of the parameters p_j so that the probabilities $P^*(j)$, $1 \le j \le m$, that Player j wins are as close to equal as possible. The values of k_j are given and remain fixed. Use the same criteria and data as in Problem 3.6.46.

Hint: There are many possible approaches to finding the critical values of an implicitly defined function of several variables. We suggest trying the following: Let at any step of the search P_1^* and P_2^* be the smallest and the largest of the quantities $P^*(j)$. Search for the values of the corresponding p_{j_1} and p_{j_2} for which $|P_2^* - P_1^*|$ is minimum. Repeat this until all $P^*(j)$ values differ from each other by less than some small amount. Since this problem will require many calls to a portion of the algorithm developed in Problem 3.6.40, it is important to write that code as efficiently as possible.

Problem 3.6.48: For the multinomial waiting time model in Example 3.5.1, suppose that some of the integers i_j are zero and that we consider the waiting time N until the various quotas of alternatives have occurred. Explain why the alternatives for which $i_j = 0$, may be combined into a single alternative. Its probability is the sum of the corresponding p_j. Supposing that has been done, let there be $m + 1$ alternatives with probabilities $p_1, \cdots, p_{m+1}$ and that $i_j \ge 1$, for $1 \le j \le m$, and $i_{m+1} = 0$.

Show that for $n \ge i_1 + \cdots + i_m = i_m^*$,

$$P\{N \le n\} = \sum \frac{n!}{k_1! \cdots k_{m+1}!} p_1^{k_1} \cdots p_{m+1}^{k_{m+1}},$$

where the summation extends over all $(m + 1)$-tuples $k_1, \cdots, k_{m+1}$ satisfying $k_1 + \cdots + k_{m+1} = n$, $k_1 \ge i_1, \cdots, k_m \ge i_m$, and $k_{m+1} \ge 0$. Show further that, in terms of the notation in Example 3.5.1, that probability may be written as

$$\sum_{k_{m+1}=0}^{n-i_m^*} \binom{n}{k_{m+1}} p_{m+1}^{k_{m+1}} (1 - p_{m+1})^{n-k_{m+1}} F(n - k_{m+1}; m; p'_1, \cdots, p'_m),$$

where $p'_j = p_j(p_1 + \cdots p_m)^{-1}$, for $1 \leq j \leq m$. Explain why the distribution of N may be computed by the same algorithm as in Example 3.5.1, but with one additional iteration that differs from the others only in the range of the summation index.

Problem 3.6.49: The random variables $X_1, \cdots, X_m$ are the counts of the occurrences of m alternatives in n multinomial trials with positive probabilities $p_1, \cdots, p_m$. Y and Z are defined by $Y = \min\{X_1, \cdots, X_m\}$, and $Z = \max\{X_1, \cdots, X_m\}$. For $r \geq 1$, the moments of Y are given by

$$E[Y^r] = \sum \frac{n!}{k_1! \cdots k_m!} p_1^{k_1} \cdots p_m^{k_m} [\min(k_1, \cdots, k_m)]^r,$$

where the summation extends over all m-tuples $k_1, \cdots, k_m$ of nonnegative integers such that $k_1 + \cdots + k_m = n$. A similar formula holds for $E[Z^r]$.

a. Set $N(m, n) = [n/m]$, the integer part of n/m, and define $P(\nu)$ by

$$P(\nu) = \sum \frac{n!}{k_1! \cdots k_m!} p_1^{k_1} \cdots p_m^{k_m},$$

where the summation extends over all m-tuples $k_1, \cdots, k_m$ of integers satisfying $k_1 \geq \nu, \cdots, k_m \geq \nu$, and $k_1 + \cdots + k_m = n$. Show that

$$E[Y^r] = \sum_{\nu=0}^{N(n,m)} \nu^r [P(\nu) - P(\nu - 1)],$$

and derive a similar formula for $E[Z_r]$.

b. Building on the ideas in Example 3.5.1, develop an efficient algorithm to compute $P(\nu)$ for $0 \leq \nu \leq n$, and the related quantities needed in the formula for $E[Z^r]$.

c. Write and test a computer program to evaluate the first four moments of Y and Z. In particular, report the numerical results for the data

$$m = 25, \quad n = 100, \quad p_j = 0.04, \quad \text{for } 1 \leq j \leq 25.$$

Problem 3.6.50: Y and Z are defined as in Problem 3.6.49.

a. Show that the probability $R(i,j) = P\{Y \geq i, Z \leq j\}$ is given for $0 \leq i \leq j \leq m$ by:

$$R(i,j) = \sum \frac{n!}{k_1! \cdots k_m!} p_1^{k_1} \cdots p_m^{k_m},$$

where the summation extends over all m-tuples $k_1, \cdots, k_m$ of integers satisfying $i \leq k_\nu \leq j$, $1 \leq \nu \leq m$, and $k_1 + \cdots + k_m = n$.

b. Develop a recursive scheme similar to that in Example 3.5.1 to compute the probabilities $R(i, j)$. Express the joint probability density $p(i, j)$ of Y and Z in terms of the $R(i, j)$ and print it out. Also print the marginal densities of Y and Z.

c. Write and test a computer program to solve this problem and report the numerical results for the data

$$m = 25, \quad n = 100, \quad p_j = 0.04, \quad \text{for } 1 \leq j \leq 25.$$

Problem 3.6.51: The question treated in this problem arises in statistical procedures for identifying the most likely alternatives in multinomial trials with m outcomes $1, \cdots, m$ with positive probabilities $p_1 \leq \cdots \leq p_m$. We consider n independent trials and for $1 \leq j \leq m$, we denote by X_j the number of times the alternative j occurs. For a given positive integer d, we wish to compute, the probability

$$\Theta(n; p_1, \cdots, p_m) = P\{X_j \leq X_m + d, \quad \text{for } 1 \leq j \leq m\},$$

that no count exceeds that of the most probable outcome by more than d. Let $E(n; m; d)$ be the set of m-tuples $k_1, \cdots, k_m$ of nonnegative integers satisfying $k_1 \leq k_m + d, \cdots, k_{m-1} \leq k_m + d$, and $k_1 + \cdots + k_m = n$.

a. Show that for $n \geq 1$,

$$\Theta(n; p_1, \cdots, p_m) = \sum_{E(n;m;d)} \frac{n!}{k_1! \cdots k_m!} p_1^{k_1} \cdots p_m^{k_m}$$

$$= \sum_{k_m=0}^{n} \sum_{E(n-k_m;m-1;d)} \frac{n!}{k_1! \cdots k_m!} p_1^{k_1} \cdots p_m^{k_m}$$

$$= \sum_{k_m=0}^{n} \binom{n}{k_m} p_m^{k_m}(1-p_m)^{n-k_m} \Theta(n-k_m; p'_1, \cdots, p'_{m-1}),$$

where the quantities p'_j are defined by $p'_j = p_j(1-p_m)^{-1}$, for $1 \le j \le m-1$.

b. In a manner analogous to that outlined in Example 3.5.1, write a computer program to evaluate the probabilities $\Theta(n; p_1, \cdots, p_m)$ recursively on m. Your code should be able to handle values of m up to 30 and obtain the desired probabilities for n up to 500. Choose appropriate data and execute your program for $d = 3, 5, 10$; $m = 15$; and $n = 250$.

References: Carson, C. C. and Neuts, M. F., "Some computational problems related to multinomial trials," *Canadian Journal of Statistics,* 3, 235 - 248, 1975, and Gupta, S. S. and Nagel, K., "On the selection and ranking procedures and order statistics from the multinomial distribution," *Sankhya,* Ser. B, 29, 1 - 34, 1967.

Problem 3.6.52: For the model in Problem 3.6.51, suppose that $p_m = p$ and $p_j = (m-1)^{-1}(1-p)$, for $1 \le j \le m-1$. For given m and d, how can one determine the value p^* of p, for which $\Theta(n; p_1, \cdots, p_m) = 0.95$? For $m = 11$, compute $p^*(d)$ for $d = 1, \cdots, 10$. Strive for high efficiency in programming the numerical solution of the implicit equations involved.

Problem 3.6.53: This problem deals with the determination of the sample size in a selection model for binomial trials. There are $m+1$ coins of which one has probability p_1 of showing heads, whereas all the others have probability $p_2 = \delta p_1$, $0 < \delta < 1$. Each coin is tossed n times and the numbers of heads $X_1, \cdots, X_{m+1}$ are recorded. The coins that yield the maximum number of heads are retained and, if there is more than one, one among them is chosen at random.

a. Prove that the probability $P^*(n)$ that the best coin (the one with probability p_1) is selected by that procedure is given by

$$P^*(n) = \sum_{r=0}^{n} \binom{n}{r} p_1^r (1-p_1)^{n-r}$$

$$\times \sum_{j=0}^{m} \frac{1}{j+1} \binom{m}{j} \left[\binom{n}{r} p_2^r (1-p_2)^{n-r}\right]^j \left[F(r-1, p_2)\right]^{m-j},$$

where $F(k, p_2)$ is defined by

$$\sum_{i=0}^{k} \binom{n}{i} p_2^i (1-p_2)^{n-i}, \quad \text{for} \quad 0 \leq k \leq n.$$

b. Write an efficient code to compute the probabilities $P^*(n)$ for successive n. The recurrence relations for the binomial probabilities should be fully exploited. Specify a value θ between 0.90 and 0.975 and stop at the smallest value of n for which $P^*(n)$ exceeds θ. Implement your code for various choices of p_1, δ, m, and θ. Discuss what the results say about the preformance of this procedure.

Hint: We suggest using two arrays to store the binomial densities with parameters p_1 and p_2 for successive n. In evaluating the first sum in the formula for $P^*(n)$, we may neglect terms for which the leftmost binomial probability is small, say, less than 10^{-9}. That avoids much computation of the inner sums. These may be computed by using a convenient recursive scheme.

Problem 3.6.54: There are $m + 1$ coins of which one has a probability p^* of showing heads. The other m coins have a common probability p. At Stage 0, one could, without any experimentation, choose a coin at random. The probability of getting the best coin is then $(m + 1)^{-1}$. If one does not stop at Stage 0, then, at the subsequent stages, each available coin is tossed ν times. The numbers of heads scored with each are noted. One then has the option of choosing the coin with the highest score or, if there are several such, to choose one of these at random. Once a coin is chosen, the procedure ends.

In addition, after each stage at which one declines to make a selection, the remaining coins are subject to independent Bernoulli trials. With probability θ, a coin either stays in the game or is removed with probability $1 - \theta$. Should all coins disappear before a selection is made, one's chances of getting the best coin also vanish.

The coins are indistinguishable, so that a decision cannot be based on information from earlier stages. At any stage, one only knows the number of surviving coins and the number of heads scored by each in the ν tosses of that stage. $P(j)$ is the probability of getting the best coin if the selection is made at Stage j.

a. For given values of m, p, p^*, θ, and ν, compute the probability $P(j)$ for $j \geq 0$. Stop computation at the index j for which $P(j)$ becomes smaller than $(m + 1)^{-1}$. Also print the probability density of the maximum number of heads scored at each stage. Implement your program for $m = 25$, $p = 0.50$, $p^* = 0.60$, $\theta = 0.10$, and $\nu = 1, 5, 10$. Discuss the qualitative implications of the numerical results.

b. Discuss whether it is worthwhile to base the selection only on the maximum reported. One could formulate a decision policy of the type "I will choose at the jth stage, if the maximum number of heads at that stage is at least r." By examining the densities of the successively reported maxima, choose r and compute the probabilities $P(j)$ for that decision rule. Give detailed justifications for your choices.

Problem 3.6.55: Integer-valued random variables N_1, N_2 have a joint density $\{p(i_1, i_2)\}$ on the indices $0 \leq n_1 \leq n_1^*$, $0 \leq n_2 \leq n_2^*$, satisfying $p(0, 0) = 0$. Develop an efficient subroutine to compute the density of the discrete random variable $Z = N_1(N_1 + N_2)^{-1}$. Z takes values in a set of fractions belonging to $[0, 1]$. Construct a table of these (reduced) fractions listed in ascending order, represented as pairs of integers and of the corresponding probabilities. Plot the density and distribution of Z. Test and implement your code for several examples. As a test case, let (N_1, N_2) be uniform on the pairs satisfying $1 \leq i_1 \leq 10$, $1 \leq i_2 \leq 10$. For that case the required density can easily be found.

Problem 3.6.56: A purchaser knows that a product is available at M different price levels. She does not know the M exact prices L_ν, $1 \leq \nu \leq M$, which we list in increasing order. N suppliers offer the commodity for sale. We assume that their prices are chosen independently and that with probability $p_j > 0$ the price L_j is charged. The prices charged by suppliers 1, $\cdots$, N are successively examined. If the offer is declined, the purchaser cannot return to that supplier. Should purchase be postponed until the final supplier N, the item *must* be bought from him.

The purchaser applies the following selection rule: She will examine the first K offers without making a purchase. If the lowest price among suppliers 1, $\cdots$,K is C, she will buy from the first supplier among $K + 1$, $\cdots$,N who charges C or less. If there is no such supplier, she has to buy from supplier N.

a. Show that the probability $P_K(i, j)$ that $C = L_i$ and that she buys at a price $L_j \le C$ is

$$P_K(i, j) = [(p_i + \cdots + p_M)^K - (p_{i+1} + \cdots + p_M)^K]$$

$$\times \frac{1 - (p_{i+1} + \cdots + p_M)^{N-K}}{1 - (p_{i+1} + \cdots + p_M)} p_j, \quad \text{for } 1 \le j \le i \le M.$$

b. Show that the probability $P_K^o(i, j)$ that $C = L_i$ and that she buys at a higher price $L_j > C$ is

$$P_K^o(i,j) = [(p_i + \cdots + p_M)^K - (p_{i+1} + \cdots + p_M)^K]$$

$$\times (p_{i+1} + \cdots + p_M)^{N-K-1} p_j, \quad \text{for } 1 \le i < j \le M.$$

c. Write an efficient code to compute for given p_r and K, the probabilities π_ν for $1 \le \nu \le M$, that the purchase price $X = L_\nu$.

d. For given p_r and L_r, write a code to find the value of K for which the expected purchase price is as small as possible.

e. For given p_r, write a code to find K for which the probability that she will pay the highest price L_M is as small as possible.

f. Set $M = 5$ and $N = 20$, choose at least three lists of p_r and L_r, and execute your code for these. Discuss the implications of your numerical answers, particularly with reference to the values of K found for the decision criteria in parts *d* and *e*.

Remark: To make this problem concrete, imagine that you want to buy gasoline at the lowest price from one of N stations along a one-way road. This problem is a simple example of the Secretary Problem, a sequential decision problem on which there is an extensive literature.

Problem 3.6.57: Suppose that $\pi(1), \cdots, \pi(n)$ is a random permutation of the integers $1, \cdots, n$ and that k satisfes $1 \le k \le n(n+1)/2$. The random variable Z_k is equal to j, $1 \le j \le n$, if and only if j is the first index for which $\pi(1) + \cdots + \pi(j) > k$. The probability density $\{p_k(\nu)\}$ of Z_k is to be computed. Note that the event $\{Z_k > j\}$ is the set of permutations for which $\pi(1) + \cdots + \pi(j) \leq k$. Denote its probability by $g(j, n; k)$.

a. Let $B(r, n; k)$ be the number of subsets of $r \geq 1$ elements of $\{1, \cdots, n\}$, for which the sum is at most k. Show that the recurrence relation

$$B(r, n; k) = B(r, n-1; k) + B(r-1, n-1; k-n),$$

holds. Specify its boundary conditions. Define $B(r, n; k)$ appropriately for $k \geq n(n+1)/2$.

b. Show that

$$g(j, n; k) = P\{Z_k > j\} = \binom{n}{j}^{-1} B(j, n; k).$$

Derive a recurrence relation for these probabilities and use it to compute the density of Z_k for given n and k.

c. Show further that, with appropriate storage in a three-dimensional array, you also have the information to print the densities of Z_k, for all $k' \leq k$ and for all $n' \leq n$. Print the density of Z_{100} for $n = 20$ and that of Z_{50} for $n = 10$.

Problem 3.6.58: The Game of Bowling: In the game of bowling, a player tries to knock over as many of 10 pins per frame as possible. Each turn consists of two shots. Any pins that remain standing after the first shot can be knocked over by the second shot. If all 10 pins are knocked over by the first shot, the player scores a *strike*. If the second shot is required to knock over all ten pins, the player scores a *spare*. In every frame (turn), the player gains one point for every pin knocked over. In addition, the number of pins knocked over by the first shot after a spare counts double. The numbers of pins, upset by the *two* shots following a strike, count double. A *game* consists of 10 frames. If the player gets a spare on the 10th frame, one more shot is made and the number of pins knocked over is added to the accumulated score. The player who scores a strike on the 10th frame gets to roll two more balls and the numbers of points scored are added to the accumulated score. The highest possible score is 300.

We shall assume that the successive frames are independent trials and (somewhat unrealistically) that the number of pins knocked over on the second shot of a frame depends only on their number and not on their configuration. The probability that a player knocks over i pins on the first ball of a frame and j on the second is denoteds by $p(i, j)$, for $i \geq 0$, $j \geq 0$, $i + j \leq 10$. The quantities $p(i, j)$ sum to one.

After the i–th frame, there are four possible situations:

1. The score is n points and there are neither strikes nor spares to remember. Denote the probability of that event by $P(n; i)$.
2. The score is n and the preceding frame resulted in a spare. Let the probability of that event be $R(n; i)$.
3. The score is n and the preceding frame, but not the one before that, was a strike. The probability of that event is $S(n; i)$.
4. The score is n and the preceding two frames were strikes. Denote the probability of that event by $V(n; i)$. Note that now each pin knocked over by the next ball is worth three points.

For each i between 1 and 10, specify the values of the index n, for which each of the probabilities $P(n; i)$, $R(n; i)$, $S(n; i)$, $V(n; i)$ can be positive. Set aside four arrays P, R, S and V, each of 301 locations and at the i–th stage of the calculation, enter $P(n; i)$, $R(n; i)$, $S(n; i)$, $V(n; i)$ into the $(n + 1)$th locations of these arrays. Initialize the arrays by noting that

$$P(n; 1) = \sum_{i+j=n} p(i, j), \qquad \text{for} \quad 0 \leq n \leq 9,$$

$$R(10; 1) = \textstyle\sum_{k=0}^{9} p(k, 10 - k), \qquad S(10; 1) = p(10, 0),$$

$$V(n; 1) = 0, \qquad \text{for all } n,$$

$$P(n; 1) = R(n; 1) = S(n; 1) = 0, \qquad \text{for all other } n.$$

By careful consideration of cases, compute the quantities $P(n; i)$, $R(n; i)$, $S(n; i)$, $V(n; i)$, first for $i = 2$ and $i = 3$ and then by a general recursion for $4 \leq i \leq 10$. The density of the final score can be computed from the arrays for $i = 10$.

Write a code to compute the probability density of the final score for any given set of probabilities $p(i, j)$. The correctness of your code should be tested by constructing special (though artificial) examples. For instance, if only the $p(i, j)$ with $i + j = k$, for some k between 0 and 9 are positive, the final score will be $10k$ with probability one. Also, if $p(10, 0) = 0$, but the player scores a spare with probability one at every frame, the final score density will exhibit special features that should be present in your numerical results. Construct other such test examples.

The parameters $p(i, j)$ are given. One may question the purpose of an algorithm that has 54 free parameters, but the logic of the program is general and in any concrete use, the user is free to add assumptions that will reduce the number of free parameters $p(i, j)$. For instance, suppose that you have specified values that reflect your skills at bowling and you wish to study the merit of increasing $p(10, 0)$ (scoring strikes) or of $p(8, 2)$ (making a spare with two pins left). You can vary these parameters in different ways, while always keeping a valid set of values for the $p(i, j)$. Do several such studies, explaining the reasons for your choices and interpreting the qualitative conclusions drawn from your numerical results.

PROBLEMS REQUIRING NUMERICAL INTEGRATION

Problem 3.6.59: The random variables U and θ are independent. U has an exponential distribution with parameter λ and θ is uniform on $[0, \pi/2]$.

a. Show that the probability distribution $H(\cdot)$ of $X = U \cos \theta$ is given by

$$H(x) = 1 - \frac{2}{\pi} \int_x^\infty \lambda e^{-\lambda u} \arccos \left[\frac{x}{u} \right] du = 1 - \frac{2}{\pi} \int_0^1 e^{-\frac{\lambda x}{v}} (1 - v^2)^{-\frac{1}{2}} dv.$$

Give a geometric explanation for the fact that H depends only on the product λx.

b. Setting $\lambda = 1$, determine the values x_k, $1 \le k \le 19$, for which $H(x_k) = 0.05k$.

Remark: With a modicum of caution, the required numerical integrations can be done using elementary methods. For the motivated reader, we suggest exploring *quadrature* methods that may be applicable.

Problem 3.6.60: The side AC of the right triangle ABC has length 1. The angle at C is the right angle. The angle θ at A is randomly chosen according to the rule

$$\theta = \frac{\pi}{4} \frac{1}{1 + X},$$

where X is exponentially distributed with parameter λ. Y is the area of the triangle ABC.

a. Use numerical integration to compute the mean and variance of Y for arbitrary values of λ. Execute your program for $\lambda = 1.5$.

b. Find the probability density of Y and code a numerical integration procedure to find $P\{Y \le y\}$ for arbitrary values of λ and y. Execute your program for $\lambda = 1.5$ and $y = 0.1k$, $1 \le k \le 5$.

c. Use Newton's method and numerical integration to find the quartiles of Y for an arbitrary λ. Execute your program for $\lambda = 1.5$. The quartiles are the numbers c_1, c_2, and c_3 for which $P\{Y \le c\}$ equals 0.25, 0.50, and 0.75. After you have found c_1, c_2, and c_3, compute $P\{Y \le c_1\}$, $P\{Y \le c_2\}$, and $P\{Y \le c_3\}$ by means of the algorithm developed in part *b*.

Problem 3.6.61: The random variables $X_1, \cdots, X_n$ are independent, normally distributed with mean $\mu = 0$ and variance σ^2. Y is independent of the $X_1, \cdots, X_n$ and has a uniform distribution on (–1, +1).

a. For a given positive constant a, give a formula for the probability

$$P_n(a; \sigma^2) = P\{Y - a \le X_j \le Y + a, \quad \text{for all } j = 1, \cdots, n\}.$$

b. Develop an algorithm to compute $P_n(a; \sigma^2)$ for given n, σ and a. Execute your program for $n = 10$, $\sigma = 2$, and $a = 1.5$.

Problem 3.6.62: The random variables X_i, $i \ge 1$, are independent, normally distributed with mean zero and variance one. Y is independent of the $\{X_i\}$ and has an exponential distribution with mean one.

a. Obtain an expression for the probability $p(n) = P\{|X_i| < Y, 1 \le i \le n\}$.

b. As efficiently as possible, compute $p(n)$, for $1 \le n \le n^*$, where n^* is the smallest n for which $p(n) \le 0.1$.

c. Interpret your numerical results as a comparison of the heaviness of the tails of the normal and exponential distributions.

Hint: In truncating the integrals, use the fact that, for $u \ge 7$, $\Phi(u) - \Phi(-u)$ is very close to one.

Problem 3.6.63: The random variables X_i, $1 \le i \le 10$, are independent, normally distributed with mean zero and variance one. Y is independent of the $\{X_i\}$ and has an exponential distribution with parameter λ.

a. Obtain a formula for the probability $p(\lambda) = P\{|X_i| < Y, \ 1 \le i \le 10\}$.

b. As efficiently as possible, find the value of λ for which $p(\lambda) = 0.5$.

c. Interpret your numerical results as a comparison of the heaviness of the tails of the normal and exponential distributions.

Problem 3.6.64: The side AB of the triangle ABC is fixed and of unit length. The angles α and β (measured in radians) at A and B are independently chosen at random between 0 and $\pi/2$. The random variable X is the area of the random triangle ABC.

a. Show that $X = [2(\cot\alpha + \cot\beta)]^{-1}$.

b. Show that the random variables $\cot\alpha$ and $\cot\beta$ have the same probability density $\phi(\cdot)$ given by

$$\phi(u) = \frac{2}{\pi}\frac{1}{1+u^2}, \quad \text{for} \quad u \ge 0,$$

and zero elsewhere.

c. Prove that the probability density $\psi(\cdot)$ of $\cot\alpha + \cot\beta$ is given by

$$\psi(v) = \frac{8}{\pi^2 v\,(v^2+4)}[\log(v^2+1) + v\arctan v], \quad \text{for} \quad v \ge 0,$$

and zero elsewhere.

d. Show that the area X has the probability density

$$\theta(u) = \frac{32u}{\pi^2(16u^2+1)}\left[\log\left(1 + \frac{1}{4u^2}\right) + \frac{1}{2u}\arctan\left(\frac{1}{2u}\right)\right], \quad \text{for } u \ge 0.$$

e. Compute a table of the density $\theta(\cdot)$ and the corresponding distribution at points $k/100$ for $1 \le k \le 500$. Plot graphs of both functions.

Hint: In part c, in order to evaluate the integral

$$\int_0^v \frac{du}{(1+u^2)[1+(v-u)^2]},$$

use the identity

$$\frac{1}{(1+u^2)[1+(v-u)^2]} = \frac{1}{v(v^2+4)}\left[\frac{v+2u}{1+u^2} + \frac{3v-2u}{1+(v-u)^2}\right],$$

which is obtained by partial fraction expansion.

Problem 3.6.65: A point ω is chosen at random in the square with vertices (0, 0), (0, 1), (1, 0), and (1, 1). The random variable X is the sum of the distances from ω to the four vertices. Compute the mean and the variance of X. This requires numerical evaluation of the integrals

$$\int_0^1\int_0^1 f(u, v)\, du\, dv, \quad \text{and} \quad \int_0^1\int_0^1 f^2(u, v)\, du\, dv,$$

where

$$f(u, v) = (u^2+v^2)^{1/2} + [(u-1)^2+v^2]^{1/2}$$

$$+ [u^2+(v-1)^2]^{1/2} + [(u-1)^2+(v-1)^2]^{1/2}.$$

These integrals should first be simplified by exploiting the symmetry of the integrand.

Problem 3.6.66: B and C are independent normal random variables with mean zero and variances σ^2 and α^2, respectively. Show that the probability p that the quadratic equation $x^2 + 2Bx + C = 0$ has two *real* roots is given by

$$p = \frac{1}{2} + 2\int_0^\infty \phi(u)\left[1 - \Phi\left(\frac{\sqrt{\alpha u}}{\sigma}\right)\right] du,$$

where $\phi(\cdot)$ and $\Phi(\cdot)$ are the standard normal density and distribution, and p depends only on the ratio $\theta = \sqrt{\alpha}\sigma^{-1}$. Efficiently evaluate p for 25 equidistant points in the interval $0.9 \le \theta \le 1.1$.

Hint: Explain why the integral may be truncated by setting the upper limit of integration at 6. As an alternative, we can use Gauss quadrature.

Problem 3.6.67: Use an appropriate numerical integration technique to find the constant C for which the function

$$\phi(u) = Ce^{-u}\sin^2 u, \quad \text{for } u > 0$$

is a probability density. Also compute the first two moments of that density. Try to use as few evaluations of transcendental functions as possible. To that end, we suggest you compute the integrals for the moments before you find C. Appropriately adjust the moments once you know C.

Problem 3.6.68: Use an appropriate numerical integration technique to find the positive constant C for which the function $\phi(u) = Ce^{-u^2}\sin^2 u$, for $u > 0$, is a probability density. Also compute the first two moments of that density as in Problem 3.6.67.

Problem 3.6.69: The independent random variables X_1 and X_2 are exponential with mean one. For a given positive constant a, find the probability distribution $G(\cdot)$ of $Y = \max[X_1 + aX_2, aX_1 + X_2]$. For $a = 0.5, 1, 2$, print a table of $G(0.01k)$, $0 \le k \le K$, where K is the smallest integer for which $G(0.01K) \ge 0.999$.

Hint: You need to distinguish between the cases $0 < a < 1$ and $a \ge 1$. Identify the set $E(x)$ of points (u_1, u_2) with positive coordinates, for which $u_1 + au_2 \le x$, $au_1 + u_2 \le x$. Compute the required double integral over $E(x)$ as efficiently as possible by partitioning $E(x)$ in a rather obvious way.

PROBLEMS ON INCLUSION - EXCLUSION

Problem 3.6.70: Evaluate the successive partial sums in the formula

$$p(m;\, k;\, n) = \binom{n}{m} \sum_{\nu=0}^{n-m} (-1)^{\nu} \binom{n-m}{\nu} \left[1 - \frac{m+\nu}{m}\right]^{k},$$

in Example 3.4.1, remembering Bonferroni's inequalities. Implement your code for $k = 1{,}000$, $n = 365$, and for successive values of m. For each successive m, stop when the upper bound is less than 0.005 or the upper and lower bounds differ by less than 0.0001, whichever comes first. Print the approximate values so obtained. Compare them to the terms of the Poisson density with $\lambda = ne^{-k/n}$.

Problem 3.6.71: Using the setup of Example 3.4.1, let $X(i)$ be the number of labels that occur exactly i times. Use a combinatorial argument to show that the probabilities $p_i(m;\, k;\, n) = P\{X(i) = m\}$ are given by

$$p_i(m;\, k;\, n) = \frac{n!k!}{m!n^k} \sum (-1)^{\nu-m} \frac{(n-\nu)^{k-\nu i}}{(\nu-m)!(n-\nu)!(k-\nu i)!(i!)^{\nu}}.$$

The summation is over all values of ν, satisfying $n \le \nu \le m$, and $i\nu \le k$. Use that formula with $i = 1$, to obtain a numerical approximation to the probability density of the number of persons in a group of 100 who do not have the same birthday as another person in the group.

Problem 3.6.72: A box contains $2n$ chips, two of which are marked j for $1 \le j \le n$. First, n chips are drawn at random and placed in a second box. From the two boxes, we now draw one chip each, compare the labels, and continue until all chips have been removed. N is the number of times pairs of chips with matching labels are drawn. The event A_j occurs if at the jth drawing, both chips bear the same number.

a. Show that, for $0 \le r \le n$, the quantity S_r arising in the inclusion - exclusion formula is given by

$$S_r = \binom{n}{r}^2 r!\, \frac{2^{-n+r}[2(n-r)]^2}{2^{-n}(2n)!} = \frac{1}{2^r r!} \frac{n(n-1)\cdots(n-r+1)}{(n-1/2)(n-3/2)\cdots(n-r+1/2)}.$$

b. It may be shown that the kth factorial moment of N converges to 2^{-k} as $n \to \infty$. This implies that N has a Poisson limit density with $\lambda = 0.5$. Write a program to evaluate the exact probability density of N for $5 \le n \le 15$. Compare it to the limiting probabilities and to the Poisson density with $\lambda_1 = n(2n-1)^{-1}$; λ_1 is the exact mean of N.

Remark: This is the simplest version of a model known as *Levene's Matching Problem.* It has applications in genetics.

Problem 3.6.73: Data from a space probe are transmitted through a chain of 10 nodes, numbered $1, \cdots, 10$. During a mission, node i fails with probability $q(i)$. Failures are independent events. If a node i fails, but nodes $i-1$ and $i+1$ remain operative, transmission bypassing node i is still possible (albeit of reduced quality). For $i = 1$ and $i = 10$, we imagine virtual nodes 0 and 11, which cannot fail. The data link fails if and only if two adjacent nodes are down.

a. Use the inclusion - exclusion formula to calculate the probability $P[q(1), \cdots, q(10)]$ that the data link survives the mission. Write a subroutine to compute that probability. Implement your code to compute P for sets of parameter values obtained by choosing the $q(i)$ independently and uniformly in (0, 0.05).

b. For given values of $q(1), \cdots, q(10)$, determine algorithmically the permutations of the nodes with the highest and the lowest failure probabilities P.

c. For each set of parameter values $q(i)$, $i = 1, \cdots, 10$, used in part *a*, perform 10,000 replications of an experiment in which the performance of the 10 nodes is simulated. For those cases where the data link fails, store all configurations that cause the failure of the link (there could be several pairs, triplets, etc.) Examine those configurations. Do they share identifiable features, such as a string of high values for the failure probabilities $q(i)$? Discuss the engineering implications, if any, of your experiment.

Hints: For part *a*, let A_i be the event that node i fails. P is then the probability that none of the events A_1A_2, A_2A_3, $\cdots$, A_9A_{10} occur. For part *b*, show that the lowest value of P is obtained if the 10 $q(i)$ are increasing (or decreasing). To find the configuration for which P is largest, start with an arbitrary configuration and systematically do pairwise interchanges to obtain progressively larger values of P.

CHAPTER 4

DISCRETE-TIME MARKOV CHAINS

4.1. MATRIX FORMULAS FOR FINITE MARKOV CHAINS

Markov chains are widely used as models for a variety of practical situations. In recent years, algorithmic methods for Markov chains with large state spaces have attracted intense research efforts. Such methods are based on matrix analysis and related numerical techniques. Most textbooks still emphasize the rare cases of Markov chains that are tractable via simple calculations. To develop algorithmic skills, we instead stress the *structural properties* of the models. In working the problems in this chapter, the reader is encouraged to do any prior analysis and calculations in matrix notation. That avoids getting mired in heavy analytic calculations and it brings out the essential simplicity of many computations for Markov chains. In addition to a large set of problems, this chapter provides a summary of some matrix-analytic results on Markov chains. We shall assume familiarity with the basic mathematical properties of finite Markov chains and with the special infinite chains discussed in introductory texts.

In this chapter, we discuss discrete-time Markov chains. The analysis of continuous-time Markov chains and of Markov renewal processes involves differential and integral equations, and therefore also the additional algorithmic techniques for such equations. Once the structural thinking that is emphasized here has become familiar, some algorithmic methods for continuous-time models will be treated in Chapter 5.

Unless otherwise noted, we consider an *irreducible* m-state Markov chain. We denote its transition probability matrix by P. Whenever we deal with reducible Markov chains, the transition probability matrices will be displayed to bring out their structure. When dealing with special models, we encourage the reader always *to display the transition matrix* P to bring out any particular *structure* it may have. Analyses of Markov chains by long strings of special equations tend to obscure where the derivations are leading. Even when the number m of states is potentially large, we suggest displaying the matrix for a smaller version of the model to gain structural insight. That is done for various examples throughout this chapter.

The n-Step Transition Probabilities: The conditional probability $P_{ij}^{(n)}$ that the Markov chain is in state j at time n, given that it started in state i at time zero, is the (i, j)-element of the matrix P^n.

If it is necessary to compute the n-step transition probabilities up to some value N of n, the most obvious approach is to form the successive matrices $P^n = P^{n-1}P$, for $2 \leq n \leq N$. If the matrix P is sparse or has a special structure, it is efficient to write a special subroutine for the product of a general matrix B by P. When the initial probability vector is α, we compute the vectors αP^n, by the recurrence $\alpha P^n = [\alpha P^{n-1}]P$. Each step then only requires a multiplication of a vector $\mathbf{b}$ by P. Sometimes we need to compute P^n for a few high values of n. The same procedure as for repeated convolutions in Section 3.2 is applied. Using the binary expansion of n greatly reduces the number of matrix multiplications.

The *spectral method* is occasionally applicable for matrices P of rather low order m. If the matrix P is *diagonalizable,* i.e., it can be written as $P = H^{-1}\Lambda H$, where Λ is a diagonal matrix with the eigenvalues of P as its diagonal elements, the powers P^n are given by $P^n = H^{-1}\Lambda^n H$, for $n \geq 0$. Usually P has some complex eigenvalues. That computation is therefore done in complex arithmetic.

Taboo Probabilities: The conditional probabilities $P_{ij,H}^{(n)}$ that the chain is in state j at time n and that, in moving from the initial state i to j, it has not visited the set H, are called *taboo probabilities*. Many questions about Markov chains, such as those on first passage time distributions, are problems about taboo probabilities. In organizing computations of such probabilities, it is useful to partition the matrix P. That is illustrated by the following examples.

Example 4.1.1: Taboo Probabilities: We rearrange the states of the Markov chain into sets, 1, containing the states in the complement H^c of H, and 2, the set H. The transition probability matrix P is then written as

$$P = \begin{array}{c} 1 \\ \\ 2 \end{array} \left| \begin{array}{cc} T(1, 1) & T(1, 2) \\ & \\ T(2, 1) & T(2, 2) \end{array} \right| .$$

If i and j are states in 1, we see that the taboo probability $P_{ij,H}^{(n)}$ is the (i, j)-element of the matrix $[T(1, 1)]^n$. Similarly, the conditional probability that the chain visits set 2 for the first time at time n in the state $j \in 2$, given that it starts in $i \in 1$ at time zero, is the (i, j)–element of the matrix

$$[T(1, 1)]^{n-1}T(1, 2).$$

For $i \in 1$, and $j \in 2$, let ϕ_{ij} be the conditional probability that, starting in i, the chain *eventually* enters set 2 in the state j. The matrix Φ with elements ϕ_{ij} is then given by

$$\Phi = \sum_{n=1}^{\infty} [T(1, 1)]^{n-1}T(1, 2) = [I - T(1, 1)]^{-1}T(1, 2).$$

The second equality holds provided the inverse exists. However, if P is an irreducible stochastic matrix, the Perron - Frobenius theorem (see Appendix 1) implies that the largest eigenvalue $\eta(1, 1)$ of $T(1, 1)$ is less than one. The inverse of $I - T(1, 1)$ therefore exists.

Clearly, the first passage time probabilities f_{ij}^{n} from state i to state j, as well as various absorption probabilities, are particular taboo probabilities. For algorithmic purposes, such probabilities are most conveniently expressed in matrix notation.

Example 4.1.2: Phase-Type Densities: Consider a finite Markov chain with a single absorbing state. Partition its transition probability matrix as

$$P = \begin{matrix} 1 \\ * \end{matrix} \begin{vmatrix} T & \mathbf{T}^{\circ} \\ 0 & 1 \end{vmatrix},$$

where T is a matrix of order $m - 1$ and $\mathbf{T}^{\circ}$ is a column vector. The single absorbing state is denoted by $*$. By the considerations in Example 4.1.1, the probabilities of reaching state $*$ at time n are the components of $T^{n-1}\mathbf{T}^{\circ}$. Starting from any state, eventual absorption into $*$ is certain, if and only if

$$\sum_{n=1}^{\infty} T^{n-1}\mathbf{T}^{\circ} = \mathbf{e}, \tag{1}$$

where $\mathbf{e}$ always stands for a column vector with all its components equal to one. Since $T\mathbf{e} + \mathbf{T}^{\circ} = \mathbf{e}$, the nonsingularity of $I - T$ is necessary and sufficient for eventual absorption into $*$, from any initial state, to be certain.

If the initial state of the chain P is chosen according to an initial probability vector (α, α_0), the probability p_n of absorption at time n is given by

$$p_0 = \alpha_0, \quad \text{and} \quad p_n = \alpha T^{n-1} \mathbf{T}^\circ, \quad \text{for } n \geq 1. \tag{2}$$

Then $\{p_n\}$ is a proper probability density. Any probability density that can be so constructed from a finite Markov chain with a single absorbing state, is a (discrete) *probability density of phase type,* or a *PH-density,* in brief. Properties of the class of *PH*-distributions are listed in Appendix 2. We see that the probability generating function of the *PH*-density $\{p_n\}$ is

$$P(z) = \alpha_0 + z\alpha(I - zT)^{-1}\mathbf{T}^\circ.$$

By differentiating and setting $z = 1$, we obtain that

$$P'(1) = \alpha(I - T)^{-1}\mathbf{T}^\circ + \alpha(I - T)^{-2} T \mathbf{T}^\circ.$$

However, $(I - T)^{-1}\mathbf{T}^\circ = \mathbf{e}$, so upon simplification, the mean μ'_1 of the *PH*-density is given by

$$\mu'_1 = P'(1) = \alpha(I - T)^{-1}\mathbf{e}.$$

A second differentiation of $P(z)$ and a similar simplification yield that

$$P''(1) = 2\alpha(I - T)^{-2}\mathbf{e} - 2\alpha(I - T)^{-1}\mathbf{e}.$$

For the variance σ^2 of the *PH*-density, we so obtain

$$\sigma^2 = P''(1) + P'(1) - [P'(1)]^2 = 2\alpha(I - T)^{-2}\mathbf{e} - \mu'_1 - \mu'^2_1.$$

Remark: Many problems in this chapter involve phase-type densities. The reader is encouraged to study the more detailed discussion in Appendix 2 before attempting their solutions.

Example 4.1.3: Absorption Probabilities: The considerations in

Example 4.1.2 are easily extended to Markov chains with more than one absorbing state. For definiteness, let there be two absorbing states. The transition probability matrix then has the generic form:

$$P = \begin{array}{c|ccc|} 1 & T & \mathbf{T}^{\circ}(1) & \mathbf{T}^{\circ}(2) \\ * & 0 & 1 & 0 \\ ** & 0 & 0 & 1 \end{array}.$$

The initial probability vector is now of the form $(\alpha, \alpha_0, \alpha'_0)$. We stress that in any specific application, the matrix may not be given in exactly that form. It is therefore useful, if only in thought, to rearrange the states so that the transient states are placed in a set 1 and the two absorbing states are distinguished. For any initial probability vector, absorption in either $*$ or $**$ is again certain if and only if $I - T$ is invertible. The probability of absorption in $*$ is then

$$P(*) = \alpha_0 + \alpha(I - T)^{-1}\mathbf{T}^{\circ}(1).$$

The corresponding probability $P(**)$ for the state $**$ is

$$P(**) = \alpha'_0 + \alpha(I - T)^{-1}\mathbf{T}^{\circ}(2).$$

From $T\mathbf{e} + \mathbf{T}^{\circ}(1) + \mathbf{T}^{\circ}(2) = \mathbf{e}$, it readily follows that $P(*) + P(**) = 1$.

The probability $P_n(*)$ of absorption at time n in the state $*$ is

$$P_0(*) = \alpha_0, \quad \text{and} \quad P_n(*) = \alpha T^{n-1}\mathbf{T}^{\circ}(1), \quad \text{for } n \geq 1.$$

Clearly, the conditional density of the absorption time, given that the chain is absorbed in $*$, is $\{[P(*)]^{-1}P_n(*)\}$. The corresponding expressions for the state $**$ and for the moments of the (conditional) absorption time densities are left to the initiative of the reader. Calculation of the moments proceeds as in Example 4.1.2.

Steady-State Probabilities: For an irreducible stochastic matrix P, there exists a *unique, positive* probability vector $\boldsymbol{\pi}$ that satisfies the equations

$$\boldsymbol{\pi} = \boldsymbol{\pi}P, \qquad \boldsymbol{\pi}\mathbf{e} = 1,$$

where $\boldsymbol{\pi}$ is called the *left-invariant* or the *steady-state* probability vector of P. Important limits are the following:

$$\lim_{n\to\infty} \frac{1}{n}\sum_{k=0}^{n} P_{ij}^{(k)} = \pi_j, \quad \text{for } 1 \le j \le m, \quad \text{and all } 1 \le i \le m,$$

and provided the Markov chain is also *aperiodic,*

$$\lim_{n\to\infty} P_{ij}^{(n)} = \pi_j, \quad \text{for } 1 \le j \le m, \quad \text{and all } 1 \le i \le m.$$

These limit theorems may be conveniently written in matrix notation as

$$\lim_{n\to\infty} \frac{1}{n}\sum_{k=0}^{n} P^k = \mathbf{e}\boldsymbol{\pi}, \quad \text{and} \quad \lim_{n\to\infty} P^n = \mathbf{e}\boldsymbol{\pi}.$$

The matrix product $\mathbf{e}\boldsymbol{\pi}$ of the column vector $\mathbf{e}$ by the row vector $\boldsymbol{\pi}$ is clearly an $m \times m$ matrix with all rows equal to $\boldsymbol{\pi}$.

If $\boldsymbol{\pi}$ is chosen as the initial probability vector of the Markov chain, then $\boldsymbol{\pi}P^n = \boldsymbol{\pi}$ for all $n \ge 0$. For that choice of initial conditions, the Markov chain is *stationary*. Qualitatively, that corresponds to the fact that the Markov chain has been going on for a long period of time (and has forgotten the specific initial state in which it started). One of the most useful interpretations of the steady-state probability π_i is as (an approximation to) the *fraction of time* the Markov chain spends in state i over a long period of time.

The Computation of the Steady-State Probabilities: This subject, a particular case of the solution to a system of linear equations, has attracted much attention, particularly in the development of procedures for very large systems. Here, we limit attention to some classical methods that are generally well-suited for the problems in a first study. For aperiodic chains, a very simple algorithm is the *power method.* It consists in computing high powers of P by means of the recurrence

$$P^{2^{n+1}} = [P^{2^n}]^2,$$

which, of course, is just successive squaring of the current power of P. For the current power P^k, we compute a test criterion such as

$$\Phi(k) = \max_i \max_j \left| (P^k)_{ij} - \bar{P}_j(k) \right|,$$

where for $1 \le j \le m$, $\bar{P}_j(k) = m^{-1}\sum_{\nu=1}^{m} (P^k)_{\nu j}$. It tests the difference between the rows of the matrix. Computation stops when the computed value of $\Phi(k)$ is sufficiently small. The power method is computationally inefficient and one can easily construct examples where its convergence is slow.

Adaptations of methods for solving linear systems of equations must, in some manner, get around the fact that $\boldsymbol{\pi}$ must satisfy $m + 1$ equations in its m unknown components. A naive method is to assign a temporary value to, say, π_1. Then, one of the equations in $\boldsymbol{\pi} = \boldsymbol{\pi}P$ is dropped and the remaining system is solved for $\pi_2, \cdots, \pi_m$. We so obtain $\boldsymbol{\pi}$ up to an unknown constant c, which is found from the normalizing equation $\boldsymbol{\pi}\mathbf{e} = 1$. That approach can fail in several ways. For example, we may assign the temporary value $\pi_1 = 1$, when the actual value of π_1 is very small. The corresponding c is then very large and the normalization can produce inaccurate results.

A Convenient Method: In computing the vector $\boldsymbol{\pi}$ for matrices P of moderate size, we usually apply the following method. In the system

$$\sum_{i=1}^{m} \pi_i = 1, \qquad \sum_{i=1}^{m} \pi_i P_{ij} = \pi_j, \quad \text{for } 1 \le j \le m,$$

the first (normalizing) equation is multiplied by P_{mj}, $1 \le j \le m - 1$, and subtracted from the equation with the corresponding index j. That leads to

$$\sum_{i=1}^{m-1} \pi_i [\delta_{ij} + P_{mj} - P_{ij}] = P_{mj}, \quad \text{for } 1 \le j \le m-1,$$

$$\pi_m = 1 - \sum_{i=1}^{m-1} \pi_i .$$

The inhomogeneous system of the first $m - 1$ equations from which π_m has been eliminated can be shown to have a unique solution (see Problem 4.4.22). That system is solved by a library routine. As accuracy checks, we always perform back substitution and we print the maximum absolute difference between the computed left- and right-hand sides of the steady-state equations.

Iterative Methods: These methods solve the equations for Markov chains by successive approximations. This is a broad subject of which we can discuss only the most salient features. Iterative methods should be considered for *Markov chains with large numbers of states,* for those with special features of the transition probability matrix that *facilitate the computation of successive iterates,* and in studying *the effect of a parameter* on numerical descriptors of Markov chains.

Iterative methods are mostly insensitive to the numerical difficulties that occasionally plague methods such as Gauss elimination. However, for some problems, they converge slowly. That is their principal drawback. Slow convergence often reflects the physical behavior of the Markov chain at hand. In using iterative methods, insight into the behavior of the chain is particularly useful.

We consider two generic problems: that of computing the invariant probability vector $\boldsymbol{\pi}$ for a given irreducible, aperiodic stochastic matrix P and the solution of a system of linear equations $\mathbf{v} = T\mathbf{v} + \mathbf{c}$, where T is a principal submatrix of P such that $I - T$ is nonsingular. Here, $\mathbf{c}$ is a given column vector.

Both problems can be solved by *successive substitutions.* For any initial probability vector $\boldsymbol{\pi}(0)$, the iterates

$$\boldsymbol{\pi}(n+1) = \boldsymbol{\pi}(n)P,$$

converge to $\boldsymbol{\pi}$. The iterate $\boldsymbol{\pi}(n)$ is just the vector of state probabilities after the nth transition. The reader is encouraged to construct some examples of Markov chains for which these iterates converge very slowly.

Moreover, because of rounding errors, it can happen that the iterates deviate from the set of probability vectors. They then eventually diverge or converge to zero. In implementing successive substitutions, we therefore recommend dividing $\boldsymbol{\pi}(n)$ by $\boldsymbol{\pi}(n)\mathbf{e}$ before computing the next iterate $\boldsymbol{\pi}(n+1)$. We call that *renormalizing the iterates.*

The vector $\mathbf{v}$ in the second problem is the limit of the iterates

$$\mathbf{v}(n+1) = \mathbf{c} + T\mathbf{v}(n).$$

The initial vector $\mathbf{v}(0)$ is arbitrary. We see that $\mathbf{v}(n) = (I + T + \cdots + T^{n-1})\mathbf{c} + T^n \mathbf{v}(0)$. Successive substitution therefore amounts to a truncation of the matrix-geometric series for the inverse of $I - T$.

The rate of convergence is significantly improved by *Gauss - Seidel iteration.* For $1 \leq i \leq m$, the iterates are now computed by

$$v_i(n+1) = c_i + \sum_{j<i} T_{ij} v_j(n+1) + \sum_{j \geq i} T_{ij} v_j(n),$$

or by the preferred formula

$$v_i(n+1) = [1 - T_{ii}]^{-1}[c_i + \sum_{j<i} T_{ij} v_j(n+1) + \sum_{j>i} T_{ij} v_j(n)].$$

Gauss - Seidel iteration is easily programmed. The next iterate is component-wise written over the preceding one, but while doing so we keep track of a the stopping criterion such as

$$\max \{ |v_i(n+1) - v_i(n)| \}.$$

Convergence of the Gauss - Seidel procedure for the computation of π is, in general, not ensured. However, cases where it does not converge are rare in practice. We have occasionally found it expedient to implement Gauss - Seidel iteration also to compute π. However, in doing so it is essential to renormalize each iterate to be a probability vector, as the components of the iterates themselves no longer sum to one.

In iterative methods we can easily exploit special features, such as sparsity or special structure, of the matrices P or T. To do so, we must write an ad hoc subroutine to perform the iteration for the problem at hand. In some problems we are interested in how a certain descriptor of a Markov model varies with a parameter θ. For such questions, iterative methods can be very useful. We solve the problem for an initial value θ_0 and iteratively compute the descriptor by varying θ. Usually, the preceding vector is then an excellent starting solution and updating the solution often requires only a few iterations.

There is a block version of Gauss - Seidel iteration that is useful in analyzing Markov chains whose transition matrices have natural block structures. Such Markov chains arise in some queueing models. A few illustrative examples are listed among the problems.

References: An excellent source on iterative numerical methods for matrix problems is Varga, R. S. *"Matrix Iterative Analysis,"* Englewood Cliffs, New Jersey: Prentice-Hall, 1962.

For a survey and comparisons of advanced numerical methods for the computation of the vector $\boldsymbol{\pi}$, see: Philippe, Bernard; Saad Youcef and Stewart, William J., "Numerical methods in Markov chains." *Operations Research,* 40, 1156 - 1179, 1992.

4.2. SOME INFINITE-STATE MARKOV CHAINS

By using special structural properties, some Markov chains with infinitely many states can be analyzed in detail. These chains arise as models for random walks, queues, dams, and inventories and are therefore important. Their analyses involve many interesting mathematical techniques, such as special equations, recurrence relations, and repeated functional iterates among others. We review the basic properties of these models. Only sketches of the proofs are given. For further details, we refer the reader to standard texts.

4.2.1: One-Dimensional Random Walk

That is the Markov chain on the integers 0, 1, $\cdots$, with the transition probabilities

$$P_{i,i-1} = q_i, \qquad P_{ii} = r_i, \qquad P_{i,i+1} = p_i,$$

for $i \geq 0$ with $q_0 = 0$. For $|i - j| \geq 2$, $P_{ij} = 0$. The parameters q_i, r_i, and p_i sum to one and we assume that $p_i > 0$, for $i \geq 0$ and $q_i > 0$ for $i \geq 1$.

Let us consider the embedded Markov chain at epochs when the state changes. The transition probability matrix P' is of the same form as P, with $p'_0 = 1$, and

$$p'_i = p_i(1 - r_i)^{-1}, \qquad r'_i = 0, \qquad q'_i = q_i(1 - r_i)^{-1}, \quad \text{for } i \geq 1.$$

The steady-state equations for P' can be solved in a somewhat explicit form. Writing π'_i for its invariant probabilities, we have the steady-state equations

$$\pi'_0 = \pi'_1 q'_1, \quad \text{and} \quad \pi'_i = \pi'_{i-1} p'_{i-1} + \pi'_{i+1} q'_{i+1}, \quad \text{for } i \geq 1.$$

By successive substitutions, we obtain that $\pi'_{i-1} p'_{i-1} = \pi'_i q'_i$, for $i \geq 1$. That in turn leads to

$$\pi'_i = \pi'_0 \frac{p'_1 \cdots p'_{i-1}}{q'_1 \cdots q'_i}, \quad \text{for } i \geq 1.$$

The quantities π'_i sum to one if and only if

$$\pi'_0 = \left[1 + \sum_{i=1}^{\infty} \prod_{r=1}^{i-1} \left(\frac{p'_r}{q'_r} \right) \frac{1}{q'_i} \right]^{-1},$$

is positive, or equivalently if the series

$$\sum_{i=1}^{\infty} \prod_{r=1}^{i-1} \left(\frac{p'_r}{q'_r} \right) \frac{1}{q'_i},$$

converges. Establishing that convergence depends on the specific parameters of the random walk.

By using an argument from the theory of Markov renewal processes, we can show that the random walk with transition probability matrix P is positive recurrent if and only if the preceding series converges and in addition, the sum $E = \sum_{k=0}^{\infty} \pi'_k (1 - r_k)^{-1}$ is finite. The steady-state probabilities π_i of P are then given by

$$\pi_i = \frac{1}{E} \pi'_i (1 - r_i)^{-1}, \quad \text{for } i \geq 0.$$

4.2.2: The Embedded Markov Chain of the *GI/M/1* Queue

A Markov chain of this type arises in queueing theory in the treatment of the *GI/M/1* queue. We discuss its general form. Let $\{a_k\}$ be a probability density with finite mean α on the nonnegative integers. To avoid uninteresting special cases, we assume that $a_0 > 0$ and $0 < a_0 + a_1 < 1$. We define the tail probabilities a'_k by $a'_k = 1 - \sum_{r=0}^{k} a_r$, for $k \geq 1$. The transition probability matrix P of the Markov chain is given by

$$P = \begin{vmatrix} a'_0 & a_0 & 0 & 0 & 0 & \cdots \\ a'_1 & a_1 & a_0 & 0 & 0 & \cdots \\ a'_2 & a_2 & a_1 & a_0 & 0 & \cdots \\ a'_3 & a_3 & a_2 & a_1 & a_0 & \cdots \\ a'_4 & a_4 & a_3 & a_2 & a_1 & \cdots \\ \cdot & \cdot & \cdot & \cdot & \cdot & \end{vmatrix} .$$

The Markov chain is *positive recurrent* if and only if the mean $\alpha > 1$. Its invariant probabilities π_i then form the *geometric* density

$$\pi_i = (1 - \xi)\xi^i, \qquad \text{for } i \geq 0,$$

where ξ is the unique solution in the interval (0, 1) of the equation $z = \sum_{k=0}^{\infty} a_k z^k$.

ξ has a noteworthy probabilistic interpretation. If the chain is started in state $i > 0$, the expected number of visits to the state $i + n$, $n > 0$, before the chain reaches any state j with $j < i$ for the first time, is ξ^n. The computation of the steady-state probabilities is simple. It suffices to solve the equation for ξ numerically by any classical method.

4.2.3: The Embedded Markov Chain of the *M/G/1* Queue

In queueing theory, a Markov chain of this type arises when we consider the numbers of customers in the system at successive departures for the single-server queue with Poisson arrivals and general independent service times. That model is the *M/G/1* queue. We shall state the main properties of the Markov chain in a general form. Here $\{a_k\}$ is a probability density on the nonnegative integers with the same properties as stated in Section 4.2.2. The transition probability matrix P of the Markov chain is of the form

$$P = \begin{vmatrix} b_0 & b_1 & b_2 & b_3 & b_4 & \cdots \\ a_0 & a_1 & a_2 & a_3 & a_4 & \cdots \\ 0 & a_0 & a_1 & a_2 & a_3 & \cdots \\ 0 & 0 & a_0 & a_1 & a_2 & \cdots \\ . & . & . & . & . & \end{vmatrix},$$

where $\{b_k\}$ is a probability density. The Markov chain is irreducible whenever $0 \le b_0 < 1$. The densities $\{a_k\}$ and $\{b_k\}$ have finite means α and β and generating functions $A(z)$ and $B(z)$, respectively.

From the steady-state equations, a direct calculation shows that the generating function $X(z)$ of the invariant probabilities $\{x_i\}$ of P is

$$X(z) = x_0 \frac{zB(z) - A(z)}{z - A(z)},$$

where x_0 remains to be found. If the equation $z = A(z)$ has a root ξ in (0, 1), the function $X(z)$ is singular at $z = \xi$ and cannot be a probability generating function. In that case, the Markov chain is transient. $z - A(z)$ vanishes only at $z = 1$ in (0, 1] if and only if $\alpha \le 1$, so that the Markov chain is recurrent if and only if $\alpha \le 1$; and *positive* recurrent only if $\alpha < 1$. In the latter case, we let $z \to 1$ and apply L'Hospital's rule to verify that

$$x_0 = [1 + \beta(1 - \alpha)^{-1}]^{-1}.$$

The computation of the invariant probabilities and other quantities of interest presents a variety of algorithmic problems. Several of these are stated at the end of this chapter.

4.2.4: Markov Chains Related to the Galton - Watson Process

With probability densities $\{a_k\}$ and $\{b_k\}$ having the same properties as in the preceding two examples, the present matrix P has the form

$$P = \begin{vmatrix} b_0 & b_1 & b_2 & b_3 & \cdots \\ a_0^{(1)} & a_1^{(1)} & a_2^{(1)} & a_3^{(1)} & \cdots \\ a_0^{(2)} & a_1^{(2)} & a_2^{(2)} & a_3^{(2)} & \cdots \\ a_0^{(3)} & a_1^{(3)} & a_2^{(3)} & a_3^{(3)} & \cdots \\ a_0^{(4)} & a_1^{(4)} & a_2^{(4)} & a_3^{(4)} & \cdots \\ . & . & . & . & \end{vmatrix} .$$

The element $a_k^{(r)}$ is the term of index r in the r-fold convolution of the sequence $\{a_k\}$. With $b_0 = 1$, $b_k = 0$, for $k \geq 1$, the Markov chain has a familiar interpretation as the Galton - Watson model for the extinction of family lines. The Markov chain $\{X_n\}$ then describes the number of individuals in the nth generation. With a different density $\{b_k\}$, the population is restarted, one time unit after each extinction, by the immigration of k new individuals.

For the Galton - Watson model, the probability P_{ij}^n that i progenitors have j descendants in the nth generation is the coefficient of z^j in the power series for $[A^{[n]}(z)]^i$. The generating functions $A^{[n]}(z)$ are recursively defined by $A^{[0]}(z) = z$, and

$$A^{[n]}(z) = A\,[A^{[n-1]}(z)] = A^{[n-1]}[A(z)], \quad \text{for } n \geq 1.$$

The probability $\phi^i(n)$ that the lineages of i progenitors have become *extinct* by the nth generation is $[A^{[n]}(0)]^i$. The sequence $\{\phi_i(n)\}$ is nondecreasing (why?) Its limit ξ^i is the probability of the *eventual* extinction of a population with i progenitors. The quantity ξ is the smallest root of the equation $z = A(z)$, in (0, 1). As in the preceding two models, we see, by considering the graph of $A(z)$, that $\xi = 1$, if and only if $\alpha \leq 1$. Otherwise, $0 < \xi < 1$.

Few quantities related to the Galton - Watson process are explicitly tractable in simple form; algorithms for their numerical computation are therefore of interest. For instance, the probability distribution of the number of generations T to extinction, starting with one individual, is given by

$$P\{T \leq n\} = A^{[n]}(0), \quad \text{for } n \geq 0.$$

When $\xi = 1$, that is a nondefective probability distribution. Its mean $E(T)$ is given by

$$E(T) = 1 + \sum_{n=1}^{\infty} [1 - A^{[n]}(0)],$$

and that series converges if and only if $\alpha < 1$. In summing that series, it is important to have bounds on the error upon truncation. Such a bound is discussed in Problem 4.4.74.

For the case of certain extinction, we are interested in the total number Z of descendants of a single progenitor (including that individual). The generating function $G(z)$ of the density of Z satisfies the equation

$$G(z) = z\sum_{k=0}^{\infty} a_k [G(z)]^k = zA[G(z)].$$

For every value z in $[0, 1]$, $G(z)$ is the *smallest* positive solution of the equation $\zeta = zA(\zeta)$. The probability distribution of Z is nondefective if and only if $\alpha \le 1$. Its mean $E(Z)$ is finite if and only if $\alpha < 1$, and is then given by $E(Z) = (1 - \alpha)^{-1}$. The corresponding density, which also arises for the class of Markov chains in Section 4.2.3, is not easily computed. Markov chains related to the Galton - Watson model present challenging algorithmic problems, several of which are listed at the end of this chapter.

4.3. THE DISCRETE MARKOVIAN ARRIVAL PROCESS

By marking the epochs of certain transitions of a finite Markov chain, we construct special discrete point processes. These have found use as models for the arrival processes to queues and for data streams in communications engineering. In this section, we present the basic properties of the *Markovian arrival process,* which is commonly referred to by its initials *MAP*. A number of problems in this chapter deal with specific *MAP* s.

We consider an irreducible $m \times m$ stochastic matrix P with invariant probability vector $\boldsymbol{\pi}$ and initial probability vector $\boldsymbol{\alpha}$. P is written as the sum of two substochastic matrices A_0 and A_1. For reasons that will soon be clear, we assume that the matrix $I - A_0$ is nonsingular.

Suppose that P_{ij} is positive. When a transition from state i to j occurs, that transition is labeled (independently of any past events) as an *arrival* with probability

$$\frac{(A_1)_{ij}}{P_{ij}},$$

and remains unlabeled with the complementary probability

$$\frac{(A_0)_{ij}}{P_{ij}}.$$

Therefore, $(A_0)_{ij}$ and $(A_1)_{ij}$ are the probabilities of a transition from i to j, respectively without and with an associated arrival. The Markovian arrival process is the point process of the arrivals (or equivalently, of the labeled transitions). The following are examples of *MAP* s:

Example 4.3.1: The Bernoulli Process: Classical Bernoulli trials arise as the trivial *MAP* in which the 1×1 stochastic matrix $P = 1$ is split into $A_0 = q = 1 - p$, and $A_1 = p$. The times of successes form the *Bernoulli process*.

Example 4.3.2: The Discrete *PH*-Renewal Process: In Example 4.1.2, set $\alpha_0 = 0$, write $A_0 = T$, $A_1 = \mathbf{T}^{\circ}\alpha$, and set $P = T + \mathbf{T}^{\circ}\alpha$. Without loss of generality, we may assume that P is irreducible. The Markov chain P corresponds to *immediately* restarting the absorbing Markov chain of Example 4.1.2 upon each absorption. The chain is instantaneously (and independently of the past) moved to a state chosen according to the probability vector α. It is clear that the time intervals between such restarts are independently, identically distributed with the *PH*-density $\{p_k\}$. That *MAP* is a *PH*-renewal process.

Example 4.3.3: The Markov-Modulated Bernoulli Process: $p = (p_1, \cdots, p_m)$ is a non-zero vector with $0 \le p_i \le 1$, and q is the vector with components $q_i = 1 - p_i$. For notational convenience, $\Delta(p)$ and $\Delta(p)$ are diagonal matrices with diagonal elements p_i and q_i. The Markov-modulated Bernoulli process has the parameter matrices $A_0 = P\Delta(q)$, and $A_1 = P\Delta(p)$. With probability p_j, a transition into the state j is marked as an arrival.

Now returning to the general discussion of the *MAP*, we see that the density $\{\phi_k\}$ of the time until the first arrival is given by

$$\phi_k = \alpha A_0^{k-1} A_1 \mathbf{e}, \qquad \text{for } k \ge 1.$$

That is a proper probability density for *any* probability vector α if and only if the series $\sum_{r=0}^{\infty} A_0^r$ converges, or equivalently if the matrix $I - A_0$ is nonsingular. The assumption stated earlier is therefore a natural one. It merely excludes cases where, for some α, the process would have no arrivals at all.

The Number of Arrivals: The random variable $N(n)$ is the number of arrivals during the time interval $(1, \cdots, n)$. $J(r)$ denotes the state of the Markov chain P at time r and $P_{ij}(k; n)$ is the conditional probability

$$P_{ij}(k; n) = P\{N(n) = k, J(n) = j \,|\, J(0) = i\}, \qquad \text{for } 0 \le k \le n.$$

We write recurrence equations for the $m \times m$ matrices $P(k; n)$ with elements $\{P_{ij}(k; n)\}$. We readily see that for $n \ge 0$, with initial conditions $P(0; 0) = I$, and $P(k; 0) = 0$ for $k \ge 1$, we have that

$$P(k; n+1) = P(k; n)A_0 + P(k-1; n)A_1, \quad \text{for } 1 \le k \le n,$$

$$P(0; n+1) = P(0; n)A_0, \quad \text{and} \quad P(n+1; n+1) = P(n; n)A_1.$$

If we introduce the matrix generating function $P^*(z; n) = \sum_{k=0}^{n} P(k; n)z^k$, the preceding recurrence relations lead to

$$P^*(z; n+1) = P^*(z; n)[A_0 + A_1 z], \quad \text{with} \quad P^*(z; 0) = I.$$

Obviously, that equation implies that

$$P^*(z; n) = [A_0 + A_1 z]^n, \qquad \text{for } n \ge 0.$$

The matrices $\{P(k; n)\}$ are a natural matrix generalization of the *binomial density*. Because, generally, the matrices A_0 and A_1 do not commute, there is no simple closed form expression for the $\{P(k; n)\}$. However, they obey recurrence relations that are easily implemented.

To calculate the mean of $N(n)$, we differentiate $P^*(z; n)$ to obtain

$$P'^*(z; n) = \sum_{k=0}^{n-1} [A_0 + A_1 z]^k A_1 [A_0 + A_1 z]^{n-k-1}.$$

Setting $z = 1$ yields

$$M_1(n) = P'^*(1; n) = \sum_{k=0}^{n-1} P^k A_1 P^{n-k-1}.$$

The element $M_{1,ij}(n)$ is the expectation $E\{N(n)I\{J_n = j\} | J_0 = i\}$, where I here denotes the indicator function. We see that

$$\boldsymbol{\pi} M_1(n) = \boldsymbol{\pi} A_1 \sum_{k=0}^{n-1} P^k,$$

so that

$$\boldsymbol{\pi} M_1(n)[I - P + \mathbf{e}\boldsymbol{\pi}] = \boldsymbol{\pi} A_1[I - P^n] + n(\boldsymbol{\pi} A_1 \mathbf{e})\boldsymbol{\pi}.$$

Using the fundamental matrix $Z = [I - P + \mathbf{e}\boldsymbol{\pi}]^{-1}$, we obtain that

$$\boldsymbol{\pi} M_1(n) = \boldsymbol{\pi} A_1[I - P^n]Z + n(\boldsymbol{\pi} A_1 \mathbf{e})\boldsymbol{\pi}.$$

A similar calculation shows that $M_1(n)\mathbf{e} = Z[I - P^n]A_1\mathbf{e} + n(\boldsymbol{\pi} A_1\mathbf{e})\mathbf{e}$.

The quantity $\boldsymbol{\pi} M_1(n)\mathbf{e} = n(\boldsymbol{\pi} A_1 \mathbf{e})$, is the expected number of arrivals when $\boldsymbol{\pi}$ is the initial probability vector. The quantity $p^* = \boldsymbol{\pi} A_1 \mathbf{e}$ is the probability that, in the steady-state version of the *MAP*, there is an arrival at an arbitrary time point.

Another useful choice of the initial probability vector is $\boldsymbol{\alpha}^\circ = \frac{1}{p^*}\boldsymbol{\pi} A_1$, which corresponds to choosing time zero at an arbitrary arrival in the stationary process. (Why is that so?)

For a general initial probability vector $\boldsymbol{\alpha}$, we get that

$$\boldsymbol{\alpha} M_1(n)\mathbf{e} = np^* + \boldsymbol{\alpha} Z[I - P^n]A_1\mathbf{e}, \quad \text{for } n \geq 0.$$

The mean number of arrivals up to time n then depends on the initial conditions. The preceding formula may be rewritten as (why?)

$$\alpha M_1(n)\mathbf{e} = np^* + [\alpha Z A_1 \mathbf{e} - p^*] + \alpha Z[\mathbf{e}\pi - P^n]A_1\mathbf{e}, \quad \text{for } n \geq 0.$$

Since $P^n \to \mathbf{e}\pi$ when P is aperiodic, this shows that $\alpha M_1(n)\mathbf{e}$ has a linear asymptote given by the first two terms on the right-hand side. Further recurrence relations for the matrix $M_1(n)$ and for the corresponding second factorial moment matrix $M_2(n)$ are stated in Problem 4.4.41.

The Intervals Between Arrivals: Let τ_k be the time between the $(k-1)$st and the kth arrivals. The random variables τ_k are positive-integer-valued. For positive integers $i_1, \cdots, i_n$, the joint density of the n random variables τ_k, $1 \leq k \leq n$, is given by

$$\phi(i_1, \cdots, i_n) = \alpha A_0^{i_1-1} A_1 A_0^{i_2-1} A_1 \cdots A_0^{i_n-1} A_1 \mathbf{e}.$$

The n intervals are, in general, dependent. The joint generating function is

$$\phi^*(z_1, \cdots, z_n) = \alpha[I - z_1 A_0]^{-1} z_1 A_1 \cdots [I - z_n A_0]^{-1} z_n A_1 \mathbf{e}.$$

By setting $z_2 = \cdots = z_{n-1} = 1$ we obtain the joint generating function

$$\phi_{1,n}^*(z_1, z_n) = \alpha[I - z_1 A_0]^{-1} z_1 A_1 [(I - A_0)^{-1} A_1]^{n-2} [I - z_n A_0]^{-1} z_n A_1 \mathbf{e},$$

of the lengths of the first and the nth interval. By routine differentiations, formulas for the first two moments and for the serial correlation of τ_1 and τ_n may be derived (see Problem 4.4.42).

Remark: We see that the special form $A_1 = \mathbf{T}^\circ\alpha$ of the matrix A_1 causes the joint generating function $\phi^*(z_1, \cdots, z_n)$ for the *PH*-renewal process to factor into the product of the marginal generating functions.

4.4. MAIN PROBLEM SET FOR CHAPTER 4

What follows is a substantial set of algorithmic problems on discrete parameter Markov chains. Some problems also treat complementary theoretical results. Most computational problems require a fair amount of prior analysis that should further the reader's facility with Markovian models. In contrast to the earlier chapters, we only rarely propose specific parameters for the implementation of computer codes. We encourage the reader increasingly to think of the computer program as a tool for the exploration of a model. The choice of interesting parameter values is essential in such an exploration. A discussion of the insight into the features of the probability model gained from a numerical study now becomes an important part of the proposed solution.

EASIER PROBLEMS

Problem 4.4.1: For a given stochastic matrix P, develop an efficient algorithm to compute the matrices P^n for $k \le n \le k + \nu$, where k is a fairly large integer and ν is small compared to k. Dimension your program so as to handle matrices P of order up to 50. Implement your program for $k = 500$, $\nu = 10$, and P given by:

$$P = \begin{array}{c|ccccccccc|} 1 & .01 & .99 & .0 & .0 & .0 & .0 & .0 & .0 & .0 \\ 2 & .0 & .0 & 1 & .0 & .0 & .0 & .0 & .0 & .0 \\ 3 & .0 & .0 & .0 & 1 & .0 & .0 & .0 & .0 & .0 \\ 4 & .0 & .0 & .0 & .0 & 1 & .0 & .0 & .0 & .0 \\ 5 & .0 & .0 & .0 & .0 & .0 & 1 & .0 & .0 & .0 \\ 6 & .0 & .0 & .0 & .0 & .0 & .0 & 1 & .0 & .0 \\ 7 & .0 & .0 & .0 & .0 & .0 & .0 & .0 & 1 & .0 \\ 8 & .0 & .0 & .0 & .0 & .0 & .0 & .0 & .0 & 1 \\ 9 & .95 & .01 & .01 & .01 & .01 & .01 & .0 & .0 & .0 \end{array} .$$

Problem 4.4.2: Consider the nine-state Markov chain with transition probability matrix P:

$$P = \begin{array}{c|ccccccccc|} 1 & .249 & .20 & .20 & .35 & .0 & .0 & .0 & .0 & .001 \\ 2 & .20 & .10 & .10 & .60 & .0 & .0 & .0 & .0 & .0 \\ 3 & .10 & .50 & .0 & .40 & .0 & .0 & .0 & .0 & .0 \\ 4 & .05 & .05 & .80 & .10 & .0 & .0 & .0 & .0 & .0 \\ 5 & .0 & .0 & .0 & .0 & .10 & .20 & .20 & .10 & .40 \\ 6 & .0 & .0 & .0 & .0 & .20 & .10 & .20 & .30 & .20 \\ 7 & .0 & .0 & .0 & .0 & .10 & .80 & .10 & .0 & .0 \\ 8 & .0 & .0 & .0 & .0 & .30 & .30 & .20 & .0 & .20 \\ 9 & .60 & .0 & .0 & .0 & .0 & .10 & .0 & .0 & .30 \end{array} .$$

a. Efficiently compute the matrices P^n for $n = 100k$, $1 \le k \le 20$. Print the matrices. What you can infer about the behavior of the Markov chain?

b. If the chain starts in State 4, compute the probabilities that by times $n = 100k$, $1 \le k \le 20$, the chain has not yet left the set of states $\{1, 2, 3, 4\}$.

c. If the first state is chosen with equal probabilities from among 1, 2, 3, 4, what is the expected time to reach the set $\{5, 6, 7, 8, 9\}$?

d. Compute the invariant probability vector $\boldsymbol{\pi}$ of P. Can you draw any qualitative conclusions?

Problem 4.4.3: Consider a stochastic matrix P of order m and a permutation $\pi_1, \cdots, \pi_m$ of the indices $1, \cdots, m$. Write a subroutine to permute the rows and columns of P according to the permutation $\pi_1, \cdots, \pi_m$ *without* using a new array to store the rearranged matrix P or any intermediate matrices.

Hint: Examine P element by element until you reach an element $P(i_1, j_1)$ that needs to be moved. Compute its new location (i_2, j_2) and store the element $P(i_2, j_2)$ in a temporary hold. Enter the quantity $-P(i_1, j_1)$ into the former location (i_2, j_2) of $P(i_2, j_2)$ and compute the new location of the old $P(i_2, j_2)$. Whenever you encounter an element that needs not be moved, change its sign to negative. Show that the process ends when there are no more positive elements in the matrix. When that happens, just change the sign of all elements and the task is completed.

Problem 4.4.4: A simple procedure to determine the *class structure* of a Markov chain with m states and with transition probability matrix P is as follows: Calculate the matrix $P_m = I + P + \cdots + P^{m-1}$. Show that the state j can be reached from state i if and only if $(P_m)_{ij}$ is positive.

Two states i and j for which $(P_m)_{ij}$ and $(P_m)_{ji}$ are both positive are said to *communicate*. A set of communicating states is called a *recurrent class*. A Markov chain in which all states communicate is an *irreducible* Markov chain. States that do not belong to a recurrent class are called *transient*. Possibly after permutation of the state indices, any finite stochastic matrix P may be written as

$$P^\circ = \begin{vmatrix} P(1) & 0 & \cdots & 0 & 0 \\ 0 & P(2) & \cdots & 0 & 0 \\ \cdots & & \cdots & & \cdots \\ 0 & 0 & \cdots & P(c) & 0 \\ T(1) & T(2) & \cdots & T(c) & T(0) \end{vmatrix},$$

where $P(1), \cdots, P(c)$ are *irreducible* stochastic matrices of orders $m_1, \cdots, m_c$, with $m_1 + \cdots + m_c \leq m$. The bottom row, which may be absent, consists of the substochastic matrix $T(0)$ of order m_0 and of matrices $T(1), \cdots, T(c)$ of dimensions $m_0 \times m_j$, $1 \leq j \leq c$. Write a subroutine that after evaluating the matrix P_m will rewrite a given stochastic matrix P into the form P°. In computing P_m, use Horner's algorithm for greater efficiency.

Remark: The procedure described in this problem is suitable for small values of m. Much more efficient procedures, based on the theory of graphs, are available to identify the class structure of a Markov chain. These are essential in dealing with large matrices.

Problem 4.4.5: The class structure of a Markov chain depends only on which elements of P are positive and not on their specific values. The algorithm described in Problem 4.4.4 can be executed more efficiently by considering the *incidence matrix* P_0 corresponding to P and by using *Boolean* arithmetic. To form P_0, we set $[P_0]_{ij} = 1$ if and only if P_{ij} is positive and zero otherwise. In Boolean arithmetic, all matrix calculations are carried out using the conventions

$$0 + 0 = 0, \quad 1 + 0 = 0 + 1 = 1, \quad 1 + 1 = 1,$$
$$0 \times 0 = 0, \quad 1 \times 0 = 0 \times 1 = 0, \quad 1 \times 1 = 1,$$

or equivalently by using the logical operations "or" and "and." Write a computer code using Boolean arithmetic to carry out the tasks set in Problem 4.4.4. Strive for efficiency by avoiding unnecessary operations; for example, to determine whether an element in a Boolean matrix product is 1, it is enough to encounter at least one product that is 1. Incorporating a test for this in your code will usually greatly reduce the number of logical operations that need be performed. Adapt *Horner's algorithm* to Boolean arithmetic to evaluate the analogue of P_m as efficiently as possible.

Problem 4.4.6: Suppose you need an algorithm *only* to test whether a given stochastic matrix is irreducible. If it is reducible, you are not interested in the details of its class structure. How would you modify the algorithm in Problem 4.4.4, and particularly that in Problem 4.4.5, to serve that purpose? Code (or modify) a subroutine in Boolean arithmetic to test for irreducibility. Strive for maximum efficiency.

Use your subroutine in a program to do the following computer experiment: Successively generate 1,000 incidence matrices, each of order 10, by choosing their elements according to Bernoulli trials with probability p, i.e., each element is either 1 with probability p or 0 with probability $1-p$. Test each incidence matrix for irreducibility and report the number of cases (out of 1,000) for which the test is positive. Carry out three such experiments, respectively with $p = 0.2, 0.5, 0.75$.

Hint: What are the intermediate matrices that you compute in implementing Horner's algorithm? Is it always necessary to carry out all $m-2$ Boolean matrix products or can you sometimes stop earlier? If you wish to explore this further, keep track of the number of (needed) Boolean matrix products performed in testing each of the 1,000 matrices in your experiments. Report the empirical frequencies of these counts for the three values of p. Does their behavior agree with what you expected?

Problem 4.4.7: Consider an irreducible Markov chain P in which each state is *periodic* with period $d > 1$. By relabeling of states, P can then be rewritten in the form

$$P^\circ = \begin{vmatrix} 0 & P(1) & 0 & 0 & \cdots & 0 \\ 0 & 0 & P(2) & 0 & \cdots & 0 \\ 0 & 0 & 0 & P(3) & \cdots & 0 \\ & & \cdots & & \cdots & \\ P(d) & 0 & 0 & 0 & \cdots & 0 \end{vmatrix},$$

where $P(1), \cdots ,P(d)$ are stochastic matrices of order m/d. The period d of such a Markov chain is clearly a divisor of the number of states m. Periodicity depends only on the incidence matrix of the Markov chain. Develop an algorithm (appropriate for small Markov chains, say, with $m \le 30$) to determine whether the chain is periodic. If it is, write its transition probability matrix in the form P°.

Problem 4.4.8: Consider an irreducible Markov chain with transition probability matrix P

$$P = \begin{array}{c} 1 \\ 2 \end{array} \left| \begin{array}{cc} T(1, 1) & T(1, 2) \\ T(2, 1) & T(2, 2) \end{array} \right|,$$

as in Example 4.1.1. In this problem, you are asked to derive matrix formulas for various probabilities and moments of interest. No computer coding is required. The initial probability vector α is of the form $[\alpha(1), 0]$ so that the chain starts in a state of 1 chosen according to the vector $\alpha(1)$.

a. Write the vector of the probabilities x_j that the set 2 is first visited in the state $j \in 2$.

b. What are the mean and variance of the time to reach the set 2?

c. What is the probability that, after the chain leaves the set 1, it returns to that set after spending exactly n units of time in 2?

d. Find the probability that the first three sojourns in 2 last n_1, n_2, and n_3 units of time respectively.

Problem 4.4.9: $F(n)$ is the probability that k consecutive successes in Bernoulli trials (with probability α of success) occur before or at the nth trial. Show that for $n > k$,

$$F(n) = F(n-1) + (1-\alpha)\alpha^k[1 - F(n-k-1),$$

with $F(n) = 0$, for $n < k$, and $F(k) = \alpha^k$.

a. Derive the probability generating function of the density $\{\phi_n\}$, where $\phi_n = F(n) - F(n-1)$. Use it to show that its mean μ and variance σ^2 are

$$\mu = \frac{1-\alpha^k}{(1-\alpha)\alpha^k} \quad \text{and} \quad \sigma^2 = \frac{1}{(1-\alpha)^2\alpha^{2k}} - \frac{2k+1}{(1-\alpha)\alpha^k} - \frac{\alpha}{(1-\alpha)^2}.$$

By a few numerical examples, show that for all but small values of k and for p close to 1, μ and σ are very large.

b. For $\alpha < k(1+k)^{-1}$, the following bound on $1 - F(n)$ is known:

$$\left| 1 - F(n) - \frac{1 - \alpha\xi}{(1-\alpha)(k+1-k\xi)} \frac{1}{\xi^{n+1}} \right| < \frac{2(k-1)\alpha}{k(1-\alpha^2)},$$

where ξ is the unique root in the interval $(1, \alpha^{-1})$ of the equation $z = 1 + (1-\alpha)\alpha^k z^{k+1}$.

Examine the quality of that approximation for $\alpha = 0.5$ and $k = 5$, by comparing the exact and approximate values of $F(n)$ for $n \leq 1000$. Also, recalling that $\mu = \sum_{n=0}^{\infty} [1 - F(n)]$, compare the exact value of μ to those obtained by replacing the terms for $n > N$ by the approximation. Note that the sum involved is a geometric series, so that it can be expressed explicitly in ξ. Make that comparison for $N = 100\nu$, $1 \leq \nu \leq 10$.

Reference: Feller, W. *"An Introduction to Probability Theory and Its Applications, Vol. I."* Third Edition; New York: J. Wiley & Sons, 1967, pp. 322 - 325.

Remark: The form of the approximation, but not the error bound, also follows from the asymptotic behavior of discrete *PH*-distributions. The quantity $\eta = \xi^{-1}$ is the Perron - Frobenius eigenvalue of the irreducible matrix T, which is displayed here for $k = 6$ and with $\beta = 1 - \alpha$.

$$T = \begin{vmatrix} \beta & \alpha & 0 & 0 & 0 \\ \beta & 0 & \alpha & 0 & 0 \\ \beta & 0 & 0 & \alpha & 0 \\ \beta & 0 & 0 & 0 & \alpha \\ \beta & 0 & 0 & 0 & 0 \end{vmatrix}.$$

For further details, see the discussion of *PH*-distributions in Appendix 2.

Problem 4.4.10: In a game of chance, the gambler wins the Grand Prize upon scoring five successes in a row. The game is announced to operate according to independent Bernoulli trials with probability 0.45 of success. Assuming that this is true, compute the probability that the gambler would win the Grand Prize in at most 100 trials.

The operator of the game has also computed that probability and finds it too large to his taste. Being far from honest, he tampers with the chance mechanism as follows: Once three consecutive successes have occurred, the success probability at subsequent trials is reduced to $p < 0.45$. As soon as a failure occurs the mechanism returns to the advertised probability 0.45. The operator wants to set p such that the probability of winning the Grand Prize in at most 100 trials is one tenth of its value under Bernoulli trials with probability 0.45. Find that value of p.

Hint: With $F(n)$ as in Problem 4.4.9, the probability in the first part is $F(100)$ with $\alpha = 0.45$ and $k = 5$. For the second part, consider a 5-state Markov chain with one absorbing state. One needs to find p so that the probability of absorption in 100 steps or less has a specific value. Should you need the 100th power of a matrix, compute it efficiently!

Problem 4.4.11: For an irreducible m-state Markov chain, the matrix P is partitioned as in Example 4.1.1. Suppose the chain is observed only when it is in the set **1**. Y_k denotes the state at the kth visit to a state in **1**.

a. Show that $\{Y_k\}$ is a Markov chain with transition probability matrix

$$P(1, 1) = T(1, 1) + T(1, 2)[I - T(2, 2)]^{-1}T(2, 1).$$

$P(1, 1)$ is called the *Schur complement* of $T(1, 1)$ with respect to P.

b. If the invariant probability vector $\boldsymbol{\pi}$ of P is partitioned as $[\boldsymbol{\pi}(1), \boldsymbol{\pi}(2)]$, obtain matrix formulas in terms of the matrices $T(1, 1)$, $T(1, 2)$, $T(2, 1)$, and $T(2, 2)$ for $[\boldsymbol{\pi}(1), \boldsymbol{\pi}(2)]$, and for the invariant probability vector $\boldsymbol{\pi}^*(1)$ of $P(1, 1)$. Following an analytic derivation, also give a probabilistic argument for these formulas.

Problem 4.4.12: For the Markov chain with transition probability matrix

$$P = \begin{array}{c|ccccc|} 1 & 0.25 & 0.20 & 0.20 & 0.35 & 0.0 \\ 2 & 0.15 & 0.10 & 0.60 & 0.15 & 0.0 \\ 3 & 0.0 & 0.0 & 0.25 & 0.50 & 0.25 \\ 4 & 0.0 & 0.25 & 0.0 & 0.75 & 0.0 \\ 5 & 0.15 & 0.10 & 0.25 & 0.0 & 0.50 \end{array},$$

and the initial probability vector (0, 0.25, 0, 0.75, 0), compute the numerical values of *a.* The probabilities of visiting the states 1, 3, 5, upon first leaving the set of states {2, 4},

b. The expected time spent in the set {2, 4}, before leaving it,

c. The probability that the state 2 is visited at the time of the third return to the set {2, 4} from a state outside that set.

Problem 4.4.13: For the Markov chain with transition probability matrix

$$P = \begin{array}{c|ccccc|} 1 & 0.25 & 0.20 & 0.20 & 0.35 & 0.0 \\ 2 & 0.15 & 0.10 & 0.60 & 0.15 & 0.0 \\ 3 & 0.0 & 0.0 & 0.25 & 0.50 & 0.25 \\ 4 & 0.0 & 0.25 & 0.0 & 0.75 & 0.0 \\ 5 & 0.15 & 0.10 & 0.25 & 0 & 0.50 \end{array},$$

first compute the steady-state probabilities $\pi_1, \cdots, \pi_5$. f_{ii}^n is the probability that, starting in the state i, the chain returns to i *for the first time* at the nth transition. For $1 \le i \le 5$, compute and print the probability densities $\{f_{ii}^n\}$ and the corresponding distributions of the return times. For each i, stop at the first index n_i^* for which

$$\frac{1}{\mu'_i}\left[\mu'_i - 1 - \sum_{k=2}^{n_i^*} [1 - \sum_{k=1}^{r-1} f_{ii}^k]\right] < 0.05,$$

where μ'_i is the mean recurrence time of the state i.

Problem 4.4.14: Consider a random walk on the integers $0, 1, \cdots, K+1$ with absorbing states 0 and $K+1$. For $1 \le i \le K$, the probabilities of moving to $i+1$ or $i-1$ are $p(i)$ and $q(i) = 1 - p(i)$ respectively and $0 < p(i) < 1$. If the walk is started in state j, $1 \le j \le K$, $\phi(j)$ is the probability of eventual absorption in state 0.

a. Express $\phi(j)$ explicitly in terms of $p(i)$, $1 \le i \le K$.

b. Discuss an efficient algorithm to find the initial state j for which $\phi(j)$ is closest to 1/2.

c. Implement your algorithm for randomly generated data as follows: Set

$K = 100$ and choose $p(i)$, $1 \le i \le 100$, independently and uniformy in $(0.1, 0.9)$.

Problem 4.4.15: The Markov chain with transition probability matrix

$$\begin{array}{c|cccc|} 1 & 1. & 0.0 & 0.0 & 0.0 \\ 2 & 0.1 & 0.6p & 0.6q & 0.3 \\ 3 & 0.2 & 0.2 & 0.5 & 0.1 \\ 4 & 0.0 & 0.0 & 0.0 & 1. \end{array}$$

is started according to the initial probability vector $(0, 0.5, 0.5, 0)$. $p = 1 - q$ is a parameter in $[0, 1]$. Is there a value of p for which the probabilities of absorption in the states 1 and 4 are equal (to 1/2)? Answer the same question for the five-state chain with transition probability matrix

$$\begin{array}{c|ccccc|} 1 & 1. & 0.0 & 0.0 & 0.0 & 0.0 \\ 2 & 0.1 & 0.6p & 0.6q & 0.0 & 0.3 \\ 3 & 0.2 & 0.2 & 0.5 & 0.1 & 0.0 \\ 4 & 0.3 & 0.1 & 0.4 & 0.0 & 0.2 \\ 5 & 0.0 & 0.0 & 0.0 & 0.0 & 1. \end{array}$$

and initial probability vector $(0.0, 0.5, 0.5, 0.0, 0.0)$. Can a value of p in $[0, 1]$ be found for which the probabilities of absorption in the states 1 and 5 are equal?

Hint: With a modest amount of analytic calculation, one can show that the probability of absorption into state 1 for the first Markov chain is

$$p_1^* = \frac{3}{4}\frac{13 - 8p}{19 - 9p},$$

so that for $p = 1/6$, $p_1^* = 1/2$. It is possible to obtain analytic expressions, also for the second Markov chain, by using symbolic computation software. We suggest solving this problem both by symbolic computation and by numerical search procedures.

Problem 4.4.16: $\{a_j\}$ is a sequence of positive constants and $a_0 = 1$. We set $A_k = a_0 + \cdots + a_k$, for $k \ge 0$. The infinite matrix P is defined by

$$P_{ij} = \frac{a_j}{A_{i+1}}, \qquad \text{for } i \geq 0, \quad 0 \leq j \leq i + 1,$$

and $P_{ij} = 0$, elsewhere.

a. Show that the steady-state probabilities π_k satisfy

$$\pi_k = \pi_0 \prod_{i=0}^{k-1} \frac{a_{i+1}}{A_i}, \qquad \text{for } k \geq 1.$$

Prove that the Markov chain P is positive recurrent if and only if

$$\sum_{k=1}^{\infty} \prod_{i=0}^{k-1} \frac{a_{i+1}}{A_i} < \infty.$$

b. Identify the density $\{\pi_i\}$ when $A_k = (1 + p)^k$, for $k \geq 0$. (That case yields a classical result for the *GI/M/1* queue).

c. What are the probabilities π_i when $a_j = 1$ for all $j \geq 0$? What if $a_j = j + 1$, for $j \geq 0$? Set $a_j = (j + 1)^{\alpha}$ for $j \geq 0$, where α is a nonnegative integer. It can be shown that the Markov chain is positive recurrent for all $\alpha \geq 0$. Simple explicit forms for $\{\pi_i\}$ are found for $\alpha = 0, 1, 2,$ only. Examine $\{\pi_i\}$ numerically for selected other values of α.

d. Set $a_j = c^j$, for $j \geq 0$, with $0 < c < 1$. Derive an analytic expression for the $\{\pi_i\}$. Examine that density numerically for selected values of c.

Reference: The Markov chain in this problem was first discussed in F. G. Foster, "A Markov chain derivation of discrete distributions," *Annals of Mathematical Statistics,* 23, 624 - 627, 1952. For other special discrete densities arising from Markov chains, see A. W. Kemp, "Steady-state Markov chain models for the Heine and Euler distributions," *Journal of Applied Probability,* 29, 869 - 876, 1992.

Problem 4.4.17: The algorithm for a geometric mixture of the convolutions of a density $\{r_k\}$ in Problem 3.6.15 can be extended to the case of a *PH* mixing density. Here are the main steps. Let $R(z)$ be the generating function of $\{r_k\}$. The generating function $P(z)$ of the *PH* mixing density is

$$P(z) = \alpha_0 + z\,\boldsymbol{\alpha}(I - zT)^{-1}\mathbf{T}^{\circ}.$$

To obtain the coefficients of $P[R(z)]$, introduce the vector-generating function $\mathbf{V}(z)$, which satisfies the equation

$$\mathbf{V}(z) = R(z)T\mathbf{V}(z) + R(z)\mathbf{T}^{\circ}.$$

By series expansion prove that the density $\{\phi_n\}$ with generating function $P[R(z)]$ is given by

$$\phi_0 = \alpha_0 + \boldsymbol{\alpha}\mathbf{v}_0, \quad \text{and} \quad \phi_n = \boldsymbol{\alpha}\mathbf{v}_n, \quad n \geq 1.$$

The vectors $\{\mathbf{v}_n\}$ with generating function $\mathbf{V}(z)$ are recursively given by

$$\mathbf{v}_0 = r_0(I - r_0T)^{-1}\mathbf{T}^{\circ},$$

$$\mathbf{v}_n = (I - r_0T)^{-1}T\sum_{k=1}^{n} r_k\mathbf{v}_{n-k} + r_n(\mathbf{I} - r_0T)^{-1}\mathbf{T}^{\circ}, \quad \text{for } n \geq 1.$$

Without necessarily writing the code, discuss how to organize this computation efficiently.

Problem 4.4.18: Find the first row of the transition probability matrix

$$P = \begin{vmatrix} P_{11} & P_{12} & P_{13} \\ 0.25 & 0.25 & 0.50 \\ 0.10 & 0.80 & 0.10 \end{vmatrix},$$

for which the function $V(\pi_1, \pi_2, \pi_3) = -2\pi_1 + 3\pi_2 + \pi_3$, of the steady-state probabilities π_1, π_2, π_3 attains its maximum and its minimum.

Hint: Verify that, expressed in P_{12} and P_{13}, the objective function is

$$V(P_{12}, P_{13}) = \frac{128P_{12} + 126P_{13} - 22}{56P_{12} + 62P_{13} + 11}.$$

Find the extreme values under the constraints $0 \leq P_{12} + P_{13} \leq 1$, $P_{12} \geq 0$, and $P_{13} \geq 0$.

Problem 4.4.19: A Search Problem: A wounded missing person is immobilized in one of N regions. Based on available information, the *a priori* probability that the person is in region i is $p_i > 0$. Whenever region i is searched and the person is there, he is found with probability $\alpha_i > 0$. With the complementary probability he is overlooked. A sequential search plan is worked out as follows: At each stage, we search the region with the highest *a posteriori* probability given that all earlier searches were negative. If that *a posteriori* probability is the same for several regions, the one with the lowest number among them is searched. For the first stage, that rule is applied to the *a priori* probabilities.

At each stage, the previous posterior probabilities are used as the current prior probabilities. These are modified to account for information of the sort "Region j was searched in vain."

a. If at any stage, the current probabilities of the various regions are $p'_1, \cdots, p'_N$, with $p'_1 + \cdots + p'_N = 1$, and region j is searched in vain, show that the *a posteriori* probabilities are given by

$$p_i^*(j) = (1 - \delta_{ij}) \frac{p'_i}{1 - p'_j \alpha_j} + \delta_{ij} \frac{p'_j (1 - \alpha_j)}{1 - p'_j \alpha_j},$$

where $\delta_{ij} = 1$ if $i = j$ and 0 otherwise.

b. A plan for a search involving up to $r > 1$ stages is a list $j_1, \cdots, j_r$ of the successive regions to be searched as long as the person has not been found. For given p_i and α_i, $1 \leq i \leq N$, write a computer code to work out a search plan with at most r stages. After r unsuccessful searches, the effort is terminated.

c. For the search plan obtained in part b, compute the probabilities $P(k)$ that the person is found at the kth stage, $1 \leq k \leq r$.

d. Implement your code for $N = 10$, $r = 20$ and

i	1	2	3	4	5	6	7	8	9	10
p_i	.10	.15	.05	.10	.05	.02	.20	.03	.20	.10
α_i	.50	.55	.85	.75	.75	.70	.40	.80	.40	.60

AVERAGE PROBLEMS

Problem 4.4.20: First Passage Densities: This problem deals with a general algorithm for the computation of the first passage densities $\{f_{ij}^n\}$ from a state i in an irreducible finite Markov chain. By f_{ij}^n we denote the probability that, starting from the state i, the Markov chain reaches the state j for the first time at time n.

Use the following notation: $T(j)$ is the matrix obtained from P by removing its jth row and column. $\mathbf{T}^\circ(j)$ and $\alpha(j)$ are respectively the column and row vectors obtained from the jth column and row of P by removing the element P_{jj}. In discussing components we retain the original indices.

a. Explain why f_{ij}^n, with $i \neq j$, is the component with index i of the vector

$$\mathbf{u}(n) = T^{n-1}(j)\mathbf{T}^\circ(j).$$

b. Explain why f_{jj}^n is given by

$$f_{jj}^1 = P_{jj}, \qquad f_{jj}^n = \alpha(j)\mathbf{u}(n-1), \quad \text{for } n \geq 2.$$

c. Derive matrix formulas for the mean M_{ij} and variance σ_{ij}^2 of the probability densities $\{f_{ij}^n\}$.

d. Write a computer program to evaluate M_{ij} and σ_{ij}^2 for a given i.

e. Write a program to compute the densities $\{f_{ij}^n\}$ for $1 \leq j \leq m$ and for a given i.

f. Implement your code for

$$P = \begin{array}{c|ccccc|} 1 & 0.25 & 0.20 & 0.20 & 0.35 & 0.0 \\ 2 & 0.15 & 0.10 & 0.70 & 0.15 & 0.0 \\ 3 & 0.0 & 0.0 & 0.25 & 0.50 & 0.25 \\ 4 & 0.75 & 0.25 & 0.0 & 0.0 & 0.0 \\ 5 & 0.15 & 0.10 & 0.25 & 0.0 & 0.50 \end{array},$$

and $i = 1$. Stop when a probability mass of at most 10^{-7} remains in the tail of each computed density.

g. Verify numerically that, when $T(j)$ is irreducible, the ratios $\{f_{ij}^{n+1}/f_{ij}^{n}\}$ tend to a constant ρ which does not depend on i or j. The constant ρ is the Perron - Frobenius eigenvalue of the matrix $T(j)$.

Problem 4.4.21: Let P be an $m \times m$ irreducible stochastic matrix with invariant probability vector $\boldsymbol{\pi}$. By $\mathbf{v}$ and $\mathbf{u}$, respectively, we denote a column and a row vector of dimension m. The inner product $\boldsymbol{\pi}\mathbf{v} \neq 0$. The vector $\mathbf{u}$ is nonnegative and differs from the zero vector. Show that the matrix $I - P + \mathbf{v}\mathbf{u}$ is nonsingular.

Hint: Give a proof by contradiction. Use the fact that, up to multiplicative constants, $\boldsymbol{\pi}$ and $\mathbf{e}$ are the only left and right invariant vectors of P. Note that, in particular, this problem establishes the nonsingularity of $I - P + \mathbf{e}\boldsymbol{\pi}$. The inverse Z of that matrix is called the *fundamental matrix* of P. It arises in various moment formulas.

Problem 4.4.22: Use the result of Problem 4.4.21 with $\mathbf{v} = \mathbf{e}$ and $\mathbf{u}$ replaced by the mth row of P to show that the system of $m - 1$ equations in the convenient method for the computation of $\boldsymbol{\pi}$, discussed in Section 4.1.1, has a unique solution.

Problem 4.4.23: P is an $m \times m$ irreducible stochastic matrix. Explain why the (i, j)–element of the matrix $V_n = I + P + \cdots + P^n$ is the conditional expected number of visits to the state j during the epochs $0, 1, \cdots, n$, given that the chain starts in state i. Establish the formula

$$V_n = [I - P^{n+1} + (n + 1)\mathbf{e}\boldsymbol{\pi}]\,[I - P + \mathbf{e}\boldsymbol{\pi}]^{-1}.$$

Derive a similar formula for the matrix

$$V_n^\circ = I + 2P + 3P^2 + \cdots + (n-1)P^n.$$

Problem 4.4.24: An irreducible m-state Markov chain is considered only when state changes occur. Transitions from a state i to itself are ignored. Let Z_k be the kth element in the sequence obtained from $\{X_n\}$, the original Markov chain, by removing all successive repetitions. For example, if the given Markov chain successively visits the states

$$1, 2, 2, 1, 1, 2, 3, 3, 3, 2, 2, 1, \cdots$$

the new sequence starts with $1, 2, 1, 2, 3, 2, 1, \cdots$.

a. Show that $\{Z_k\}$ is also an irreducible m-state Markov chain. Obtain its transition probability matrix P'. If Δ is the diagonal matrix with diagonal elements P_{jj}, write P' in a concise formula by using the matrix Δ.

b. How are the invariant vectors π' of P' and π of P related?

Problem 4.4.25: We define a Markov chain that describes the queue length immediately after departures during the busy period of the $M/G/1$ queue of capacity N. Its transition probability matrix $P(N)$ is displayed for $N = 4$. Its structure is the same for general N; only the order is larger.

$$P(4) = \begin{vmatrix} 1 & 0 & 0 & 0 & 0 \\ a_0 & a_1 & a_2 & a_3 & a_4^\circ \\ 0 & a_0 & a_1 & a_2 & a_3^\circ \\ 0 & 0 & a_0 & a_1 & a_2^\circ \\ 0 & 0 & 0 & a_0 & a_1^\circ \end{vmatrix},$$

where the a_k are the terms of a discrete probability density. The entries in the column with index N are defined by $a_k^\circ = \sum_{r=k}^{\infty} a_r = 1 - \sum_{r=0}^{k-1} a_r$, for $k \ge 1$. For the $M/G/1$ queue, $\{a_k\}$ is the probability density of the number of arrivals during a service time, which has the probability distribution $H(\cdot)$. With a Poisson arrival process of rate λ, a_k is given by

$$a_k = \int_0^\infty e^{-\lambda u} \frac{(\lambda u)^k}{k!} \, dH(u), \quad \text{for } k \ge 0.$$

a. Evaluate the probability density $\{a_k\}$ explicitly for $H(\cdot)$, the *Erlang distribution* of order m

$$H(x) = \int_0^x e^{-\mu u} \frac{(\mu u)^{m-1}}{(m-1)!} \mu \, du, \quad \text{for } x \geq 0,$$

and for the *hyperexponential distribution*

$$H(x) = 1 - pe^{-\mu_1 x} - (1-p)e^{-\mu_2 x}, \quad \text{for } x \geq 0,$$

where $0 < p < 1$ and $0 < \mu_2 < \mu_1$.

b. If $H(\cdot)$ has mean μ'_1, the quantity $\rho = \lambda\mu'_1$ is called the *traffic intensity* of the unbounded queue. Construct an input file of parameters for several Erlang and the hyper-exponential distributions. After reading in the file, call a subroutine to find the value of λ for which $\rho = 0.80$.

c. For N = 5, 10, 15, 20, 25, compute the invariant probability vector $\boldsymbol{\pi} = [\pi(0), \cdots, \pi(N)]$ of the matrix $P(N)$ and interpret your numerical results.

Problem 4.4.26: The Classical Random Walk: Consider the classical symmetric random walk on the integers in which a point moves from its present position i, with equal probability, to $i + 1$ or to $i - 1$.

a. Show that the conditional probability $P_{00}(n)$, that starting in 0, the random walk is at 0 at time n, is given by

$$P_{00}(2n) = \binom{2n}{n} \left(\frac{1}{2}\right)^{2n}, \quad \text{for } n \geq 0,$$

and $P_{00}(2n+1) = 0$, for $n \geq 0$. By using Stirling's formula, show that

$$P_{00}(2n) \approx \frac{1}{\sqrt{\pi n}},$$

where, as usual, the symbol $\approx$ indicates that the ratio of the left- and right-hand sides tends to one as $n \to \infty$. Examine the quality of that approximation by comparing $P_{00}(2n)$ and the right-hand side for n up to 100. Print the exact and approximate values and the relative error.

b. Let $\{f_{i0}(n)\}$ be the first passage density from state i to state 0. Denote the generating function of that density by $F_i(z)$. Give simple, insightful arguments to establish the following equalities (no calculations are required!)

$$F_{-i}(z) = F_i(z), \quad \text{for } i \geq 1,$$

$$F_i(z) = [F_1(z)]^i, \quad \text{for } i \geq 1, \qquad F_0(z) = zF_1(z).$$

Show that $F_1(z)$ satisfies $F_1(z) = z/2 + z/2[F_1(z)]^2$, and that the solution which is a probability generating function is

$$F_1(z) = \frac{1 - \sqrt{1 - z^2}}{z}.$$

c. By expanding $F_0(z)$ in a power series, show that $f_{00}(2n + 1) = 0$, and

$$f_{00}(2n) = \frac{1}{2n - 1}\binom{2n}{n}\left(\frac{1}{2}\right)^{2n} = \frac{1}{2n - 1}P_{00}(2n), \quad \text{for } n \geq 1.$$

Verify that the mean of the density $\{f_{00}(n)\}$ is infinite. Compute the 99th percentile n_{99}^* of that density. Is it surprising that n_{99}^* is not all that large?

d. To see the significance of the infinite mean, write a subroutine to simulate 1,000 independent replications of the first passage from state 1 to state 0. For each replication, record the duration and the state i^* with the largest abscissa reached by the random walk. Summarize your results informatively. Discuss the qualitative conclusions of that experiment.

Reference: Feller, W. "*An Introduction to Probability Theory and Its Applications, Vol. I.*" Third Edition; New York: J. Wiley & Sons, 1967, Chapter 3.

Problem 4.4.27: The deviation of some physical process from its norm, which corresponds to state 0, is described by a random walk on the integers $-M, -M+1, \cdots, M-1, M$. Whenever the walk is in state i, the probability $p(i)$ of a step to the right is given by

$$p(i) = \frac{1}{2} + \frac{1}{2}\left[-\frac{i}{M}\right]^{\frac{1}{\alpha}}, \quad \text{for} \quad -M \le i \le 0,$$

$$p(i) = \frac{1}{2} - \frac{1}{2}\left[\frac{i}{M}\right]^{\frac{1}{\alpha}}, \quad \text{for} \quad 0 \le i \le M.$$

The probability $q(i)$ of a step to the left is $1 - p(i)$. The positive parameter α serves as a control parameter. For $\alpha \to 0$, the random walk approaches the simple symmetric random walk with reflecting barriers at M and $-M$. For $\alpha \to \infty$, the walk becomes constrained to states -1, 0, $+1$.

If it is increasingly costly to set a value of α, we may want to find the smallest α for which certain specifications are met. As a basis for reference, we use the limiting case $\alpha = 0$. For that case, it is easily verified that the steady-state probabilities $\pi(i; 0)$ are $\pi(i; 0) = (2M)^{-1}$, for $-M < i < M$, and $\pi(-M; 0) = \pi(M; 0) = (4M)^{-1}$. The mean recurrence time $T^*(0)$ of the state 0 is then $2M$. For the purposes of this problem, we consider only values of M between 25 and 200. If we define M_1 by $M_1 = [(9M)/10]$, where $[\cdot]$ denotes integer part. We agree that a *large deviation* is a visit to a state either to the right of M_1 or to the left of $-M_1$. For the simple random walk, the steady-state fraction of time that the system experiences a large deviation is

$$\phi(0) = \frac{2(M - M_1) - 1}{2M}.$$

The terms $\pi(i; \alpha)$, $T^*(\alpha)$, and $\phi(\alpha)$ are the corresponding quantities for the random walk with parameter α. The objective of forcing the random walk back to zero can be expressed by several mathematical criteria. For each of these criteria, the reader is asked to compute the smallest α for which it is met. For the optimal α-value for each criterion, the numerical values of the other criteria should also be computed.

Criterion A: The steady-state probability $\pi(M; \alpha) + \pi(-M; \alpha)$ should be less that 10^{-6}.

Criterion B: The mean recurrence time $T^*(\alpha)$ of state 0 should not exceed $0.5T^*(0)$.

Criterion C: The fraction of time $\phi(\alpha)$ that the system experiences a large deviation should not exceed $0.01\phi(0)$.

Criterion D: Suppose we start the simple random walk (corresponding to $\alpha = 0$) in state 0. Let $F(n; 0)$ be the probability that the walk returns to zero in at most n steps. $n^*(0)$ is the smallest n for which $F(n; 0) \geq 0.99$. Set $n^*(\alpha) = [n^*(0)/10]$. We want to determine α such that for the corresponding return time distribution $F(n; \alpha)$, $F[n^*(\alpha); \alpha] \geq 0.99$.

Each criterion has an interesting physical meaning. For example, Criterion D requires that the time $n^*(\alpha)$ by which nearly all deviations from zero have been corrected should be at most one tenth of what it is for the uncontrolled walk. The reader should give similar interpretations for the other criteria.

Write computer subroutines to compute all necessary probabilities and means to find the smallest α for each criterion. Also compute the numerical values of the other criteria for those α. Implement your code for at least three choices of M between 25 and 200. Give qualitative discussions of the various criteria. Which one requires the largest α and is therefore most stringent? Describe imaginative scenarios in which you would be guided by each criterion. How sensitive are the optimal αs to the parameter M?

Hint: In all computations, exploit the symmetry about state 0 of the transition probabilities. That results in major simplifications in the evaluation of the steady-state probabilities, the mean return time of state 0 and the recurrence time distribution of state 0. This problem involves much computation. Exploit special structure to achieve an efficient algorithm.

Problem 4.4.28: Consider a two-state Markov chain with transition probability matrix

$$P = \begin{array}{c} 1 \\ 2 \end{array} \begin{vmatrix} p_1 & 1 - p_1 \\ 1 - p_2 & p_2 \end{vmatrix},$$

with the initial probability vector $(c, 1 - c)$. K_1 and K_2 are positive integers. The random variable T is the number of transitions until either state 1 is visited K_1 times in succession or until state 2 is visited K_2 times at successive transitions, whichever comes first. If the required run of K_1 visits to state 1 occurs first, we say that state 1 wins. Otherwise, state 2 wins.

In numerical computations for this problem, set

$$c = \frac{1 - p_2}{2 - p_1 - p_2}.$$

That choice corresponds to the stationary Markov chain. Next, construct a finite Markov chain with two absorbing states. These correspond to the alternatives that state 1 or state 2 wins. That Markov chain can be used to compute many quantities of interest for this waiting time model.

Specificaliy, develop formulas and implement them algorithmically for the following items:

1. The probability that state 1 wins
2. The mean time until absorption
3. The conditional mean time until absorption given that state 1 (resp. state 2) wins
4. The conditional distributions of the time until absorption, given that state 1 (resp. state 2) wins.

This waiting time problem can serve as a model for a sequential test of the hypothesis H_0: $p_1 > p_2$, against the alternative H_1: $p_1 \leq p_2$. The probability that state 2 wins when the hypothesis H_0 holds is called the probability *of Type I error*. Print a table and generate a three-dimensional plot of the probability of Type I error as a function of p_1 and p_2.

Problem 4.4.29: The states of the Markov chain with

$$P = \begin{vmatrix} p_0 & 1 - p_0 \\ 1 - p_1 & p_1 \end{vmatrix},$$

are zero and one. $0 < p_0 < 1$ and $0 < p_1 < 1$. Write X_k for the state visited at time $k \geq 0$ and choose the initial state according to the steady-state probabilities $(\pi, 1 - \pi)$, where $\pi = (1 - p_1)(2 - p_0 - p_1)^{-1}$.

a. Show that the correlation coefficient γ of X_0 and X_1 equals $p_0 + p_1 - 1$. Verify that p_0 and p_1 are uniquely determined if we specify π and γ with $0 < \pi < 1$ and $-1 < \gamma < +1$. That is, we may choose π and γ as the parameters of the chain. Note that $\gamma = 0$ if and only if the chain corresponds to a sequence of independent Bernoulli trials.

Y_n is the number of indices j among $0, \cdots, n$ for which $X_j = 0$, or the amount of time spent in state 0. Show that $E(Y_n) = (n+1)\pi$. Derive a formula for the variance $\text{Var}(Y_n)$.

b. $Z_n(0)$ and $Z_n(1)$ are the numbers of *runs* in states 0 and 1 during the times $0, \cdots, n$. They are the numbers of uninterrupted strings of zeros or ones generated by the chain in its first $n+1$ time epochs. Develop and code recurrence relations to compute the joint densities of the pairs $\{Y_n, Z_n(0)\}$ and $\{Y_n, Z_n(1)\}$. Is it necessary to compute *two* joint densities?

c. Fix π and study how the joint densities in part *b* vary with γ. Execute your code for $n = 99$ and $\pi = 0.25, 0.40, 0.50$, with $\gamma = \pm 0.10k$, for $k = 0, \cdots, 9$. If possible, plot three-dimensional graphs of the joint densities to visualize how they depend on γ. Also, compute marginal densities, correlation coefficients, and other descriptors that provide further insight. Draw qualitative conclusions from your numerical results.

Problem 4.4.30: The stochastic matrix P depends on a parameter p. Specifically,

$$P(p) = \begin{array}{c|ccccc|} 1 & P_{11} & P_{12} & P_{13} & P_{14} & P_{15} \\ 2 & 0.05 & 0.0 & 0.80 & 0.0 & 0.15 \\ 3 & 0.0 & 0.10 & 0.0 & 0.90 & 0.0 \\ 4 & 0.15 & 0.0 & 0.10 & 0.10 & 0.65 \\ 5 & 0.25 & 0.0 & 0.20 & 0.10 & 0.45 \end{array},$$

where the elements in the first row are given by

$$P_{11} = 0.05p + 0.75q, \quad P_{12} = 0.15p + 0.05q, \quad P_{13} = 0.25p + 0.15q,$$

$$P_{14} = 0.05q, \quad P_{15} = 0.55p, \qquad \text{with } 0 \le p = 1 - q \le 1.$$

a. Compute the invariant probability vector $\pi(p)$ of $P(p)$ for $p = 0.05k$, $0 \le k \le 20$, and plot its components as functions of p. First use a non-iterative method to compute $\pi(p)$ for the successive k values.

b. As an alternative, first compute $\pi(0)$ as in part a. For $k \ge 1$, compute

$\pi(k)$ by successive substitutions. Use the preceding vector as a starting solution. In the iterative solution, stop when the maximum absolute difference between successive iterates becomes smaller than 10^{-6}.

c. Discuss the processing time and accuracy of both approaches.

Problem 4.4.31: Again consider the decision model in Problem 2.3.37. A *sequential* procedure for the selection of the better coin is as follows: A positive integer r is specified. Both coins are repeatedly tossed together. $Z_1(m)$ and $Z_2(m)$ are their numbers of successes after m tosses. We stop after τ trials, where

$$\tau = \min\left\{ m : m \le \left[\frac{N}{2}\right];\ |Z_1(m) - Z_2(m)| = r \right\},$$

and use the coin with most successes in the remaining $N - 2\tau$ trials.

a. $\psi_r(j)$ is the probability that Coin 1 is selected after j tosses. Show how, for a given r, $\psi_r(j)$, $1 \le j \le [N/2]$, may be computed. Discuss how the results may be used to avoid computation for uninteresting cases in part *b*.

b. Develop an efficient algorithm for the density and the mean of the total number K of successes in the N trials.

c. Determine r for which $E(K)$ is largest. As in Problem 2.3.38, set $L(r)$ to one less than the 25th percentile of the distribution of K. Determine r for which $L(r)$ is largest.

Remark: It is interesting to solve this problem and the two related ones in Chapter 2 for the same values of N, $p(1)$, and $p(2)$. For large N and for all three problems, information from early stages of the algorithm can be used to avoid computing for uninteresting cases.

Problem 4.4.32: The Galton - Watson Process: A Tractable Case: A *modified geometric probability density* $\{r_k\}$ with parameters p_0 and p is defined by

$$r_0 = q_0, \quad \text{and} \quad r_k = p_0\, q^{k-1} p, \quad \text{for } k \ge 1.$$

$0 < p_0 = 1 - q_0 < 1$, $0 < p = 1 - q < 1$. Its probability generating function $P(z)$ is given by $P(z) = q_0 + p_0 pz(1 - qz)^{-1}$.

We form the densities whose generating functions $P_n(z)$ are the functional iterates of $P(z)$, i.e., $P_{n+1}(z) = P[P_n(z)]$, for $n \geq 0$, with $P_0(z) = z$. $P_n(z)$ is the generating function of the number of descendants of an individual in the nth generation of a Galton - Watson process.

a. Verify that $P_n(z)$ corresponds to a modified geometric density with parameters $p_0(n)$ and $p(n)$ recursively by

$$p_0(n+1) = p_0 p_0(n)[1 - q_0 q(n)]^{-1}, \quad \text{and} \quad p(n+1) = pp(n)[1 - q_0 q(n)]^{-1},$$

for $n \geq 1$. By a direct calculation, show that $p_0(n)$ decreases. Argue that $q_0(n)$ tends to 1 or to $q_0 q^{-1}$, depending on whether the mean $p_0 p^{-1}$ of $\{r_k\}$ is at most 1 or exceeds 1. The probability $q_0(n)$ tends to the smallest positive root of the equation $z = P(z)$.

b. If $p_0 < p$, extinction is certain. The mean number $E(T)$ of generations to extinction is then finite. Show that $E(T) = \sum_{n=1}^{\infty} p_0(n)$. As $P_n(z)$ is convex increasing, its graph lies above its tangent at $z = 1$. Use this to show that

$$p_0(n) < P'_n(1) = \left[\frac{p_0}{p}\right]^n, \qquad \text{for } n \geq 1.$$

The series for $E(T)$ therefore converges at least as fast as the geometric series of ratio $p^{-1}p_0$.

c. This present Galton - Watson process is unusually tractable analytically. Use the properties in parts *a* and *b* to study the effect of variability in the number of offspring on the behavior of the process. First set $p = (1 - \delta)p_0$, where $0 < \delta < 1$. Then, successively set $p = 0.1 + 0.2\nu$, $0 \leq \nu \leq 4$. For $\delta = 0.1$, compute and plot the density of the time T to extinction. Calculate $E(T)$ so that the relative error in the computed value is at most 0.01. For successive n, compute and plot the 90th percentile of the distribution of the number of descendants in the nth generation.

d. Next, set $p = (1 + \delta)p_0$, where $0 < \delta < 1$ and successively set $p = 0.1 + 0.2\nu$, $0 \leq \nu \leq 4$. Compute and plot the extinction probability $q_0(n)$ for successive generations until $q_0 q^{-1} - q_0(n) < 0.001$. For successive n, compute and plot the 90-th percentile of the distribution of the size of the nth generation.

Hint: For a discrete random variable N, the 90th percentile is the smallest index i for which $P(N \le i) \ge 0.90$. For the modified geometric distribution, that index is easily found. What do your numerical results tell you about the behavior of the process?

Problem 4.4.33: Write an efficient program to simulate the behavior of the Galton - Watson process in Problem 4.4.32. Execute it for 20 realizations and this for each of the parameter values specified there. Interpret the behavior of the simulation runs in light of the numerical results for Problem 4.4.32.

Problem 4.4.34: Modify the model of Problem 4.4.32 as follows: One time unit after an extinction, new individuals are introduced. The number of new individuals has a Poisson distribution of mean λ. The modified process is an irreducible, aperiodic Markov chain. Show that, when $p_0 < p$, that chain is positive recurrent.

a. Establish that the generating function $\Pi(z)$ of its steady-state probabilities is given by

$$\Pi(z) = \pi_0 \left[1 + \sum_{n=0}^{\infty} \{\exp[-\lambda(1 - P_n(z))] - \exp[-\lambda(1 - P_n(0))]\} \right],$$

where

$$\pi_0 = \left[1 + \sum_{n=0}^{\infty} \{1 - \exp[-\lambda(1 - q_0(n))]\} \right]^{-1}.$$

Using results from Problem 4.4.32, write a program to evaluate π_0 to high precision.

b. Verify that, if $P(z)$ is the generating function of a modified geometric density, the generating function $\exp[-\lambda(1 - P(z))]$ may be written as

$$\exp[-\lambda(1 - P(z))] = \prod_{r=0}^{\infty} \exp[-\lambda p_0 p q^{r-1}(1 - z^r)].$$

Interpret each factor on the right as a Poisson density on an appropriate support. Use that formula to evaluate the density with generating function $\exp[-\lambda(1 - P(z))]$. Also, adapt your code to compute the steady-state probabilities π_ν. Execute your code with $\lambda = 1$ and for the parameter values in part *c* of Problem 4.4.32.

Hint: Also see Problems 3.6.26 and 3.6.27.

Problem 4.4.35: The transition probability matrix $P(K)$ is of the form:

$$P(K) = \begin{vmatrix} b_0(K) & b_1(K) & b_2(K) & b_3(K) & b_4(K) & \cdots \\ a_0 & a_1 & a_2 & a_3 & a_4 & \cdots \\ 0 & a_0 & a_1 & a_2 & a_3 & \cdots \\ 0 & 0 & a_0 & a_1 & a_2 & \cdots \\ . & . & . & . & . & \end{vmatrix}.$$

The sequence $\{a_k\}$ is a probability density on the nonnegative integers with finite variance σ^2. To avoid trivial cases, we assume that $a_0 > 0$ and $a_0 + a_1 < 1$. The generating function of $\{a_k\}$ is $A(z)$; its mean is $\alpha < 1$. The terms of the density $\{b_k(K)\}$ are zero, except for some $b_K(K)$, $K \geq 1$, which is 1. Under these conditions, the Markov chain with transition probability matrix $P(K)$ is irreducible, aperiodic, and positive recurrent.

a. Show that the generating function $X(z; K)$ of the invariant probability density $\{x_i(K)\}$ of $P(K)$ is given by

$$X(z; K) = \frac{1 - \alpha}{K + 1 - \alpha} \frac{z^{K+1} - A(z)}{z - A(z)}.$$

Verify that its mean $L(K)$ is

$$L(K) = \frac{K}{2} + \frac{K\sigma^2}{2(1 - \alpha)(K + 1 - \alpha)}.$$

b. Show that $X(z; K)$ may be rewritten as

$$X(z; K) = \frac{1 - \alpha}{K + 1 - \alpha} + \frac{K}{K + 1 - \alpha} \frac{z - z^{K+1}}{K(1 - z)} Y(z).$$

The probability generating function

$$Y(z) = (1-\alpha)\left[1 - \alpha\frac{1-A(z)}{\alpha(1-z)}\right]^{-1},$$

does not depend on K. What is the corresponding probability density? Develop an efficient algorithm for its computation. Also identify $[K(1-z)]^{-1}(z - z^{K+1})$ as the generating function of the uniform density on the integers $1, \cdots, K$.

c. First choosing $\{a_k\}$ to be the Poisson density with mean $\alpha = 0.80$, and next as the density with $a_0 = 0.92$, and $a_{10} = 0.08$, find the smallest value(s) of K for which the mean $L(K) \geq 30$. Also compute the 95th percentiles of the corresponding densities $\{x_i(K)\}$.

d. For the same data as in part c, find the smallest value(s) of K for which $\sum_{i=0}^{60} x_i(K) \leq 0.95$, and compute the means $L(K)$ of the corresponding densities $\{x_i(K)\}$.

Problem 4.4.36: The $M/G/1$ Markov Chain: Initially, a box contains i coins. Every morning, one coin is paid for that day's activities and at closing time, $R_n \geq 0$ coins are added to the box. Z_n denotes the amount in the box just after the nth coin has been paid out. The random variables $\{Z_n\}$ are recursively defined by $Z_{n+1} = Z_n - 1 + R_n$, whenever the right hand side is nonnegative. Whenever at the end of a day, the box is empty, a munificent friend adds a random number of coins so that business can continue the next day. The probability density of those amounts is $\{b_k\}$. The amounts are independent of any past history. The probability density $\{b_k\}$ has a finite mean β and generating function $B(z)$.

The intakes R_n are independent, identically distributed with probability density $\{a_k\}$ on the nonnegative integers. To avoid uninteresting cases, assume that $a_0 > 0$ and $a_0 + a_1 < 1$. The generating function of $\{a_k\}$ is $A(z)$ and the corresponding (finite) mean is α.

$\{Z_n\}$ is a Markov chain on the nonnegative integers with

$$P = \begin{vmatrix} b_0 & b_1 & b_2 & b_3 & b_4 & \cdots \\ a_0 & a_1 & a_2 & a_3 & a_4 & \cdots \\ 0 & a_0 & a_1 & a_2 & a_3 & \cdots \\ 0 & 0 & a_0 & a_1 & a_2 & \cdots \\ \cdot & \cdot & \cdot & \cdot & \cdot & \end{vmatrix}.$$

For the positive recurrent chain, $X(z)$ is the generating function of the sequence $\{x_i\}$ of steady-state probabilities of P.

a. Verify that $X(z)$ is given by

$$X(z) = x_0 \frac{zB(z) - A(z)}{z - A(z)}.$$

Refer to a text on Markov chains for a proof of the following: The Markov chain is recurrent if and only if $\alpha \le 1$ and *positive* recurrent only if $\alpha < 1$.

b. Use L'Hospital's rule to show that $x_0 = [1 + \beta(1-\alpha)^{-1}]^{-1}$.

c. Explain why computing the quantities x_i by the recurrence $x_{i+1} = a_0^{-1}[\, x_i - x_0 b_i \; - \sum_{\nu=1}^{i} x_\nu a_{i-\nu+1}\,]$, for $i \ge 0$, is likely to suffer from loss of significance. An equivalent recursion with better numerical properties is obtained as follows: Introduce the notation

$$\hat{a}_\nu = 1 - \sum_{r=0}^{\nu} a_r, \quad \hat{b}_\nu = 1 - \sum_{r=0}^{\nu} b_r, \quad \text{for } \nu \ge 0,$$

and note that

$$\sum_{\nu=0}^{\infty} \hat{a}_\nu z^\nu = \frac{1 - A(z)}{1 - z}, \qquad \sum_{\nu=0}^{\infty} \hat{b}_\nu z^\nu = \frac{1 - B(z)}{1 - z}, \quad \text{for } 0 \le z < 1.$$

Next, derive the equation

$$X(z) = \frac{1 - A(z)}{1 - z} X(z) + x_0 + x_0 \left\{ z \, \frac{1 - B(z)}{1 - z} - \frac{1 - A(z)}{1 - z} \right\},$$

and after series expansion and noting that $\hat{A}_0 = 1 - A_0$, verify that

$$x_i = a_0^{-1} [\, x_0 \hat{b}_{i-1} + \sum_{\nu=1}^{i-1} x_\nu \hat{A}_{i-\nu} \,], \quad \text{for } i \geq 1.$$

d. Write a subroutine to compute the density $\{x_i\}$ for given $\{a_k\}$ and $\{b_k\}$. Test the code on several examples of your choice. To avoid very long-tailed densities, use only examples where $\alpha \leq 0.99$.

Reference: Section 1.4 of Neuts, M. F. *"Structured Stochastic Matrices of M/G/1 Type and Their Applications."* New York: Marcel Dekker Inc., 1989.

The next three problems elaborate on the model in Problem 4.4.36. Each is motivated by a question related to the $M/G/1$ queue. For the information of those familiar with queueing theory, we explain that question in a remark after each problem.

Problem 4.4.37: The following is a result on the factorial moments of the probability density $\{x_i\}$: Provided that $\alpha < 1$ and the $(N+1)$st moments of the probability densities $\{A_k\}$ and $\{B_k\}$ exist, the Nth moment of the probability density $\{x_i\}$ exists and the nth *factorial* moments $X^{(n)}(1-)$ are recursively given by $X^{(0)}(1-) = 1$, and

$$X'(1-) = \frac{1}{2(1-\alpha)} \left\{ A''(1-) + x_0[2B'(1-) + B''(1-) - A''(1-)] \right\},$$

$$X^{(n)}(1-) = \frac{1}{(n+1)(1-\alpha)} \left\{ \sum_{\nu=2}^{n+1} \begin{bmatrix} n+1 \\ \nu \end{bmatrix} A^{(\nu)}(1-) X^{(n-\nu+1)}(1-) \right.$$

$$\left. + x_0 \, [B^{(n+1)}(1-) + (n+1)B^{(n)}(1-) - A^{(n+1)}(1-)] \right\},$$

for $2 \leq n \leq N$, where $A^{(\nu)}(1-)$ and $B^{(\nu)}(1-)$ are the νth factorial moments of $\{A_k\}$ and $\{B_k\}$, respectively. Use this result in a subroutine to compute the first three *central* moments of the density $\{x_i\}$. Test your code on the following two examples. For each, choose three sets of admissible parameters.

1. $\{a_k\}$ and $\{b_k\}$ are Poisson densities with means α and β, respectively.

2. $\{a_k\}$ and $\{b_k\}$ are negative binomial densities corresponding to

$$A(z) = [p_1(1 - z(1 - p_1)^{-1}]^{m_1}, \quad \text{and} \quad B(z) = [p_2(1 - z(1 - p_2)^{-1}]^{m_2},$$

with means $\alpha = m_1(1 - p_1)(p_1)^{-1}$ and $\beta = m_2(1 - p_2)(p_2)^{-1}$, respectively.

Remark: This deals with the first three central moments of the stationary queue length in an $M/G/1$ queue where the *first* service time of a busy period may have a different distribution than the others. The first example deals with *constant* service times; the second with *Erlang* service times of orders m_1 and m_2.

Problem 4.4.38: For the model in Problem 4.4.36, define $\{b_k\}$ by

$$b_k = 0, \qquad \text{for } 0 \le k < j^*,$$

$$= a_{j^*+k}, \qquad \text{for } k \ge j^*.$$

Notice that for $j^* = 0$, the first two rows of the matrix P are identical. As we increase j^*, the elements in the first row are successively shifted one place to the right. Compute the steady-state probabilities for $j^* = 0$ and find the smallest index i^* for which

$$\sum_{i=0}^{i^*} x_i \ge 0.995.$$

The object of this problem is to determine the *largest* value of j^* for which the corresponding steady-state probabilities $\{x'_i\}$ will satisfy

$$\sum_{i=0}^{i^*} x'_i \le 0.90.$$

Examine the probability generating function $X(z;j^*)$ analytically to see if you can express how the steady-state probabilities depend on j^*. Next, write an efficient code to solve this problem for the same $\{a_k\}$ as in Problem 4.4.37.

Remark: Suppose we have an $M/G/1$ queue with an unlimited supply of stored jobs. Whenever the queue becomes empty, we add j^* such customers at the beginning of the next busy period. The index i^* is an indicator of the long queues experienced in the original queue. We wish to add as many stored jobs as indicated without unduly increasing the fraction of time that long queues are experienced. The problem is one way of formalizing that practical question.

Problem 4.4.39: Using the first example data in Problem 4.4.37, examine how the steady-state probabilities of the matrix P vary with the parameter β of the Poisson distribution $\{b_k\}$. Compute tables and plot graphs of the mean and variance of the density $\{x_i\}$ as well as of its 90th and 99th percentiles. Discuss their significance. Examine analytically how the generating function $X(z;\beta)$ depends on β. That should significantly reduce the algorithmic effort.

Remark: Consider the $M/D/1$ queue with constant service times of length 1. At the arrival of a customer to an empty queue there is a constant setup time that augments the service time of that customer. This corresponds to the proposed model with $\beta > \alpha$. The object of the problem is to examine the effect of the setup time on the steady-state queue lengths. Specific analytic relations can be derived. They should be used to avoid brute force computation.

Reference: For ample details, see Section 1.4 of Neuts, M. F. *"Structured Stochastic Matrices of M/G/1 Type and Their Applications."* New York: Marcel Dekker Inc., 1989.

Problem 4.4.40: Consider a sequence of dependent Bernoulli trials in which the probability of success at a trial depends on the outcomes of the preceding two trials. Specifically, if the preceding two outcomes are 00, 01, 10, or 11, the conditional probabilities of success at the next trial are p_1, p_2, p_3, and p_4. The symbols 0 and 1 indicate failure and success, respectively.

a. Show that the sequence of successes is a *MAP*. Set up its coefficient matrices A_0 and A_1 and determine the initial probability vector π corresponding to the stationary version of the process. What is the probability p^* of a success at any given trial in that process?

b. Write a computer program to evaluate for $0 \le k \le n$, the probabilities that in that stationary version, k successes occur in n trials. For various choices of the parameters p_1, p_2, p_3, and p_4 for which p^* is the same, compare these to the binomial density with parameter p^*.

Problem 4.4.41: For the *MAP* with coefficient matrices A_0 and A_1, differentiate in $P^*(z; n+1) = P^*(z; n)[A_0 + A_1 z]$, to obtain the recurrence relations

$$M_1(n + 1) = M_1(n)P + P^n A_1,$$

and

$$M_2(n + 1) = M_2(n)P + 2M_1(n)A_1, \quad \text{for } n \geq 0,$$

for the first and second factorial moment matrices $M_1(n)$ and $M_2(n)$. From the preceding equation, deduce that

$$\pi M_2(n)\mathbf{e} = 2\sum_{k=1}^{n-1} \pi M_1(k)A_1\mathbf{e},$$

and use the formula for $\pi M_1(k)$, given in Section 4.3, to derive an explicit formula for the variance $\sigma^2(n)$ of the number $N(n)$ of arrivals up to time n in the stationary *MAP*. Show that the graph of $\sigma^2(n)$ has a linear asymptote as $n \to \infty$ and derive expressions for its coefficients. Write a program to compute $\sigma^2(n)$ for given coefficient matrices A_0 and A_1.

Problem 4.4.42: For the *MAP* with coefficient matrices A_0 and A_1, twice differentiate the joint generating function

$$\phi^*_{1,n}(z_1, z_n) = \alpha[I - z_1 A_0]^{-1} z_1 A_1 P^{n-2} [I - z_n A_0]^{-1} z_n A_1 \mathbf{e},$$

of the lengths τ_1 and τ_n to obtain formulas for their means $E(\tau_1)$ and $E(\tau_n)$, variances $\mathrm{Var}(\tau_1)$ and $\mathrm{Var}(\tau_1)$, and their correlation coefficient $\rho_{1,n}$.

Note the simplifications in these formulas for $\alpha = \pi$. With the initial probability $\alpha^\circ = (\pi A_1 \mathbf{e})^{-1} \pi A_1$, time zero corresponds to an arbitrary arrival in the stationary version.

Using the initial probability vectors π and α°, write a computer program to evaluate the means, variances and the serial correlations of the random variables $\{\tau_n\}$ for n up to a specified value. What happens to the serial correlations as $n \to \infty$? Can you prove this? Can you provide precise information of the rate of the convergence? Implement your code for various choices of the *MAP* and discuss the qualitative significance of your numerical results.

Problem 4.4.43: Runs of Arrivals: Consider the *MAP* with coefficient matrices A_0 and A_1.

a. Explain why setting the initial probability vector $\boldsymbol{\alpha}$ equal to

$$\boldsymbol{\alpha}^* = (\boldsymbol{\pi} A_0 A_1 \mathbf{e})^{-1} \boldsymbol{\pi} A_0 A_1 \mathbf{e}$$

corresponds to choosing time zero at an arbitrary arrival that is preceded by a time point at which there is no arrival. Next, consider an arbitrary run of arrivals at successive time points in the stationary version of the *MAP*.

b. Show that the probability density $\{\phi_k\}$ of the length R of an arbitrary run of arrivals is $\phi_k = \boldsymbol{\alpha}^* A_1^{k-1} A_2 \mathbf{e}$, for $k \geq 1$. Explain why $\{\phi_k\}$ is a *PH*-density. Derive formulas for its first two moments.

Consider the stationary two-state *MAP* with coefficient matrices

$$A_0 = \begin{vmatrix} p_{11}q_1 & p_{12}q_2 \\ p_{21}q_1 & p_{22}q_2 \end{vmatrix}, \qquad A_1 = \begin{vmatrix} p_{11}p_1 & p_{12}p_2 \\ p_{21}p_1 & p_{22}p_2 \end{vmatrix},$$

where $q_1 = 1 - p_1$ and $q_2 = 1 - p_2$. That is the two-state Markov-modulated Bernoulli process. In the statistical literature, it is also know as the *Markov - Bernoulli* process.

c. Give an explicit expression for the probability p^* that there is an arrival at an arbitrary time.

d. Write a program to compute the density $\{\phi_k\}$ for that *MAP* and compare it to the corresponding density for Bernoulli trials with probability p^*. Select various parameter sets for the Markov-modulated Bernoulli process for which you understand its qualitative behavior well. Do the numerical results obtained agree with your understanding of the process?

e. Derive a formula for the joint density for the lengths R_1 and R_2 of two arbitrary *successive* runs of arrivals. Obtain a measure of the dependence between R_1 and R_2. Compute the joint density for the same examples used in part *b*. Again, compare the numerical results to those for Bernoulli trials with probability p^*.

Problem 4.4.44: Various models for Bernoulli trials in which a success at a given trial modifies the success probability for a random number of subsequent trials are particular cases of discrete *MAP*s. Set up the following two models as *MAP*s and calculate as explicitly as possible the probability $P(k;n)$ of k successes in n trials. For each, also obtain expressions for the conditional probability $P^{\circ}(k;n)$ of k successes in $\{1, \cdots, n\}$ given that there is a success at time zero. Give a formula for the probability p^* of a success at an arbitrary time.

a. Each ordinary success (occurring with probability p) is followed by a geometrically distributed number of trials with success probability p_1. The probability that there are r such trials is $g_r = (1-\theta)^{r-1}\theta$, for $r \geq 1$. After these special trials, the success probability reverts to p.

b. Each ordinary success is followed by m trials with success probability p_1 after which the success probability reverts to p.

c. Perform informative numerical computations to illustrate how these models differ from ordinary Bernoulli trials.

Reference: With $p_1 = 0$, the model in part b leads to a special distribution, known as *Dandekar's modified binomial distribution.* If X is the number of successes in n trials and the probability of success at the *first* trial is p, then

$$P\{X \leq k\} = (1-p)^{n-km} \sum_{j=0}^{k} \binom{n-km+j-1}{j} p^j,$$

for $0 \leq k \leq [n/m]$. See Norman L. Johnson, Samuel Kotz and Adrienne Kemp, *"Univariate Statistical Distributions,"* (Second Edition), New York: John Wiley & Sons, p. 435, 1992.

Problem 4.4.45: The following is a simple scheme to generate a sequence of *dependent* Bernoulli trials. Consider the *MAP* with coefficient matrices

$$C_0 = \begin{vmatrix} \alpha & \beta & 0 \\ q\beta & q\alpha & 0 \\ 0 & 0 & 0 \end{vmatrix}, \qquad C_1 = \begin{vmatrix} 0 & 0 & 0 \\ 0 & p\alpha & p\beta \\ 0 & \beta & \alpha \end{vmatrix},$$

where p and α are probabilities and $q = 1-p$, $\beta = 1-\alpha$. The initial probability vector $\boldsymbol{\gamma} = (0, 1, 0)$.

a. Verify that the invariant probability vector $\boldsymbol{\theta}$ of the stochastic matrix $C = C_0 + C_1$ is $1/2(q, 1, p)$ and that $\boldsymbol{\theta} C_1 \mathbf{e} = p$. The stationary version of the *MAP* is therefore a sequence of dependent Bernoulli trials with individual probability p of success.

b. Verify that

$$\boldsymbol{\gamma} C^{n-1} C_1 \mathbf{e} = p, \quad \text{for } n \geq 1. \tag{1}$$

Therefore, also with the initial probability vector $\boldsymbol{\gamma}$, we obtain (dependent) Bernoulli trials with probability p of success. In contrast to the sequence in part a, that sequence is not stationary.

To verify the equality (1), show that the eigenvalues of C are 1, α, and $\alpha - \beta$, with the corresponding pairs of left and right eigenvectors

$$\begin{array}{lll} 1, & (q, 1, p) & (1, 1, 1)^T, \\ \alpha, & (1, 0, -1) & (p, 0, -q)^T, \\ \alpha - \beta, & (-q, 1, -p) & (1, -1, 1)^T. \end{array}$$

Write the spectral decomposition $C = H^{-1} \Lambda H$ of the matrix C and use it to evaluate the matrix C^n explicitly. By a final simple calculation, you should obtain the equality (1).

c. Notice that for $\alpha = 1$, we obtain a degenerate case of the *MAP* in which states 1 and 3 are never visited. The *MAP* then reduces to ordinary Bernoulli trials. What is the behavior of the *MAP* for $\alpha = 0$?

d. For $p = 1/2$, compute a table of the probabilities of the 32 possible strings of zeros and ones (failures and successes) that can result from the first five trials. Compute these tables using the initial probability vectors $\boldsymbol{\theta}$ and $\boldsymbol{\gamma}$ and for $\alpha = 0.99, 0.75, 0.50$, and 0.25. See if the effect of the dependence and/or stationarity of the sequence is apparent in your numerical results.

e. For the same α as in part d and for the initial vectors $\boldsymbol{\gamma}$ and $\boldsymbol{\theta}$, compute and compare tables of the probabilities $P(k; n)$ of k, $0 \leq k \leq n$, successes in n trials. Implement your code for $p = 0.25, 0.50, 0.75$, and $n = 20$.

f. We describe the construction of a random sequence of zeros and ones. Explain why it is stochastically equivalent to the sequence generated by the *MAP* with the initial probability vector $\boldsymbol{\gamma}$.

Starting from $N_0 = 0$, we form a discrete renewal process $\{N_k\}$ with geometric lifetimes, i.e., the random variables $N_k - N_{k-1}$, $k \geq 1$, have the geometric density $h_r = \alpha^{r-1}(1-\alpha)$, $r \geq 1$. If $N_1 > 1$, we perform independent Bernoulli trials with probability p at times $1, \cdots, N_1 - 1$. Points where successes occur are marked with a 1, the others with a 0. At time N_1, we perform a Bernoulli trial with probability p. If a success occurs, all points $N_1, \cdots, N_2 - 1$ are labeled 1; otherwise are all marked 0. At the times $N_2, \cdots, N_3 - 1$, we perform Bernoulli trials with probability p and mark them as for the first interval. That construction is continued indefinitely. For the intervals $N_{2v}, \cdots, N_{2v+1} - 1$, we mark all points depending on the outcome (1 or 0) at time N_{2v}. In the intervals $N_{2v-1}, \cdots, N_{2v} - 1$, we mark as for the first interval.

Also discuss how the scheme should be modified to obtain the stationary sequence corresponding to the initial vector $\boldsymbol{\theta}$.

Problem 4.4.46: The following simple scheme generates a sequence of *dependent* Bernoulli trials with probability 1/2. Consider the *MAP* with

$$C_0 = \begin{vmatrix} \alpha & \beta & 0 \\ 0 & p\alpha & p\beta \\ 0 & 0 & 0 \end{vmatrix}, \qquad C_1 = \begin{vmatrix} 0 & 0 & 0 \\ q\beta & q\alpha & 0 \\ 0 & \beta & \alpha \end{vmatrix},$$

where p and α are probabilities and $q = 1 - p$, $\beta = 1 - \alpha$. The initial probability vector $\boldsymbol{\gamma} = (0, 1, 0)$.

a. For the probability vector $\boldsymbol{\theta}$ in Problem 4.4.45, verify that $\boldsymbol{\theta} C_1 \mathbf{e} = 1/2$. Therefore, the stationary *MAP* consists of dependent Bernoulli trials of probability 1/2.

b. Use the spectral decomposition of the matrix C in Problem 4.4.45 to prove that for all $n \geq 1$,

$$\boldsymbol{\gamma} C^{n-1} C_1 \mathbf{e} = \frac{1}{2} + \frac{1}{2}(q - p)(\alpha - \beta)^n.$$

This shows that, with either $p = 1/2$ or $\alpha = 1/2$, the success probability at any n is 1/2, also with the initial probability vector $\boldsymbol{\gamma}$,

c. Discuss a construction analogous to that in part *f* of Problem 4.4.45 for the present model. What effect do changes in p and/or α have on the typical behavior of the random sequence?

Remark: Problems 4.4.45 and 4.4.46 were suggested by Problem 10201 in *The American Mathematical Monthly,* 99, p. 163, 1992, proposed by Gunnar Blom.

Problem 4.4.47: As Example 4.1.1, the irreducible transition probability matrix P is partitioned as

$$P = \begin{array}{c} 1 \\ \\ 2 \end{array} \left| \begin{array}{cc} T(1,1) & T(1,2) \\ & \\ T(2,1) & T(2,2) \end{array} \right| .$$

Its invariant probability vector $\boldsymbol{\pi}$ is accordingly written as $[\boldsymbol{\pi}(1), \boldsymbol{\pi}(2)]$. The matrix obtained by replacing $T(1,2)$ and $T(2,2)$ by zero matrices is denoted by D_1. The matrix D_2 is similarly constructed by replacing the two left blocks by zero matrices. Immediately after each transition in the chain P, a Bernoulli trial with positive probability p is performed. At each success, we record whether the chain is in 1 or in 2.

a. Show that, if the initial probability vector is $\boldsymbol{\alpha}$, the time τ_1 until the chain is *recorded* for the first time as being in 1, has a *PH*-density with representation $[\boldsymbol{\alpha}, qP + pD_2]$.

b. Show that the times when the chain is *recorded* as being in the set 1 form a *MAP* and write down its coefficient matrices $C_0(p)$ and $C_1(p)$. Show that the probability $\phi(p)$ that, at an arbitrary time, the chain is *recorded* as being in 1 is $p\boldsymbol{\pi}(1)\mathbf{e}$. Suppose that at an arbitrary time (in steady state) the chain is recorded as being in 1. What is the probability density of the time until it is again so recorded?

HARDER PROBLEMS

Problem 4.4.48: A random walk on the integers $0, 1, \cdots, K+1$ has absorbing states 0 and $K+1$. For $1 \leq i \leq K$, the probabilities of moving to $i+1$ or $i-1$ are $p(i)$ and $q(i) = 1 - p(i)$, respectively, with $0 < p(i) < 1$.

The walk is started in state j, $1 \leq j \leq K$. We say that the walk has a *reversal* if a step to the right is followed by a step to the left, or vice versa.

a. How would you compute the probability that at time n, the walk has not been absorbed and has experienced k reversals up to that time?

b. How would you evaluate the probability density of the number R of reversals before absorption?

c. Implement your algorithm for randomly generated data: Set $K = 21$, $j = 11$, $n = 100$, and choose the $p(i)$, $1 \leq i \leq 100$, independently and at random in the interval (0.3, 0.6).

Hint: Define appropriate probabilities for which simple recurrence relations are available. You will need a three-dimensional storage array or you can use two-dimensional arrays efficiently.

Problem 4.4.49: The Range of a Random Walk: For the random walk in Problem 4.4.48 with initial state j, $P(i;\, i_1, i_2;\, n)$ is the conditional probability that at time n, the chain is in the state i and i_1 and i_2 are respectively the left and rightmost states visited up to that time.

a. Establish recurrence relations for the probabilities $P(i;\, i_1, i_2;\, n)$. Discuss how, by making good use of storage arrays, the probabilities for successive n can efficiently be updated.

b. The difference Y_n between the right- and leftmost states visited is called the *range* of the random walk at time n. $R(i;\, k;\, n)$ is the conditional probability that, at time n the walk is in state i and the difference between the right- and the leftmost states visited up to n is k. Discuss how the probabilities $R(i;\, k;\, n)$ are related to the $P(i;\, i_1, i_2;\, n)$. What is the interpretation of the conditional probabilities $P(i;\, 0, i_2;\, n)$ and $P(i;\, i_1, K+1;\, n)$?

c. Compute and print the probability densities of the range of the random walk for $1 \leq n \leq 50$. Implement your algorithm for randomly generated data as follows: Set $K = 21$, $j = 11$, and choose the $p(i)$, $1 \leq i \leq 100$, independently and at random in the interval (0.4, 0.6).

Problem 4.4.50: A random walk on the integers i, with $0 \leq i \leq a$, $a > 1$, is performed as follows: From a position i with $0 < i < a$, the walk moves with probability 1/2 to a position chosen at random from $0, 1, \cdots, i-1$ or with probability 1/2 to a position chosen at random from $i+1, \cdots, a$.

From 0 or a, the next position is picked at random from among the other points.

a. Write a subroutine to generate the transition probability matrix $P(a)$ of the Markov chain that describes this random walk.

b. For $3 \le a \le 50$, compute the invariant probability vector $\pi(a)$ of $P(a)$. By examining the steady-state equations, it is possible to obtain a simple recurrence (or a somewhat more involved explicit formula). For selected values of a, plot these probabilities by placing the masses $\pi_j(a)$ at the points j/a in the interval [0, 1]. For reference, overlay these plots with a graph of the function

$$\phi(v) = \frac{1}{\pi} v^{-1/2}(1 - v)^{-1/2}, \quad \text{for } 0 < v < 1.$$

We may think of the random walk as moving on the points with abscissae j/a, $0 \le j \le a$ in [0, 1].

c. Modify your computer program to allow for different transition probabilities from states 0 or a. For example, upon reaching 0 the chain may jump to some state $i_0 \ne 0$, while from a it may move to a state $i_1 \ne a$. There are many ways of modifying the behavior of the walk at the boundary states; just select one and encode it into your program. For the same values of a as in part *b* again compute the steady-state probabilities and draw the corresponding graphs for the same a as chosen earlier.

d. The purpose of the comparison with the function $\phi(\cdot)$ is the following: The continuous analogue of the model is the random walk on the interval (0, 1) in which a point moves with probability 1/2 to a random point in the interval, either to the left or to the right of its present position. If $\phi(\cdot)$ denotes that stationary probability density of that Markov process on (0, 1), a heuristic argument (made rigorous in the theory of Markov processes on general state spaces) yields that $\phi(\cdot)$ is symmetric about 1/2 and satisfies the integral equation

$$\phi(y) = \int_0^y \frac{\phi(u)}{2(1 - u)}\, du + \int_y^1 \frac{\phi(u)}{2u}\, du,$$

for $0 < y < 1$. Differentiation with respect to y yields the differential equation

$$\frac{\phi'(y)}{\phi(y)} = \frac{2y-1}{2y(1-y)}.$$

Upon integration and using the fact that $\phi(\cdot)$ is a probability density, the stated expression for $\phi(\cdot)$ is obtained. That density is known as the *arc-sine law*. It plays an important role in probability theory. Verify and complete the analytic calculations sketched here.

Problem 4.4.51: The stationary probabilities of the Markov chain described in Problem 4.4.50 may be written as fractions with a common denominator. For example, the vector $\boldsymbol{\pi}(5)$ is given by

$$\boldsymbol{\pi}(5) = \frac{1}{126}[25, 20, 18, 18, 20, 25].$$

Write a computer code that solves the steady-state equations of the stochastic matrix $P(a)$ as efficiently as possible (using the symmetry of the solution) and *in exact fractional form.*

Hint: The steady-state equations can be solved by Gauss elimination in which all successive tableaus contain only integers. In pivoting and in ultimately writing the solution in reduced fractions, one must often determine the least common multiple of sets of integers. The code for the solution of this problem mainly requires meticulous planning.

Problem 4.4.52: Write a computer program to simulate the behavior of the random walk described in Problem 4.4.50, allowing for different specifications of the transition probabilities from the boundary states 0 and a, i.e., allow the user to specify different transition probabilities from 0 to the other states and also from a to the other states. Also simulate the continuous Markov process on (0, 1) that is described there. Explain how the simulated behavior of these Markov chains is reflected in the steady-state probabilities computed in Problem 4.4.50.

Problem 4.4.53: The integer-valued random variables $\{X_n\}$, $n \geq 1$, are independent, identically distributed with common density $\{p(k)\}$, $0 \leq k \leq m$, and $p(m) > 0$. The jth random variable is a *record* if and only if $X_j > \max(X_1, \cdots, X_{j-1})$. Obviously, there can be at most m records in the infinite sequence $\{X_n\}$. Set $X_0 = 0$. Z_i is the value of the ith record.

Show that the random variables $\{Z_i\}$ form a Markov chain with states $1, \cdots, m$. The state m is absorbing. Show that elements of the transition probability matrix R of that chain on and below the diagonal are zero. Compute the probabilities $r(\nu)$, $1 \le \nu \le m$, that there are exactly ν records in the infinite sequence $\{X_n\}$. Execute your program for the binomial density on $0, \cdots, 50$. Examine a sufficiently large number of values of the parameter p, so that you can draw sound qualitative conclusions about the number of records seen for increasing values of p.

Problem 4.4.54: Consider the same model as in Problem 4.4.53. This problem deals with the density of the number $M(n)$ of records among $X_1, \cdots, X_n$. Its computation is more involved. Set $Y_n = \max(X_1, \cdots, X_n)$. $L(n)$ is the index of the last record among $X_1, \cdots, X_n$. Note that $Y_n = X_{L(n)}$, provided that Y_n is positive. If $Y_n = 0$, we set $M(n) = L(n) = 0$. This last case is easy to handle since

$$P\{M(n) = 0\} = P\{Y_n = 0\} = [p(0)]^n, \quad \text{for} \quad n \ge 1.$$

a. Verify that the joint probability density of the triple $\{M(n), Y_n, L(n)\}$ may be recursively computed as follows: $P\{M(n) = \nu, Y_n = j, L(n) = r\}$ depends on the indices n, ν, j, and r. The index n is specified by the user. The index ν is usually the simplest on which to perform a recursion. The number of records is typically much smaller than n (for some special cases one can show that it is of the order of magnitude of $\log n$). Recurrence relations on ν are the most straightforward and easiest to code. First show that

$$P\{M(n) = 1, Y_n = j, L(n) = r\} = p_0^{r-1} p_j (p_0 + \cdots + p_j)^{n-r},$$

for $1 \le r \le n$ and $1 \le j \le m$. Compute these quantities up to the specified value of n. Next, show that for $\nu > 1$:

$$P\{M(n) = \nu, Y_n = j, L(n) = r\}$$

$$= P\{M(r-1) = \nu, Y_{r-1} \le j - 1\} p_j (p_0 + \cdots + p_j)^{n-r},$$

for $\nu \le r \le n$ and $\nu \le j \le m$. From the joint probability density of $\{M(n), Y_n, L(n)\}$, the marginal densities of $M(n)$, Y_n, and $L(n)$ are routinely computed. The marginal density of Y_n can easily be evaluated directly. It can serve as an accuracy check on the other computations.

b. Program the recursive algorithm to obtain the marginal densities of $M(n)$ and $L(n)$ for $n = 25, \cdots, 75 = n^*$, and for the same binomial density as in Problem 4.4.53. Organize your program to make efficient use of memory storage. For large values of m and n^*, that is important.

Problem 4.4.55: The transition probability matrix P of an *irreducible* Markov chain is written in the partitioned form

$$P = \begin{array}{c|cc|} 1 & T(1, 1) & T(1, 2) \\ 2 & T(2, 1) & T(2, 2) \end{array}.$$

a. The invariant probability vector $\boldsymbol{\pi}$ of P is accordingly partitioned as $[\boldsymbol{\pi}(1), \boldsymbol{\pi}(2)]$. Express $\boldsymbol{\pi}(1)$ and $\boldsymbol{\pi}(2)$ in terms the matrices $T(i, j)$, $1 \leq i, j \leq 2$.

b. Verify that $\boldsymbol{\pi}(1)T(1, 2)\mathbf{e} = \boldsymbol{\pi}(2)T(2, 1)\mathbf{e}$, and offer an intuitive explanation for that equality.

c. We wish to start the Markov chain at the epoch of an arbitrary entry into the set 1 in steady state. Explain why that is accomplished by choosing the initial probability vector as $[\boldsymbol{\alpha}^*(1), \mathbf{0}]$, where

$$\boldsymbol{\alpha}^*(1) = \frac{\boldsymbol{\pi}(2)T(2, 1)}{\boldsymbol{\pi}(2)T(2, 1)\mathbf{e}}.$$

d. Use the significance of the vector $[\boldsymbol{\alpha}^*(1), \mathbf{0}]$, to show that the generating function of the duration T^* of an arbitrary sojourn in 1 is given by:

$$\phi(u) = u\boldsymbol{\alpha}^*(1)[I - uT(1, 1)]^{-1}T(1, 2)\mathbf{e}.$$

Derive the following formulas for the moments $\mu'_1 = E(T^*)$ and $\mu'_2 = E(T^{*\,2})$ of T^*:

$$\mu'_1 = \frac{\boldsymbol{\pi}(1)\mathbf{e}}{\boldsymbol{\pi}(1)T(1, 2)\mathbf{e}}, \qquad \mu'_2 = \frac{2\boldsymbol{\pi}(1)[I - T(1, 1)]^{-1}\mathbf{e}}{\boldsymbol{\pi}(1)T(1, 2)\mathbf{e}} - \mu'_1.$$

e. With the same initial conditions, consider the durations T_1^* and T_n^* of the first and the nth sojourns in set 1, $n \geq 2$. Show that their joint probability generating function $\phi(z_1, z_2)$ is

$$\phi(z_1, z_2) = z_1 \boldsymbol{\alpha}^*(1)[I - z_1 T(1, 1)]^{-1}[K(1)]^{n-2}$$

$$\times T(1, 2)[I - T(2, 2)]^{-1} z_2 T(2, 1)[I - z_2 T(1, 1)]^{-1} T(1, 2)\mathbf{e}.$$

The matrix $K(1)$, of the same dimension as the number of states in **1**, is defined by $K(1) = T(1, 2)[I - T(2, 2)]^{-1} T(2, 1)[I - T(1, 1)]^{-1}$. Note that $\boldsymbol{\pi}(1)K(1) = \boldsymbol{\pi}(1)$. Verify that the matrix

$$L(1) = [I - T(2, 2)]^{-1} T(2, 1)[I - T(1, 1)]^{-1} T(1, 2),$$

is stochastic and has the invariant probability vector

$$\boldsymbol{\alpha}^*(2) = [\boldsymbol{\pi}(1)T(1, 2)\mathbf{e}]^{-1} \boldsymbol{\pi}(1)T(1, 2).$$

e. Show that the cross-moment $E(T_1^* T_n^*)$ is given by

$$E(T_1^* T_n^*) = [\boldsymbol{\pi}(1)T(1, 2)\mathbf{e}]^{-1} \boldsymbol{\pi}(1)[I - T(1, 1)]^{-1}[K(1)]^{n-1}\mathbf{e}$$

$$= [\boldsymbol{\pi}(1)T(1, 2)\mathbf{e}]^{-1} \boldsymbol{\pi}(1)[L(1)]^{n-1}[I - T(1, 1)]^{-1}\mathbf{e}.$$

Use this and preceding formulas to compute the correlation coefficients of T_1^* and T_n^* for successive values of n.

f. Test your code for the Markov chain with transition probability matrix

1	0.25	0.10	0.0	0.65
2	0.05	0.80	0.0	0.15
3	0.0	0.10	0.90	0.0
4	0.15	0.10	0.10	0.65

and with **1** = {1, 2} and **2** = {3, 4}.

Problem 4.4.56: Consider the Markov chain in Problem 4.4.25 for $\rho = 0.80$ and for the data sets given there. Set $N = 20$ and study the sojourn times in the set of states $\{16, \cdots, 20\}$ by using the results in Problem 4.4.55. Compare your numerical results for Erlang service times of several orders m and for hyperexponential distributions of different coefficients of variation. What qualitative conclusions about the behavior of the queue can you draw from these examples?

Remark: Note that the set 1 is now $\{16, \cdots, 20\}$. The notation in Problem 4.4.55 needs adjusting.

Problem 4.4.57: This problem deals with the numbers $N_2, \cdots, N_m$ of visits to the states $2, \cdots, m$ *between returns* to the state 1 in an irreducible m-state Markov chain with transition probability matrix P. We write P as

$$P = \begin{vmatrix} P_{11} & \mathbf{a} \\ \mathbf{T}^{\circ} & T \end{vmatrix},$$

where T is a matrix of order $m - 1$, $\mathbf{T}^{\circ}$ a column vector, and $\mathbf{a}$ a row vector. Let $\Delta(\mathbf{z})$ be a diagonal matrix with diagonal elements $z_2, \cdots, z_m$. $\Delta_1(i)$ is a diagonal matrix whose only nonzero element is a 1 at (i, i).

a. Show that the joint probability generating function $F(\mathbf{z})$ of the random variables $N_2, \cdots, N_m$ is $F(\mathbf{z}) = P_{11} + \mathbf{a}\Delta(\mathbf{z})[I - T\Delta(\mathbf{z})]^{-1}\mathbf{T}^{\circ}$.

b. Writing $K(\mathbf{z}) = [I - T\Delta(\mathbf{z})]^{-1}$, show that

$$\frac{\partial}{\partial z_i} K(\mathbf{z}) = K(\mathbf{z}) T \Delta_1(i) K(\mathbf{z}),$$

and

$$\frac{\partial^2}{\partial z_i \partial z_j} K(\mathbf{z}) = K(\mathbf{z}) T \Delta_1(i) K(\mathbf{z}) T \Delta_1(j) K(\mathbf{z}) + K(\mathbf{z}) T \Delta_1(j) K(\mathbf{z}) T \Delta_1(i) K(\mathbf{z}).$$

Express the first two partial derivatives of $F(\mathbf{z})$ in terms of $K(\mathbf{z})$.

c. By setting $z_k = 1$ for $2 \le k \le m$, and simplifying, show that

$$M(1;\, i) = E(N_i) = \mathbf{a}(I - T)^{-1}\Delta_1(i)\mathbf{e}.$$

Note that $M(1;\, i)$ is just the component with index i of the vector $\mathbf{a}(I - T)^{-1}$.

d. Verify that the variance $V(1;\, i)$ of N_i is given by

$$V(1;\, i) = 2\mathbf{a}(I - T)^{-1}\Delta_1(i)(I - T)^{-1}\Delta_1(i)\mathbf{e} - M(1;\, i) - [M(1;\, i)]^2,$$

and the covariance $R(1;\, i, j)$ of N_i and N_j with $i \neq j$, by

$$R(1;\, i, j) = \mathbf{a}(I - T)^{-1}\Delta_1(i)(I - T)^{-1}\Delta_1(j)\mathbf{e}$$

$$+ \mathbf{a}(I - T)^{-1}\Delta_1(j)(I - T)^{-1}\Delta_1(i)\mathbf{e} - M(1;\, i)M(1;\, j).$$

e. Write a subroutine to evaluate the quantities $M(1;\, i)$, $V(1;\, i)$ and $R(1;\, i, j)$ for all values of i and j. Organize the subroutine so that it can be called for an arbitrary irreducible stochastic matrix P with any state designated as state 1. Use the fact that the first two moments are easily computed in terms of elements of the inverse of $I - T$.

f. Test your subroutine for a number of stochastic matrices of your choice. Discuss any qualitative conclusions that can be drawn from the values of these first and second moments.

Hint: In simplifying the matrix formulas, remember that $(I - T)^{-1}\mathbf{T}^{\circ} = \mathbf{e}$ and that $\Delta_1(i)\Delta_1(j) = \delta_{ij}\Delta_1(i)$, where δ_{ij} is the Kronecker delta.

Remark: An interesting application of this problem is the study of the sojourns of the embedded Markov chain of an $M/G/1$ queue with finite capacity in its higher states. For application, see Problem 4.4.58.

Problem 4.4.58: For the $M/G/1$ queue of finite capacity K, let N_i, $1 \leq i \leq K$, be the number of times a departure leaves behind i customers during a *busy period* of the queue. Use the results of Problem 4.4.57 to compute the means EN_i and the covariance matrix $\{E(N_i - EN_i)(N_j - EN_j)\}$. Select three different service time distributions $H(\cdot)$ with mean one and successively set $\lambda = 0.5, 0.75, 1.0$. For all nine cases, implement your algorithm with $K = 15$ and $K = 20$. Compare the numerical results and discuss what they tell about the behavior of the queue.

Remark: A little caution is needed since the busy period starts in state 1 and ends in state 0. The count N_1 must be increased by 1 to account for the initial visit to state 1.

The following five problems deal with the computation of the Perron - Frobenius eigenvalue η of an irreducible nonnegative matrix A. The notation of Appendix 1 is used throughout.

Problem 4.4.59: The Power Method: This is a naive method suitable for aperiodic matrices of low order that are not close to periodic. It has the same drawbacks as the power method for the computation of the steady-state probabilities of an irreducible finite Markov chain. For an aperiodic matrix A of order m, it is easily shown that A^n is positive for $n \geq m$. Recalling the limit property

$$A^n = \eta^n \mathbf{v}\mathbf{u} + o(\eta^n),$$

we see that

$$\frac{A_{1j}^n}{A_{11}^n} \rightarrow \frac{u_j}{u_1} \quad \text{and} \quad \frac{A_{i1}^n}{A_{11}^n} \rightarrow \frac{v_i}{v_1}.$$

By successively forming the powers A^{2^k} of A until the values of the ratios remain nearly constant, we obtain the components of the eigenvectors $\mathbf{u}$ and $\mathbf{v}$ up to multiplicative constants. The computed vectors are normalized by $\mathbf{ue} = 1$ and $\mathbf{uv} = 1$. Computed values of η and a partial accuracy check are obtained from the equalities

$$\eta = \frac{(\mathbf{u}A)_i}{u_i}, \quad \text{and} \quad \eta = \frac{(A\mathbf{v})_i}{v_i}.$$

The choice of the diagonal element A_{11}^n is arbitrary. If it is very large or very small compared to the other diagonal elements, another choice is preferable. What changes are then needed in the preceding discussion?

Write a subroutine to implement the power method. Test your code on a number of examples such as matrices with large positive elements, nearly periodic matrices, and others with potential numerical problems. One such difficulty arises when η is significantly larger than one. Some or all elements of A^n then grow rapidly. A similar problem arises when η is very small. A partial remedy is to rescale A^n to ensure that $A^n_{11} = 1$. How would you implement that in your code? What can happen if you chose the wrong diagonal element?

Problem 4.4.60: Elsner's Method: This method is based on a variational property of the Perron - Frobenius eigenvalue. For a *positive* row vector **x** we form the ratios

$$\frac{[\mathbf{x}A]_j}{x_j} = C_j(\mathbf{x}), \qquad \text{for } 1 \le j \le m.$$

The Perron - Frobenius eigenvalue η then satisfies

$$\max_{\mathbf{x}} \min_{1 \le j \le m} C_j(\mathbf{x}) = \eta = \min_{\mathbf{x}} \max_{1 \le j \le m} C_j(\mathbf{x}).$$

The idea is to construct a sequence of vectors $\mathbf{x}_n$ for which the gap between the largest and the smallest ratios $C_j(\mathbf{x}_n)$ tends to zero. That is accomplished by an iterative scheme due to Elsner. The proof of its convergence is lengthy and quite technical.

Define $\mathbf{x}_0$ to be an arbitrary positive vector with $\mathbf{x}_0\mathbf{e} = 1$. For successive n, ν_n, and μ_n are indices for which

$$C_{\nu_n}(\mathbf{x}_n) \le C_j(\mathbf{x}_n) \le C_{\mu_n}(\mathbf{x}_n), \quad \text{for } 1 \le j \le m.$$

For a fixed constant α , $0 < \alpha < 1$, we define the quantity d_n by

$$d_n = \frac{C_{\nu_n}(\mathbf{x}_n) - A_{\nu_n \nu_n}}{C_{\nu_n}(\mathbf{x}_n) - A_{\nu_n \nu_n} + \alpha[C_{\mu_n}(\mathbf{x}_n) - C_{\nu_n}(\mathbf{x}_n)]},$$

and we construct the vector $\mathbf{x}_{n+1}$ by setting

$$[\mathbf{x}_{n+1}]_j = \frac{[\mathbf{x}_n]_j}{1 - (1 - d_n)[\mathbf{x}_n]_{\nu_n}} \qquad \text{for } j \ne \nu_n,$$

$$= \frac{d_n[\mathbf{x}_n]_j}{1-(1-d_n)[\mathbf{x}_n]_{\nu_n}} \quad \text{for } j = \nu_n.$$

Elsner proved that

$$\lim_{n\to\infty} C_{\mu_n}(\mathbf{x}_n) = \eta = \lim_{n\to\infty} C_{\nu_n}(\mathbf{x}_n), \quad \text{and} \quad \lim_{n\to\infty} \mathbf{x}_n = \mathbf{u}.$$

a. Write and test a code to compute η and $\mathbf{u}$ by Elsner's procedure. Set the parameter $\alpha = 0.75$. Stop at the first n for which

$$C_{\mu_n}(\mathbf{x}_n) - C_{\nu_n}(\mathbf{x}_n) < 10^{-7}.$$

If you have a graphics unit, visualize the effect of the successive steps by plotting a graph of the $C_j(\mathbf{x}_n)$ for $1 \le j \le m$. See how that graph evolves as n increases.

b. The parameter α affects the convergence rate in a complicated manner. On empirical grounds, a choice between 0.5 and 0.75 has been recommended. For the matrix A in one of your test examples and with the same initial vector $\mathbf{x}_0$, implement the algorithm with $\alpha = 0.05k$, $1 \le k \le 19$. For each k, report the number of iterations until the stopping criterion is satisfied.

Problem 4.4.61: Quasi-Stationary Densities: For an irreducible m-state Markov chain, P is partitioned as

$$P = \begin{array}{c|cc|} 1 & T(1,1) & T(1,2) \\ 2 & T(2,1) & T(2,2) \end{array}.$$

The $m_1 \times m_1$ matrix $T(1, 1)$ is irreducible and aperiodic. Its Perron - Frobenius eigenvalue and the corresponding left and right eigenvectors are η, $\mathbf{u}$, and $\mathbf{v}$. The eigenvectors are normalized as in Appendix 1.

a. Show that the limit property

$$[T(1,1)]^n = \eta^n \mathbf{vu} + o(\eta^n), \quad \text{as } n \to \infty,$$

implies that for any nonnegative row vector $\boldsymbol{\alpha} \neq \mathbf{0}$,

$$[\boldsymbol{\alpha}T(1, 1)^n \mathbf{e}]^{-1}\boldsymbol{\alpha}T(1, 1)^n \rightarrow \mathbf{u}, \quad \text{as } n \rightarrow \infty,$$

What does that limit formula tell us? The probability vector $\mathbf{u}$ is called the *quasi-stationary density* of set 1.

b. If the row vector $(\mathbf{u}, 0)$ is chosen as the initial probability vector of P, what is the probability density of the time to the first visit to the set 2? Can you now explain the probabilistic significance of η? For any probability vector $(\boldsymbol{\alpha}, 0)$, the time to the first exit from **1** has a *PH*-density. When $T(1, 1)$ satisfies the assumptions of this problem, what is η for that *PH*-density?

Problem 4.4.62: Coincidences: Consider two independent, irreducible, aperiodic Markov chains with transition probabilities P_1 and P_2. The chains have m_1 and m_2 states and their invariant probability vectors are $\boldsymbol{\pi}(1)$ and $\boldsymbol{\pi}(2)$. A *coincidence* occurs whenever both chains are in state 1.

a. Explain why the times Z_k between successive coincidences are independent and identically distributed. By a simple argument show that

$$E(Z_k) = \frac{1}{\pi_1(1)\pi_1(2)}.$$

b. Compute the probability density of Z_k for

$$P_1 = \begin{vmatrix} 0.2 & 0.1 & 0.2 & 0.5 \\ 0.1 & 0.0 & 0.9 & 0.0 \\ 0.0 & 0.2 & 0.4 & 0.4 \\ 0.3 & 0.1 & 0.1 & 0.5 \end{vmatrix}, \qquad P_2 = \begin{vmatrix} 0.2 & 0.4 & 0.4 \\ 0.3 & 0.1 & 0.6 \\ 0.1 & 0.3 & 0.6 \end{vmatrix}.$$

Hint: Partition the matrices P_1 and P_2 as

$$P_1 = \begin{vmatrix} a(1) & \mathbf{b}(1) \\ \mathbf{c}(1) & T(1) \end{vmatrix}, \qquad P_2 = \begin{vmatrix} a(2) & \mathbf{b}(2) \\ \mathbf{c}(2) & T(2) \end{vmatrix},$$

and show that $Z_k - 1$ is the absorption time in the Markov chain with transition probability matrix

$$\begin{array}{c|cccc|}
* & 1 & 0 & 0 & 0 \\
1^1 & a(1)\mathbf{c}(2) & a(1)T(2) & \mathbf{b}(1)\otimes \mathbf{c}(2) & \mathbf{b}(1)\otimes T(2) \\
1^2 & a(2)\mathbf{c}(1) & \mathbf{c}(1)\otimes \mathbf{b}(2) & a(2)T(1) & T(1)\otimes \mathbf{b}(2) \\
2 & \mathbf{c}(1)\otimes \mathbf{c}(2) & \mathbf{c}(1)\otimes T(2) & T(1)\otimes \mathbf{c}(2) & T(1)\otimes T(2)
\end{array} .$$

In the above, 1^1 is the set of states when the first chain is in state 1 and the second is not; 1^2 is similarly defined with the roles of the Markov chains reversed. The set 2 consists of the pairs (j, j') where j and j' differ from 1. The initial probability vector is $[a(1)a(2), a(2)\mathbf{b}(1), a(1)\mathbf{b}(2), \mathbf{b}(1)\otimes \mathbf{b}(2)]$.

Reference: For further details and applications, see Neuts, M. F., "An algorithm for the distribution of the time between coincidences of two independent *PH*-renewal processes," *Zastosowania Matematyki,* 20, 1 - 13, 1988, and the references cited therein. When the density of Z_k has a long tail, it is more useful to compute its geometric asymptote.

Problem 4.4.63: Consider the Markov chain in Problem 4.4.25 for $\rho = 0.80$ and for the data sets given there. Set $N = 20$ and consider the sojourn times in the set of states $\{16, \cdots, 20\}$. The generating function $\phi(u)$ of the density of an arbitrary sojourn time is given in part d of Problem 4.4.55. Verify that the corresponding density is given by

$$\phi_k = \alpha^*(1)[T(1, 1)]^{k-1}T(1, 2)\mathbf{e}, \quad \text{for } k \geq 1.$$

Note that the principal submatrix $T(1,1)$ corresponding to the row and column indices $\{16, \cdots, 20\}$ is irreducible. The density $\{\phi_k\}$ therefore has a geometric asymptote, i.e.,

$$\phi_k \approx \alpha^*(1)\mathbf{v} \cdot \mathbf{u}T(1,2)\mathbf{e}\, \eta^{k-1} \quad \text{as } k \to \infty,$$

where $\mathbf{u}$ and $\mathbf{v}$ are respectively the left and right eigenvectors corresponding to the Perron - Frobenius eigenvalue η of $T(1, 1)$. The eigenvectors are normalized by $\mathbf{ue} = \mathbf{uv} = 1$. For the data in Problem 4.4.25, compute the exact probability densities $\{\phi_k\}$. Compare their tails to the approximation given by the asymptotic formula. Discuss the significance of your examples. What information does η provide?

Problem 4.4.64: For the finite capacity $M/G/1$ queue of Problem 4.4.25, select one of your data sets, preferably one with a hyperexponential service time. As in Problem 4.4.63, study the asymptotic formula for the sojourn time above a certain queue length, but now consider the sets $\{K+1, \cdots, 20\}$, for $5 \leq K \leq 18$. Examine how the quantity η and the multiplicative constant in the asymptotic formula vary with K.

Discuss the significance of your numerical examples. You should find that the Perron - Frobenius eigenvalue decreases as K increases. What happens to the multiplicative constant? Think about how the queue behaves and see whether that behavior is reflected in the items you have computed.

Problem 4.4.65: The states of a Markov chain are the integers $\{0, 1, 2, \cdots, m\}$ where m may be infinite. At time zero, the chain is started in state 1. State 0 is the only absorbing state of the chain. The transition probability matrix P is written as

$$P = \begin{array}{c|ccc|} 0 & 1 & 0 & 0 \\ 1 & \mathbf{T}^{\circ}(1) & T(1,1) & T(1,2) \\ 2 & \mathbf{T}^{\circ}(2) & T(2,1) & T(2,2) \end{array},$$

where set 1 consists of the k states $\{1, 2, \cdots, k\}$. The random variable $M \geq 1$ is the index of the highest state visited by the chain before absorption in state 0.

a. Show that the probability $\psi_k = P\{M \leq k\}$, $k \geq 1$, is given by the first component of the column vector $\psi_k = [I - T(1,1)]^{-1}\mathbf{T}^{\circ}(1)]$.

b. By successively enlarging the dimension of $\mathbf{T}^{\circ}(1)$ and $T(1,1)$, compute the probability distribution, the density, and the mean of M. Implement your code for the embedded Markov chain of the finite-capacity $M/G/1$ queue in Problem 4.4.25. Obtain numerical results for the data given there. What do the results tell you?

Problem 4.4.66: Use the approach in Problem 4.4.65 to compute the probability distribution and the density of the maximum queue length immediately after departures in the busy period of the *unbounded* $M/G/1$ queue. Evaluate the probabilities $\psi_k = P\{M \leq k\}$ until either ψ_k exceeds 0.99 or until k reaches 75. Implement your code for data where $\rho = \lambda\mu'_1$ is less than, equal to, or greater than one, and interpret your numerical results. Note that, when $\rho > 1$, M has a defective probability distribution. How should the procedure be modified if the chain is started in a state $i > 1$?

Problem 4.4.67: Consider an irreducible m-state Markov chain with transition probability matrix P. The Markov chain operates during a geometrically distributed time T; i.e., at every time $n \geq 0$, a Bernoulli trial with probability p of success is performed. As long as there are successes, the Markov chain is allowed one more transition. At the time of the first failure, the process ends.

Let $\Delta(\mathbf{z})$ be an $m \times m$ diagonal matrix with diagonal elements $z_1, \cdots, z_m$. The initial state is chosen according to the probability vector $\boldsymbol{\alpha}$. $K(\mathbf{z}) = K(z_1, \cdots, z_m)$ is the joint probability generating function of the numbers $N_1, \cdots, N_m$ of visits to the various states during the life of the Markov chain. By convention, we include the first but not the final state in the counts.

a. Show that $K(\mathbf{z}) = \boldsymbol{\alpha}[I - p\Delta(\mathbf{z})P]^{-1}q\Delta(\mathbf{z})\mathbf{e}$. (If we include the last but not the first state in the counts, we obtain the same formula, but with $pP\Delta(\mathbf{z})$ inside the square brackets).

b. By setting $z_j = 1$ for all $j \neq i$, verify that the number N_i of visits to the state i during the life of the Markov chain has a discrete PH-density. Obtain a representation for that density.

c. Introduce the matrix $L(\mathbf{z})$ defined by $[I - p\Delta(\mathbf{z})P]L(\mathbf{z}) = q\Delta(\mathbf{z})$. Differentiate twice in that equation to obtain matrix formulas for $E(N_i)$, $E[N_i(N_i - 1)]$, and $E(N_iN_j)$, $i \neq j$ of the counts N_i. Write all formulas in matrix form only! It is useful to introduce $\Delta_1(i)$, a diagonal matrix whose only nonzero element is a 1 at (i, i). For example, the partial derivative of $L(\mathbf{z})$ with respect to z_i, at $z_1 = \cdots = z_m = 1$, is given by

$$L_1(i) = (I - pP)^{-1}q\Delta_1(i)(I - pP)^{-1},$$

so that $E(N_i)$ may be written as $E(N_i) = \boldsymbol{\alpha}(I - pP)^{-1}\Delta_1(i)\mathbf{e}$. That way we conveniently represent some simple operations. For example, $\Delta_1(i)\mathbf{e}$ is just an m-vector with a single one as the ith component and all others zero. Formulas for the second moments are more involved but the suggested notation shows immediately and in modular fashion how they can be programmed.

d. Write and implement a program for the means EN_i and the (symmetric) covariance matrix with elements $E[N_i - EN_i][N_j - EN_j]$. Also print the matrix of the *correlations* of the random variable N_i and N_j. Interpret the numerical results for some Markov chains whose behavior you understand well, such as random walks. Are the moment matrices helpful in understanding their behavior? Take one of your example matrices P and do the computations for increasing values of p. What do the numerical results tell you?

Problem 4.4.68: A Discrete-Time Queue: Consider a service system operating in discrete time. It has a waiting room with K places and one server. Any job that is admitted for service has a constant processing time of a units. Jobs are served in the order in which they are admitted; those arriving when the waiting room is full are lost.

Jobs arrive according to a Bernoulli process. At every time unit there is an arrival with probability p. The waiting time of an admitted job is the length of time between its arrival and the epoch at which its service starts. Arrivals that find the system empty start service immediately. The service operator announces that every customer having to wait more than b units is served free. In addition, the customer receives C_1 dollars. Any customer who waits at most b units pays the server C dollars. Our objective is to determine the capacity K for which the expected steady-state revenue per (arriving) customer is maximum.

Consider a discrete-time Markov chain with states 0 and (i, j), $1 \leq i \leq K + 1$, $1 \leq j \leq a$. The chain is in state 0 if immediately after a time epoch, the system is empty. State (i, j) is visited if and only if after a time epoch, there are i customers in the system and the customer in service still requires j units of service time. To avoid ambiguity in the accounting of customers, assume that upon completion of a service the customer departs immediately, the next service starts immediately, and a space in the waiting room becomes available. In particular, if at time n the waiting room is full and a service is completed, any job arriving at that time is admitted.

Display the transition probability matrix P for small values of K and a, say, $K = 3$ and $a = 4$. Use the special structure of P to compute the steady-state probability vector π as efficiently as possible. (Use an iterative method.) Express the steady-state probability density $\{w(k)\}$ of the waiting time of a customer admitted at an arbitrary time in terms of π and derive the expected revenue obtained from such a customer. Write a code to compute these items and to find the optimal value of K.

Implement your code for $a = 10$, $b = 40$, $C = 10$, $C_1 = 25$ and for 11 equidistant values of p in the interval $[0.07, 0.08]$. For each K, also compute the steady-state probability that an arriving customer is admitted.

Problem 4.4.69: For the queueing model in Problem 4.4.68, suppose that no money is involved. However, there is a requirement that fewer than 5% of the admitted customers have to wait more than b time units. Develop a program to find the largest K for which that constraint is satisfied. In some cases such a K may not exist. How would you identify these cases?

Implement your code for $a = 10$, $b = 40$, and for the same p-values as in Problem 4.4.68. For each K, compute the steady-state probability that an arriving customer is admitted.

Problem 4.4.70: The $m \times m$ stochastic matrix P is irreducible. Its invariant probability vector is π. Establish that, for Re $\xi > 0$, the matrix

$$\Psi(\xi) = (I - e^{-\xi}P)^{-1}(I - P)$$

may be written as a power series and that the first few coefficients A_n of that series are

$$A_0 = I - e\pi, \quad A_1 = e\pi - ZP, \quad A_2 = -\frac{3}{2}e\pi + \frac{1}{2}ZP + (ZP)^2, \tag{1}$$

where $Z = (I - P + e\pi)^{-1}$. (See Problem 4.4.21.)

Next, derive a general recurrence relation for the matrices A_n. By equating coefficients of powers of ξ in an appropriate equality, show that $A_0 = I - e\pi$, and that for $n \geq 1$,

$$A_n = \sum_{r=0}^{n-1} \frac{(-1)^{n+1-r}}{(n+1-r)!}[e\pi - (n+1-r)ZP]A_r.$$

Show further that for $n \geq 1$, A_n is of the form

$$A_n = a_n e\pi + \sum_{k=1}^{n} b_{nk}(ZP)^k,$$

where the coefficients a_n and b_{nk} satisfy the recurrence relations

$$b_{n1} = \frac{(-1)^n}{n!}, \quad \text{and} \quad b_{nk} = \sum_{r=k-1}^{n-1} \frac{(-1)^{n-r}}{(n-r)!} b_{r,k-1}, \quad \text{for } 2 \leq k \leq n,$$

and

$$a_n = \sum_{k=1}^{n-1} (b_{n+1,k+1} + b_{n,k}) - \frac{(-1)^n}{n!} - \sum_{r=1}^{n-1} (n-r)\frac{(-1)^{n+1-r}}{(n+1-r)!}a_r, \quad \text{for } n \geq 2.$$

Explicitly calculate the coefficients for $n \leq 2$, to check the expressions in (1). Write a computer program to compute the coefficients b_{nk}, $1 \leq k \leq n$, for n up to 50.

For a selection of 10×10 stochastic matrices and for $\xi = 0.1$, compute the function $\Psi(\xi)$ directly and compare the matrices obtained to the sum of the first 11 terms of the power series for $\Psi(\xi)$.

CHALLENGING PROBLEMS

Problem 4.4.71: We have two coins whose probabilities p_1 and p_2 of coming up heads are of the forms

$$p_1 = \frac{1}{2} + \delta, \qquad p_2 = \frac{1}{2} - \delta,$$

with $0 < \delta < 1/2$. One of these coins is to be used in a large number n of tosses. We are interested in having as many heads show up, but do not know which coin has the higher probability.

We use one of the coins, chosen at random, until the difference V_k between the numbers of heads and tails reaches either K or $-K$. K is an integer between 1 and n. If that occurs at some time $k < n$, we continue using the first coin for all remaining tosses if K is reached; otherwise all remaining tosses are done with the other coin. It is clear that if K is reached at the kth toss, where $k \geq K$, we have scored $(K + k)/2$ heads. Likewise, if $-K$ is reached first at time k, we have scored $(k - K)/2$ heads. For any given K, at the possible decision times k, $k + K$ must be even.

a. Denote by A the event that K is reached first, by B the event that $-K$ is reached first, and by C the event that during all n trials, V_k remains strictly between K and $-K$. Let N be the total number of heads scored. Develop an algorithm to compute the probabilities

$$P\{N=\nu, A\}, \quad P\{N=\nu, B\}, \quad P\{N=\nu, C\},$$

if the better, respectively the worse coin, is used initially. Also, for each coin, compute the expected number $E(N)$ of heads scored. Tabulate and plot these probabilities for $n = 50$, $\delta = 0.05$, and for various values of K. Discuss the qualitative conclusions you draw for your results.

b. For given n and δ, develop an algorithm to find the value of K for which the expectation $E(N)$ is largest. Implement your algorithm for $n = 50$ and $\delta = 0.05$.

Hint: For whichever coin is used first, let ϕ_k be the probability that K is reached for the first time at the kth toss. Let ψ_k be the corresponding probability for $-K$. There is a simple recursive scheme to compute the quantities ϕ_k and ψ_k for $1 \le k < n$. Clearly, ϕ_k is also the probability that the remaining $n - k$ tosses will be made with the initial coin, whereas ψ_k is the probability that these tosses will be made with the other coin. For the desired probabilities, we need to accumulate a weighted sum of shifted binomial densities. The weights are the ϕ_k and ψ_k. The binomial densities are shifted to account for the number of heads scored up to time k.

Problem 4.4.72: For the partitioned Markov chain in Problem 4.4.55, a *cycle* is the time between two successive entries into set **1**.

a. Specifying the initial probability vector as in part *c* of Problem 4.4.55, derive the joint probability generating function of the times X_1 and X_2 spent in sets **1** and **2** during the cycle.

b. Derive formulas for the means, variances, and the correlation coefficient of X_1 and X_2.

c. Develop an efficient algorithm to compute the joint probability density of X_1 and X_2.

d. Implement your code for the examples in Problems 4.4.25 and 4.4.55.

Remark: The joint density is an example of a bivariate *PH*-density. For its computation, use matrix recursions recursions similar to those for univariate *PH*-densities. A three-dimensional plot of that density could be highly informative, particularly for the examples related to the finite-capacity $M/G/1$ queue.

Problem 4.4.73: A discrete-time Markov chain has a single absorbing state and m transient states. Its transition probability matrix P is partitioned as

$$P = \begin{vmatrix} T(1,1) & T(1,2) & \mathbf{T}^\circ(1) \\ T(2,1) & T(2,2) & \mathbf{T}^\circ(2) \\ 0 & 0 & 1 \end{vmatrix},$$

and with the initial probability vector $[\alpha(1), \alpha(2), \alpha_0]$. The square matrices $T(1, 1)$ and $T(2, 2)$ are of orders $m_1 > 0$ and $m_2 > 0$ respectively. The matrices $T(1, 2)$ and $T(2, 1)$ are of dimensions $m_1 \times m_2$ and $m_2 \times m_1$. The representation of the *PH*-distribution of the time until absorption is irreducible, so that there are no superfluous transient states.

We denote the sets consisting of the first m_1 and of the next m_2 transient states by **1** and **2**. We are interested in the joint density $\phi(n_1, n_2)$ of the numbers of visits N_1 and N_2 to sets **1** and **2**, prior to absorption. The components of the vectors $\mathbf{p}_1(n_1, n_2)$ and $\mathbf{p}_2(n_1, n_2)$ are the probabilities that after n_1 visits to **1** and n_2 visits to **2**, the chain is in a specific state of the sets **1** or **2**. The initial state is included in the count.

a. Show that

$$\mathbf{p}_1(1, 0) = \boldsymbol{\alpha}(1), \quad \text{and} \quad \mathbf{p}_2(0, 1) = \boldsymbol{\alpha}(2),$$

$$\mathbf{p}_1(n_1, 0) = \mathbf{p}_1(n_1 - 1, 0)T(1, 1), \quad \text{and} \quad \mathbf{p}_2(n_1, 0) = \mathbf{0}, \quad \text{for } n_2 \geq 1,$$

$$\mathbf{p}_1(0, n_2) = \mathbf{0}, \quad \text{and} \quad \mathbf{p}_2(0, n_2) = \mathbf{p}_2(0, n_2 - 1)T(2, 2), \quad \text{for } n_2 \geq 2,$$

and for $n_1 \geq 1$, $n_2 \geq 1$,

$$\mathbf{p}_1(n_1, n_2) = \mathbf{p}_1(n_1 - 1, n_2)T(1, 1) + \mathbf{p}_2(n_1 - 1, n_2)T(2, 1),$$

$$\mathbf{p}_2(n_1, n_2) = \mathbf{p}_1(n_1, n_2 - 1)T(1, 2) + \mathbf{p}_2(n_1, n_2 - 1)T(2, 2).$$

In terms of the vectors $\mathbf{p}_1(n_1, n_2)$ and $\mathbf{p}_2(n_1, n_2)$ the probabilities $\phi(n_1, n_2)$ are given by $\phi(0, 0) = \alpha_0$, and $\phi(n_1, n_2) = \mathbf{p}_1(n_1, n_2)\mathbf{T}^\circ(1) + \mathbf{p}_2(n_1, n_2)\mathbf{T}^\circ(2)$, for $n_1 \geq 1$ or $n_2 \geq 1$.

b. By direct calculation, or by using generating functions, show that the marginal densities of N_1 and N_2 are *PH*-densities and obtain representations for them. (For the corresponding result for continuous-time Markov chains, see Problem 5.2.36.)

c. Write a subroutine to compute the marginal densities until a probability of at most 10^{-5} remains in the tails. Suppose that this requires computation up to the indices n_1^* and n_2^* respectively. Next, evaluate the probabilities $\phi(n_1, n_2)$ fcr $0 \le n_1 \le n_1^*$, $0 \le n_2 \le n_2^*$. (Specify an array of 200×200 for the storage of these probabilities. In the recursive computation use storage efficiently.)

d. Write a subroutine, using the procedure developed in Problem 3.6.54 to obtain the probability density of the ratio $N_1(N_1 + N_2)^{-1}$. Be sure to include a test that returns an error message if the subroutine is called with parameters for which $\phi(0, 0) > 0$.

e. This algorithm has a variety of applications; here is one: Consider the queue length during a busy period of an $M/M/1$ queue of capacity K. If the arrival and service rates are λ and μ, respectively, write the transition probability matrix P of the queue length *at arrival times* in terms of the quantity $p = \mu(\lambda + \mu)^{-1}$. Note that this is a Markov chain with states $0, \cdots, K - 1$. Verify that its steady-state probabilities $\pi(j)$ are of the form $\pi(j) = c[(1-p)p^{-1}]^j$, where c is a normalizing constant. We find the smallest value j^* of j for which $\pi(0) + \cdots + \pi(j) \ge 0.5$. Next, we study the joint density of N_1, N_2, the numbers of times that, during a busy period, an arrival encounters a queue length $\le j^*$ or $> j^*$, respectively.

Use caution in initializing this problem; a busy period starts with a transition from 0 to 1. The initial state is 1, but the transition at time zero must be counted in N_1. The busy period ends with the first return to state 0. Also compute the density of $N_1(N_1 + N_2)^{-1}$. Implement your code for values of p about the value 1/2. Explain what your numerical results reveal about the behavior of the queue. Specifically study the case $K = 20$.

Problem 4.4.74: Referring to Problem 4.4.36, consider the case where, upon emptiness, the box is replenished with a random number of coins according to the probability density $\{b_k\}$. The density $\{a_k\}$ is as defined in Problem 4.4.36. If the coin initially contains $T_0 = i > 0$ coins, the operation can continue for at least that many days. Let the number of coins at the end of the ith day be T_1. If $T_1 = 0$, the box is replenished the next day. If T_1 is positive, again the operation can go on for that many days. In either case, T_2 is the number of coins in the box when all coins available on the morning of day $T_1 + 1$ have been used. Successively define the random variable T_n in the same manner. Explain why the sequence $\{T_n\}$ is a Markov chain on the nonnegative integers.

a. Show that its transition probability matrix R is

$$R = \begin{vmatrix} b_0 & b_1 & b_2 & b_3 & \cdots \\ a_0^{(1)} & a_1^{(1)} & a_2^{(1)} & a_3^{(1)} & \cdots \\ a_0^{(2)} & a_1^{(2)} & a_2^{(2)} & a_3^{(2)} & \cdots \\ a_0^{(3)} & a_1^{(3)} & a_2^{(3)} & a_3^{(3)} & \cdots \\ a_0^{(4)} & a_1^{(4)} & a_2^{(4)} & a_3^{(4)} & \cdots \\ \cdot & \cdot & \cdot & \cdot & \end{vmatrix},$$

where $\{a_k^{(\nu)}\}$ is the ν-fold convolution of $\{a_k\}$.

b. Show that when $\alpha < 1$, the generating function $U(z) = \sum_{i=0}^{\infty} u_i z^i$ of the stationary probabilities $\{u_i\}$ of the stochastic matrix R is given by

$$U(z) = u_0 \{ 1 + \sum_{\nu=0}^{\infty} \{B\ [A_{(\nu)}(z)] - B\ [A_{(\nu)}(0)]\}\ \},$$

where the generating functions $A_{(\nu)}(z)$, $\nu \geq 0$, are recursively defined by

$$A_{(0)}(z) = z, \qquad A_{(1)}(z) = A(z) = \sum_{k=0}^{\infty} a_k z^k,$$

$$A_{(\nu)}(z) = A\ [A_{(\nu-1)}(z)] = A_{(\nu-1)}\ [A(z)], \quad \text{for } \nu \geq 2,$$

and $B(z) = \sum_{k=0}^{\infty} b_k z^k$.

c. Verify that u_0 is given by $u_0 = \{ 1 + \sum_{\nu=0}^{\infty} [1 - B\ [A_{(\nu)}(0)]\ ]\ \}^{-1}$.

d. By considering the graph of the function $B[A_\nu(z)]$ on the interval $[0, 1]$, show that if $0 < a_0 + a_1 < 1$, the inequality $B[A_\nu(z)] > 1 - \beta\alpha^\nu$ holds for $\nu \geq 1$. From those inequalities deduce that $u_0 > x_0$, where x_0 is the stationary probability of state 0 in the model of Problem 4.4.36. Further show that the series $\sum_{\nu=0}^{\infty} [1 - B\ [A_{(\nu)}(0)]\ ]$ converges at least as fast as a geometric series of ratio $\alpha < 1$. That result is useful in computing u_0.

e. Differentiate the expression for $U(z)$ to show that the mean of the density $\{u_i\}$ is $u_0\beta(1-\alpha)^{-1}$.

f. The computation of the quantities u_i for $i \geq 1$ is more involved. We recursively compute the sequences $\{\psi_r(\nu), r \geq 0\}$ by use of the formulas

$$\psi_r(0) = b_r, \qquad \text{for } r \geq 0,$$

$$\psi_r(\nu) = \sum_{k=0}^{\infty} \psi_k(\nu - 1)\, a_r^{(k)}, \qquad \text{for } r \geq 0,\ \nu \geq 1,$$

and by summing term by term over ν. We also accumulate the sum

$$S_N = \sum_{\nu=0}^{N} \sum_{i=1}^{\infty} \psi_i(\nu),$$

which tends to $u_0^{-1} - 1 = \sum_{\nu=0}^{\infty} [1 - B\ [A_{(\nu)}(0)]\ ]$, as $N \to \infty$. We stop at the index N for which the computed S_N is sufficiently close to its limit value which is readily expressed in terms of u_0.

g. Develop a carefully written computer code to evaluate the sequence $\{u_i\}$. Implement your code for the example data in Problem 4.4.36, as well as for other data of your choice. Compare the numerical values of the terms in the sequences $\{x_i\}$ and $\{u_i\}$ for the same data. Can you explain why these densities are so different?

Problem 4.4.75: The combinatorial game in this problem has applications in computer operations. It is analytically very involved, but can be studied in detail by an algorithmic approach. Initially, there are r individuals, who independently select an integer between 1 and n at random. The r integers are examined and all individuals who are the only ones to have picked their particular number leave the group. Let us call such individuals *singletons*. Any remaining persons again pick an integer between 1 and n at random and the singletons depart. The process stops when all persons have left.

a. X_j is the number of persons remaining after the jth stage. The sequence $\{X_j\}$ is a Markov chain with states $r, r-1, \cdots, 0$. Clearly, $X_0 = r$. If $P_{ik} = P\{X_{j+1} = k \mid X_j = i\}$, use the inclusion - exclusion formula to show that for $i \geq k$,

$$P_{ik} = \sum_{\nu=i-k}^{\min(i,n)} (-1)^{\nu-i+k} \binom{\nu}{i-k} \binom{n}{\nu} \binom{i}{\nu} \nu!\, n^{-i} (n-\nu)^{i-\nu}$$

$$= \sum_{\nu=i-k}^{\min(i,n)} (-1)^{\nu-i+k} \frac{n!i!}{(\nu - i + k)!(i - k)!(n - \nu)!(i - \nu)!} n^{-i} (n - \nu)^{i-\nu},$$

and $P_{ik} = 0$, for $i < k$.

b. Write a subroutine to compute and store the upper triangular matrix $P = \{P_{ik}\}$ as efficiently and accurately as possible.

c. M is the number of stages in the game, the absorption time into state 0 starting from state r. Write a subroutine to compute the probabilities $P\{M = t \mid X_0 = r\}$ for $t = 0, \cdots, t_0$, where t_0 is the smallest index for which $P\{M > t_0 \mid X_0 = r\} < 10^{-5}$.

d. Show that the mean times μ_i to absorption into zero, starting from state i, satisfy the system of linear equations

$$\mu_i = 1 + \textstyle\sum_{\nu=1}^{i} P_{i\nu}\, \mu_\nu, \quad \text{for } 1 \le i \le r.$$

Solve that system recursively to obtain the mean number of stages $\mu_r = E(M)$ of the game. Write a program to compute the means μ_i. Execute your program for $r = 20$, $n = 20$, and for $r = 20$, $n = 2$. Verify that for $n = 2$, μ_r is given by

$$\mu_r = \sum_{j=1}^{r} \frac{2^{j-1}}{j}.$$

Use that formula as a check of your program for $n = 2$. Discuss the numerical difficulties that may arise when r and n are large or if the P_{ik} is not computed with care.

Hint: Let A_h be the event that person h is a singleton. The intersection $A_{j_1}A_{j_2} \cdots A_{j_\nu}$ where the indices $j_1, \cdots, j_\nu$ are distinct and chosen from among i available labels of the persons has probability

$$\binom{n}{\nu} \nu! \frac{1}{n^\nu} \left[\frac{n - \nu}{n}\right]^{i-\nu}.$$

The ν indices in the intersection can be chosen in $\binom{i}{\nu}$ ways, so that, for

$i - k \leq \nu \leq \min(i, n)$, the sums S_ν in the inclusion - exclusion formula are given by

$$S_\nu = \binom{i}{\nu}\binom{n}{\nu}\nu!\frac{1}{n^\nu}\left[\frac{n-\nu}{n}\right]^{i-\nu}.$$

The stated formulas are now easily obtained.

Reference: Dym, H. and Luks, E. M., "On the mean duration of a ball and cell game: A first passage problem." *Annals of Mathematical Statististics,* 37, 517 - 521, 1966.

The following four problems deal with Markov chains related to the same basic model. They describe the random motion of a particle that, after geometrically distributed upward steps, is returned to the left of its current position. The downward steps are made according to different conditional probability densities. These problems explore how the random walk depends on these conditional densities. Quite different arguments need to be invoked for the various cases. The conditions for positive recurrence are also different. We suggest that the reader consider these models together and try to gain an intuitive understanding of why they are so dissimilar. In all cases, the construction of an algorithm is interesting and somewhat challenging. These problems offer nice tests of the algorithmic maturity attained by the study of this book.

Reference: Neuts, M. F. "An interesting random walk on the nonnegative integers." *Journal of Applied Probability,* 31, 48 - 58, 1994.

Problem 4.4.76: Let $\{r_k\}$ be a probability density on the positive integers with generating function $R(z)$ and with mean $\rho \geq 1$ (when it exists). Let p be a parameter satisfying $0 < p < 1$ and $q = 1 - p$. Consider the Markov chain on $\{0, 1, 2, \cdots\}$ with the transition probability matrix P defined by

$$P_{ij} = \frac{p}{i+1}, \quad \text{for } i \geq 0,\ 0 \leq j \leq i,$$

$$= q\, r_{j-i}, \quad \text{for } i \geq 0,\ j > i.$$

Note that the chain is irreducible and aperiodic and denote its invariant probability density (if it exists) by $\{\pi_i\}$ with generating function $\Pi(z)$.

a. Use the steady-state equations to derive the integral equation

$$(1 - z)[1 - qR(z)]\, \Pi(z) = \pi_0 - p \int_0^z \Pi(u)\, du,$$

for $0 \le z < 1$. Take derivatives of both sides of that equation and solve the resulting differential equation to show that

$$\Pi(z) = \Pi_0(z) \exp\left[-qp^{-1} \int_z^1 \Theta_1(u)\, du\right],$$

where $\Pi_0(z) = p[1 - qR(z)]^{-1}$, the generating function of a geometric mixture of the convolutions of $\{r_\nu\}$ and $\Theta_1(u)$ is defined by:

$$\Theta_1(u) = \left[\frac{1 - R(u)}{1 - u}\right] \Pi_0(u).$$

b. Show that $\Pi(z)$ is a valid generating function if and only if the integral

$$\int_0^1 \frac{1 - R(u)}{1 - u}\, du,$$

converges. That integral may be rewritten as

$$\int_0^1 (1 - u)^{-1} \sum_{k=1}^{\infty} r_k (1 - u^k)\, du = \sum_{k=1}^{\infty} r_k \sum_{j=1}^{k} \frac{1}{j},$$

and the sum and the integral converge or diverge together. Use the Euler - Mascheroni formula $\lim_{k \to \infty} [\sum_{j=1}^{k} 1/j - \log k] = \gamma$, where γ is the Euler's constant, to prove that the integral converges if and only if $\sum_{k=1}^{\infty} \log k \; r_k < \infty$.

The mean of $\{\pi_k\}$ is finite if and only if the mean ρ of $\{r_k\}$ is finite. It is then given by $\pi^* = \Pi'(1-) = 2\rho q p^{-1}$. In that case, $\Theta(u) = \rho^{-1}\Theta_1(u)$ is a probability generating function. The Markov chain is then clearly positive recurrent.

The generating function of the sequence $\{r'_k\}$ with terms $r'_k = 1 - \sum_{\nu=0}^{k} r_\nu$, $k \geq 0$ is $(1-u)^{-1}[1 - R(u)]$. We recall from renewal theory that when ρ is finite, $\{\rho^{-1} r'_k\}$ is the delay density corresponding to $\{r_k\}$.

c. If $\theta_{1,k}$ is the coefficient of u^k in the power series for $\Theta_1(u)$, verify that $\Pi(z)$ may be written as

$$\Pi(z) = \Pi_0(z) \exp\left[-qp^{-1} \sum_{k=1}^{\infty} k^{-1} \theta_{1,k-1}(1 - z^k)\right]$$

$$= \Pi_0(z) \prod_{k=1}^{\infty} \exp\left[-q\,(kp)^{-1} \theta_{1,k-1}(1 - z^k)\right].$$

Note the second factor is the generating function of the density of a sum of the form $\sum_{k=1}^{\infty} kV_k$, where, for $k \geq 1$, the V_k are independent Poisson random variables with means $\psi'_k = q\,(kp)^{-1}\theta_{1,k-1}$. For a procedure to compute the corresponding density, see Problems 3.6.26 and 3.6.27.

If σ^2 denotes the variance of $\{r_k\}$, verify that the mean of the density $\{\theta_k = \rho^{-1}\theta_{1,k}\}$ is given by

$$\theta^* = \frac{q\rho}{p} + \frac{\sigma^2 + \rho^2 - \rho}{2\rho}.$$

The following is a numerical procedure for evaluating the steady-state probabilities when θ^* is of moderate size. Write

$$\Theta(u) = qR(u)\Theta(u) + p\left[\frac{1 - R(u)}{\rho(1-u)}\right],$$

and expand in power series to obtain the following recursive equations for the density $\{\theta_k\}$:

$$\theta_0 = p\rho^{-1}, \qquad \theta_n = q\sum_{k=1}^{n} r_k \theta_{n-k} + p\rho^{-1} r'_n, \quad \text{for } n \geq 1.$$

Note that the density $\{\phi_k\}$ with transform $\Pi_0(z)$ and with mean $q\rho p^{-1}$ is similarly computed recursively by:

$$\phi_0 = p, \qquad \phi_n = q\sum_{k=1}^{n} r_k \phi_{n-k}, \quad \text{for } n \geq 1.$$

d. Use the procedure of Problem 3.6.26 and the given recurrence relations to write a general code for the computation of the density $\{\pi_i\}$. Implement your code for a number of choices of $\{r_k\}$ and values of p. Vary p and discuss its effect on the computed density.

e. The following are a few cases for which $\Pi(z)$ is obtained explicitly. For these, the π_k can be derived in closed form:

For $R(z) = z$,

$$\Pi(z) = \left[\frac{p}{1 - qz}\right]^2,$$

and for $R(z) = z^2$, with $\beta = \sqrt{q}$,

$$\Pi(z) = \left[\frac{1 - \beta}{1 - \beta z}\right]^{\frac{3+\beta}{2}} \left[\frac{1 + \beta}{1 + \beta z}\right]^{\frac{3-\beta}{2}}.$$

If $R(u) = u[\rho - u(\rho - 1)]^{-1}$ (the geometric density with mean ρ on the positive integers)

$$\Pi(z) = [\rho - z(\rho - 1)]\left[\frac{p}{\rho - z(\rho - p)}\right]^{\frac{\rho q + \rho - p}{\rho - p}}.$$

Verify the stated expressions and expand the generating function $\Pi(z)$ in power series. Write separate codes to evaluate these densities and use them to check the correctness of your general code.

Remark: The condition for positive recurrence is very weak. The density $\{r_k\}$ may be very heavy-tailed. The path behavior of the Markov chain can then be very erratic, with long upward surges followed by possibly large drops as one or more downward transitions occur. For their instructional merits, we recommend visual simulations of the paths of this Markov chain, using various p-values and some heavy-tailed densities $\{r_k\}$. For such densities, the computation of the steady-state probabilities $\{\pi_k\}$ presents corresponding difficulties. The sequence $\{\theta_{1,k}\}$ then converges slowly, so that a large number of convolutions, each requiring many terms, is needed to compute a satisfactory approximation to the probabilities $\{\pi_i\}$.

Problem 4.4.77: Let $\{r_k\}$ be a probability density on the positive integers with generating function $R(z)$ and with mean $\rho \geq 1$ (when it exists). Let p be a parameter satisfying $0 < p < 1$ and $q = 1 - p$. Consider the Markov chain on the set $\{0, 1, 2, \cdots\}$ with the transition probability matrix P defined by

$$P_{ij} = p\binom{i}{j}(1-\beta)^{i-j}\beta^j, \quad \text{for } 0 \leq j \leq i,$$

$$= q\, r_{j-i}, \qquad \text{for } i \geq 0,\ j > i,$$

where β is a constant satisfying $0 < \beta < 1$. Denoting the invariant probabilities by π_i, write the steady-state equations

$$\pi_i = q\sum_{k=0}^{i-1} \pi_k r_{i-k} + p\sum_{\nu=i}^{\infty} \pi_\nu \binom{\nu}{i}(1-\beta)^{\nu-i}\beta^i, \quad \text{for } i \geq 0,$$

where the first sum is zero for $i = 0$.

a. Show that the generating function $\Pi(z)$ of $\{\pi_i\}$ satisfies the equation

$$\Pi(z) = p\,[1 - qR(z)]^{-1}\Pi(1 - \beta + \beta z).$$

Upon iteration in that equation, show that, provided the infinite product converges

$$\Pi(z) = \prod_{k=0}^{\infty} \Pi_0[1 - \beta^k + \beta^k z],$$

where $\Pi_0(z) = p\,[1 - qR(z)]^{-1}$. By a comparison argument, the infinite product converges for all $|z| \leq 1$, if and only if the product

$$\prod_{k=0}^{\infty} \Pi_0[1 - \beta^k] = \pi_0$$

converges, or equivalently, if and only if $\sum_{k=0}^{\infty} [1 - \Pi_0(1 - \beta^k)] < \infty$. The following argument shows that, whenever the mean ρ is finite, that series converges: If ρ is finite, so is the mean of the density $\{\phi_k\}$ with generating function $\Pi_0(z)$. Since

$$1 - \Pi_0(z) = (1 - z)\sum_{n=0}^{\infty} z^n \sum_{\nu=n+1}^{\infty} \phi_\nu,$$

we obtain upon setting $z = 1 - \beta^k$ that

$$1 - \Pi_0(1 - \beta^k) = \beta^k \sum_{n=0}^{\infty} (1 - \beta^k)^n \sum_{\nu=n+1}^{\infty} \phi_\nu < \beta^k \sum_{n=0}^{\infty} \sum_{\nu=n+1}^{\infty} \phi_\nu = \beta^k q p^{-1} \rho,$$

and convergence of the sum follows by comparison.

b. For finite ρ, calculate the mean of the density $\{\pi_i\}$ and note that it depends only on p and ρ. Write an efficient code to compute the density $\{\pi_i\}$. Note that you will need to form successive convolutions of the probability densities with generating functions $\Pi_0[1 - \beta^k + \beta^k z]$. Once you have computed the density $\{\phi_\nu\}$ corresponding to $\Pi_0(z)$, these densities can be evaluated very efficiently. Notice that, for large k, the factors in the infinite product for $\Pi(z)$ place nearly all mass at zero. Use this to construct a stopping criterion for your algorithm. Implement your code for some selected densities $\{r_k\}$ and, for each, for various values of p and β. Discuss the effects of these parameters on the computed probabilities.

c. For $R(z) = z + (1 - z)\log(1 - z)$, which corresponds to the density $r_k = [k(k-1)]^{-1}$, $k \geq 2$, which has an infinite mean, verify that the Markov chain is positive recurrent. (Use a comparison test.) The mean of $\{\pi_i\}$ is now clearly infinite. Write a code to compute the first 5,000 terms of that density.

Problem 4.4.78: In a further variant of the models in Problems 4.4.76 and 4.4.77, we define the transition probability matrix P by

$$P_{i0} = p(1 - \beta)^i,$$

$$P_{ij} = p\beta(1-\beta)^{i-j}, \quad \text{for } 1 \le j \le i,$$

$$= q\, r_{j-i}, \quad \text{for } i \ge 0,\ j > i,$$

for all $i \ge 0$. The parameter β satisfies $0 \le \beta < 1$. For the invariant probabilities $\{\pi_i\}$, we obtain the equations

$$\pi_i = q \sum_{k=0}^{i-1} \pi_k r_{i-k} + p\beta \sum_{\nu=i}^{\infty} \pi_\nu (1-\beta)^{\nu-i}, \quad \text{for } i \ge 1,$$

$$\pi_0 = p \sum_{\nu=0}^{\infty} \pi_\nu (1-\beta)^{\nu} = p\,\Pi(1-\beta),$$

where $\Pi(z)$ is the generating function of $\{\pi_i\}$. Both equations lead to the following expression for $\Pi(z)$:

$$\Pi(z) = \frac{(1-\beta)(1-z)\pi_0}{p\beta z - (z-1+\beta)[1-qR(z)]}.$$

That equation is similar to that for the steady-state probabilities for the queue length in the $M/G/1$ queue. It is discussed by the same argument. We shall sketch it and ask you to verify the details. If $\Pi(z)$ is a probability generating function, the stated expression must be analytic in the unit disk. The zeros in the unit disk of the numerator are the roots of the equation

$$z = 1 - \beta + \beta p z\,[1 - qR(z)]^{-1} = 1 - \beta + \beta z\,\Pi_0(z),$$

and the right hand side of that equation is a probability generating function. By considering the graph of the right-hand side on [0, 1], we see that there is a root at some point in (0, 1) if the mean ξ of the right side is greater than one. Then, the chain cannot be positive recurrent. If $\xi \le 1$, the only root is at $z = 1$. When equality holds, that root is double and $\pi_0 = 0$. That corresponds to the null-recurrent case. By appealing to Rouché's theorem, it may be verified that there cannot be other roots in $|z| \le 1$. The chain is therefore positive recurrent if and only if $\xi < 1$. We easily verify that

$$\xi = \beta(1 + p^{-1} q \rho).$$

By L'Hospital's rule, we obtain that if $\xi < 1$, $\pi_0 = p - q\beta(1-\beta)^{-1}\rho$.

a. Show that the expression for $\Pi(z)$ may be rewritten as

$$\Pi(z) = \Pi_0(z)(1-\xi)\left[1 - \xi \frac{1 - z\Pi_0(z)}{(1 + p^{-1}q\rho)(1-z)}\right]^{-1},$$

where $\Pi(z)$ is the product of the generating functions $\Pi_0(z)$ and that of a geometric mixture of the convolutions of the delay density corresponding to $z\Pi_0(z)$. The latter is the generating function of $\{\phi_k\}$, shifted one place to the right.

b. In addition, verify that the mean π^* of $\{\pi_i\}$ is finite, if and only if $\{r_k\}$ has a finite *second* moment and derive an expression for π^* in terms of p, β, and the mean and variance of $\{r_k\}$.

c. Use the final expression for $\Pi(z)$ to derive a stable recurrence relation for the steady-state probabilities $\{\pi_i\}$. Write a code, using the suggestions given in the preceding problems, to evaluate the steady-state probabilities for this model. Again, select a density $\{r_k\}$ and vary the parameters p and β to gain insight into the dependence of the $\{\pi_i\}$ on the values of these parameters.

Problem 4.4.79: To illustrate how strongly the steady-state probabilities depend on the choice of the conditional density of the downward transitions, consider the transition probability matrix P in which for all $i \geq 0$,

$$P_{i0} = p\left[1 - \sum_{\nu=1}^{i} \beta^{\nu}\right],$$

$$P_{ij} = p\beta^{i-j}, \quad \text{for } 1 \leq j \leq i,$$

$$= q\, r_{j-i}, \quad \text{for } i \geq 0,\ j > i.$$

Since the P_{i0} must be nonnegative for all $i \geq 1$, it is necessary that $0 \leq \beta \leq 1/2$. Similar derivations as in the preceding problems lead to the following equations for the generating function $\Pi(z)$ of the invariant probabilities:

$$p\beta\,\Pi(\beta) = (1-\beta)\pi_0 - p(1-2\beta),$$

$$\Pi(z) = \pi_0 + qR(z)\Pi(z) + p\beta z(\beta - z)^{-1}[\Pi(\beta) - \Pi(z)].$$

Elimination of $\Pi(\beta)$ from these equations yields that

$$\Pi(z) = \frac{\beta(1-z)\pi_0 - pz(1-2\beta)}{p\beta z + (\beta - z)[1 - qR(z)]}.$$

We get π_0 by requiring that, when the chain is positive recurrent, $\Pi(z)$ be an analytic function. The denominator vanishes at the roots of

$$z = \beta + \beta z\,\Pi_0(z). \tag{1}$$

The case $2\beta = 1$ is a particular case of the model in Problem 4.4.78. The Markov chain is then positive recurrent, if and only if $p > \rho(1+\rho)^{-1}$. By L'Hospital's rule, we see that $\pi_0 = p - q\rho$.

a. For the case $2\beta < 1$, since the right-hand side in (1) is an increasing, convex function that equals $2\beta \le 1$ at 1, there is exactly one root η, satisfying $\beta < \eta < 1$, in (0,1]. For $\Pi(z)$ to be analytic, that singularity at $z = \eta$ must be removable so that the numerator must also vanish at that point. This leads to

$$\pi_0 = \frac{p\eta(1-2\beta)}{\beta(1-\eta)}.$$

For $2\beta < 1$, the chain is always positive recurrent. With equality, the mean ρ must be finite and p must exceed a certain critical value.

b. Verify that the formula for $\Pi(z)$ may be rewritten as follows to show that it is a proper probability generating function. By substitution for π_0 and using equation (1) for $z = \eta$, we successively obtain that

$$\Pi(z) = \frac{1-2\beta}{1-\eta}\Pi_0(z)\frac{z-\eta}{z - \beta - \beta z\,\Pi_0(z)}$$

$$= \frac{1-2\beta}{1-\eta}\Pi_0(z)\frac{z-\eta}{z-\eta-\beta[z\Pi_0(z)-\eta\Pi_0(\eta)]}$$

$$= \frac{1-2\beta}{1-\eta}\Pi_0(z)\left[1-\beta\frac{[z\Pi_0(z)-\eta\Pi_0(\eta)]}{z-\eta}\right]^{-1}.$$

The rightmost fraction may be expanded as

$$\frac{z\Pi_0(z)-\eta\Pi_0(\eta)}{z-\eta} = \frac{1}{z-\eta}\sum_{k=0}^{\infty}\phi_k(z^{k+1}-\eta^{k+1})$$

$$= \sum_{k=0}^{\infty}\phi_k\sum_{r=0}^{k}z^r\eta^{k-r} = \sum_{r=0}^{\infty}z^r\sum_{k=r}^{\infty}\phi_r\eta^{k-r}.$$

The coefficient $c_r = \sum_{k=r}^{\infty}\phi_r\eta^{k-r}$ of z^r in the final series is clearly positive and

$$\sum_{r=0}^{\infty}\sum_{k=r}^{\infty}\phi_r\eta^{k-r} = (1-\eta)^{-1}[1-\eta\Pi_0(\eta)].$$

$\Pi(z)$ may now be written in the final form

$$\Pi(z) = \frac{1-2\beta}{1-\eta}\Pi_0(z)\left[1-\beta\sum_{r=0}^{\infty}c_r z^r\right]^{-1}.$$

To validate the geometric expansion of the last factor, we verify that

$$\beta(1-\eta)^{-1}[1-\eta\Pi_0(\eta)] < 1.$$

However, that inequality is equivalent to $1 > \eta + \beta - \beta\eta\Pi_0(\eta) = 2\beta$, since η satisfies (1).

c. Normalize the sequence $\{c_r\}$ to be a probability density and check that $\Pi(z)$ is the product of the generating functions $\Pi_0(z)$ and that of a geometric mixture of the successive convolutions of the normalized $\{c_r\}$. Obtain a stable numerical recurrence for the coefficients of $\Pi(z)$. Since no moment restrictions are placed on the density $\{r_k\}$, this recurrence convergences slowly when that density is heavy-tailed.

d. Determine under what conditions the mean π^* of the density $\{\pi_i\}$ is finite and derive an expression for π^*. Carefully outline all the necessary algorithmic steps for the computation of $\{\pi_i\}$, starting with the evaluation of the root η. Select a probability density $\{r_k\}$ and parameters p and β for which $\pi^* = 25$.

(That, in itself, is an interesting question. Choose some parametric family, say the negative binomial on the positive integers, and perform a numerical search for appropriate parameter values such that $\pi^* = 25$.) Next, compute the density $\{\pi_i\}$ for that case, but by using a general purpose code.

Problem 4.4.80: Suppose that $\{r_k\}$ is a *PH*-density on the positive integers with representation $[\alpha, T]$ and mean ρ. The parameter p satisfies $0 < p < 1$. $R(z)$ is the generating function of $\{r_k\}$. In Problem 4.4.76, the densities $\{\phi_k\}$ and $\{\theta_k\}$, respectively, with the generating functions $\Pi_0(z) = p[1 - qR(z)]^{-1}$, and

$$\Theta(u) = \left[\frac{1 - R(u)}{\rho(1 - u)}\right] \Pi_0(u)$$

play a role. The present problem deals with the efficient computation of those densities when $\{r_k\}$ is a *PH*-density.

a. Show that $\{\theta_k\}$ is the density of the absorption time in the Markov chain with transition probability matrix

$$P = \left| \begin{matrix} T + q\mathbf{T}^{\circ}\alpha & p\mathbf{T}^{\circ}\gamma T & \rho^{-1}p\mathbf{T}^{\circ} \\ 0 & T & \mathbf{T}^{\circ} \\ 0 & 0 & 1 \end{matrix} \right|,$$

where $\gamma = \rho^{-1}\alpha(I - T)^{-1}$. The initial probability vector is $[q\alpha, p\gamma T, p\rho^{-1}]$.

b. Construct a recursive scheme to compute the densities $\{\phi_k\}$ and $\{\theta_k\}$ *together*. Write a code to implement that procedure as well as an algorithm to determine the value of p for which the 95th percentile of the density $\{\theta_k\}$ is ten times that of the density $\{r_k\}$. The 95th percentile is the smallest index for which the cumulative probability distribution exceeds 0.95.

Implement your code for the negative binomial density with generating function $\beta z^{\nu}[1-(1-\beta)z]^{-\nu}$, which is a *PH*-density for ν a positive integer. (Construct its representation with care.) Obtain numerical results for $\nu = 5$ and for several values of β. Print both densities $\{\phi_k\}$ and $\{\theta_k\}$ for the value of p which you have found.

Problem 4.4.81: This problem deals with a probability density arising in the study of certain queues and of the Galton - Watson process. In queueing theory, it is the density of the number of customers served during a busy period. In the Galton - Watson process, it is the probability law of the total number of offspring in the entire line of descent of an individual.

Initially, a box contains one coin. By paying that coin, an activity is carried out that results in the receipt of a random number $X_1 \geq 0$ of coins. These coins are added to the box and are used one at a time to pay for future activities. The operation continues as long as there are coins. The nth activity yields X_n coins. Provided the box has not become empty, let Z_n denote the contents of the box after the nth coin has been used. The random variables $\{Z_n\}$ are recursively defined by $Z_{n+1} = Z_n - 1 + X_n$, where X_n is the revenue yielded by the nth coin. The process ends when the box becomes empty.

The successive yields X_n are independent, identically distributed random variables with probability density $\{a_i\}$. To avoid uninteresting cases, we assume that $a_0 > 0$ and $a_0 + a_1 < 1$. The generating function of the density $\{a_i\}$ is $A(z)$ and its (finite) mean is α.

The random variable $Y \geq 1$ is defined as the time until the box becomes empty, or equivalently as the total number of coins that will be used in the process. We are interested in the probability density $\{g_k\}$ of Y. Note that the probabilities $\{g_k\}$ do not necessarily sum to one; there could be a positive probability that the game lasts forever.

a. Show that the probability generating function $G(z)$ of $\{g_k\}$ satisfies the equation $G(z) = z\sum_{i=0}^{\infty} a_i[G(z)]^i = zA[G(z)]$, and that for every value z in $[0, 1]$, $G(z)$ is the *smallest* positive solution of the equation $\xi = zA(\xi)$.

b. Show that $\{g_k\}$ is a proper probability density if and only if $\alpha \leq 1$, and that $E(Y)$ is finite if and only if $\alpha < 1$. Then $E(Y) = (1-\alpha)^{-1}$.

The rigorous proof of the statements in parts *a* and *b* requires some care and may be omitted at first reading. It can be found in books on queues or branching processes. The remainder of this problem deals with the algorithmic aspects of the computation of $\{g_k\}$. That computation can be organized in several ways. For each, you are asked to write the necessary code efficiently and accurately. Discuss the merits of each approach.

c. Equate the coefficients of z^k in both sides of the equation in *a* to show that

$$g_1 = a_0, \qquad g_{k+1} = \sum_{i=1}^{k} a_i g_k^{(i)}, \quad \text{for } k \geq 1,$$

where $g_k^{(i)}$ is the term with subscript k in the i-fold convolution of $\{g_k\}$. Show that these formulas permit a recursive computation of the density $\{g_k\}$. Discuss the storage requirements of that recursive computation and explain why it best suited for cases where $a_i = 0$, for $i > K$, where K is a small integer.

d. Successive Substitution: Let T be an operator on number sequences that places a zero in front of the first element of the sequence. So,

$$T(\mathbf{x}) = [0, x_1, x_2, \cdots].$$

Let $\mathbf{g}$ be the sequence $\{0, g_1, g_2, \cdots\}$ and $\mathbf{g}^{(i)}$ the i-fold convolution of $\mathbf{g}$. We agree that $\mathbf{g}^{(0)}$ is the sequence $\varepsilon = 1, 0, 0, \cdots$. Explain why the equation in *a* may also be written as the convolution equation

$$\mathbf{g} = a_0 T(\varepsilon) + a_1 T(\mathbf{g}) + a_2 T(\mathbf{g}^{(2)}) + \cdots$$

That equation suggests various schemes of numerical solution by successive substitutions. Consider the scheme wherein $\mathbf{g}(0) = \mathbf{0}$, the sequence with all zero terms, and successively evaluate the sequences

$$\mathbf{g}(n+1) = \sum_{i=0}^{\infty} a_i T[\mathbf{g}^{(i)}(n)], \quad \text{for } n \geq 0.$$

Show that these sequences converge monotonically in the sense that

$$\sum_{k=0}^{N} g_k(n+1) \geq \sum_{k=0}^{N} g_k(n), \quad \text{for all } N \geq 0.$$

Discuss the numerical implementation of that recursive scheme and show that it is suitable to evaluate exactly any finite number N of terms of the sequence $\mathbf{g}$. The number of terms that contain all but a negligible amount of the probability mass in the density $\mathbf{g}$ is not known in advance. Compute the mean m_1 and the standard deviation σ_1 of the density $\mathbf{g}$ and set $N = [m_1 + 3\sigma_1]$. In most cases, the first N terms of $\mathbf{g}$ contain all but a negligible amount of the probability mass.

e. If $\alpha \leq 1$, the recursive scheme part *d* converges if $\mathbf{g}(0)$ is any probability density on the positive integers. Convergence is, in general, not monotone. Implement the iterative solution by choosing various initial sequences. Define and implement appropriate stopping criteria. The support of repeated convolutions grows rapidly, but most of the terms in the tails of sequences are very small. Retaining such terms adds to the computational effort, without contributing much to the accuracy of the terms of interest. In implementing the recursive schemes in parts *d* and *e*, propose and implement several ways of *trimming* the intermediate sequences. Notice that it is not advisable merely to drop small terms in the tails of the successively computed probability densities. Specifically in part *e*, the intermediate sequences $\mathbf{g}(n)$ should remain valid probability densities (at least to machine accuracy); otherwise the convergence of the procedure could be jeopardized.

The codes developed for this problem should be tested on many examples. Appropriate choices of test sequences $\{a_i\}$ can reveal programming or numerical errors in the code. We suggest choices such as $a_0 = 0.9$, $a_5 = 0.1$, for which the sequence $\mathbf{g}$ can be expected to have special features.

Problem 4.4.82: In Problem 4.4.81, let $a_i = \theta(1-\theta)^i$, for $i \geq 0$. Then g_k is explicitly given by

$$g_k = \frac{1}{k}\binom{2k-2}{k-1}\alpha^{k-1}(1+\alpha)^{1-2k}, \quad \text{for } k \geq 1,$$

with $\alpha = \theta^{-1}(1-\theta)$. Write a subroutine to compute the quantities g_k. It can be used to test the algorithm for the general case.

Problem 4.4.83: The terms of the r-fold convolution of the density $\{g_k\}$ in Problem 4.4.82 are given by

$$g_k^{(r)} = \frac{r}{k}\binom{2k-r-1}{k-1}\alpha^{k-r}(1+\alpha)^{r-2k}, \quad \text{for } k \geq r \geq 1.$$

Derive a recurrence relation on k for the $g_k^{(r)}$. When $\alpha \leq 1$, the sequence $\{g_k^{(r)}\}$ is a proper probability density. For $\alpha > 1$, verify that the sum of the terms of the sequence $\{g_k^{(r)}\}$ is α^{-r}. Use that fact to specify a stopping criterion in the computation of the density.

CHAPTER 5

CONTINUOUS-TIME MARKOV CHAINS

5.1. SOME MATRIX FORMULAS FOR FINITE MARKOV CHAINS

We start with a summary of basic matrix-analytic results for continuous-time Markov chains. We assume familiarity with the definitions and basic properties of such chains with finitely many states. As in Chapter 4, we emphasize the use of structural properties in the matrix calculations.

The Markov chain has m states $\{1, \cdots, m\}$. Whenever it is in state i, it remains there for an exponentially distributed time whereupon, independently of the past, it moves to a *different* state j. That state is chosen in a multinomial trial with probabilities depending on i. The parameter of the exponential *sojourn time* in a state i is traditionally denoted by $-Q_{ii}$. Q_{ii} is a therefore a negative quantity. For $i \neq j$, the probability P^*_{ij} of moving to state j at the end of a sojourn in i is denoted by $[-Q_{ii}]^{-1}Q_{ij}$. Clearly, the probabilities P^*_{ij} sum to one over j for every $1 \leq i \leq m$. Therefore,

$$\sum_{i \neq j} Q_{ij} = -Q_{ii}.$$

The transition probabilities of a continuous-time Markov chain are parameterized by the $m \times m$ matrix Q. Its diagonal elements Q_{ii} are *nonpositive*. Its off-diagonal elements Q_{ij}, $i \neq j$, are nonnegative. Moreover, all row sums of Q are zero. In matrix notation, $Q\mathbf{e} = \mathbf{0}$. The matrix Q is called the *infinitesimal generator* or, briefly, the *generator* of the chain. The notation and some terminology for continuous-time Markov chains was suggested by the theory of differential equations. These play an important role in their theory.

Zero Rows: If $Q_{ij} = 0$, for $1 \leq j \leq m$, this signifies that the chain cannot leave state i. Generators with one or more zero rows arise in absorption time problems.

The process $\{X(t), t \geq 0\}$ describes the states visited at the values t of a parameter commonly referred to as time. The probability vector α with $\alpha_i = P\{X(0) = i\}$ specifies the initial conditions of the chain. The sequence

$\{Y_n, n \geq 0\}$ of the successive states visited by the chain form a discrete-time Markov chain. It is called the *embedded Markov chain*. Its transition probability matrix P^* has zero diagonal elements. Its off-diagonal elements P^*_{ij}, with $i \neq j$, are as defined earlier. The *class structure* of a continuous-time Markov chain is completely determined by that of its embedded chain. For example, the Markov chain is irreducible if and only if the embedded chain is irreducible.

The Probabilities $P_{ij}(t)$

The conditional probability $P_{ij}(t)$ that the Markov chain is in state j at time t, given that it started in state i at time zero, is the (i,j)-element of the matrix $P(t)$. A classical property states that the matrix $P(t)$ satisfies systems of differential equations known as the *Chapman - Kolmogorov equations:*

$$P'(t) = P(t)Q, \qquad P'(t) = QP(t),$$

with the initial condition $P(0) = I$. In terms of the *matrix exponential function,* whose properties are summarized in Appendix 1, the matrix $P(t)$ is given by

$$P(t) = \exp(Qt), \qquad \text{for } t \geq 0.$$

Computation of the Matrix $\exp(Qt)$

There are a variety of algorithms to compute the exponential of a matrix. Any method for solving differential equations can be adapted for that purpose. Here, we describe some methods particularly well-suited for Markov chains. Before doing so, we discuss the important choice of the *time scale.*

Choosing the Time Unit: Suppose we want to compute the matrix $\exp(Qt)$ (or some related function) over some interval $(0, t^*)$. The mean sojourn times in the various states i are given by $(-Q_{ii})^{-1}$. If all the parameters $-Q_{ii}$ have small values (as compared, say, to one) the Markov chain has few state changes per unit of time. Correspondingly, the elements of $\exp(Qt)$ change very slowly. Alternatively, when all $-Q_{ii}$ values are large, the chain changes states rapidly. At least initially, the elements of $\exp(Qt)$ then vary rapidly. Sometimes, the $-Q_{ii}$ values differ by several orders of magnitude. In such cases, the Markov chain exhibits both slow and rapid transitions. The differential equations for $P(t)$ are then said to be *stiff.* A numerically accurate solution then calls for specialized numerical techniques. To obtain informative numerical results, even for well-behaved Markov chains with few states, we must choose the time unit in a physically meaningful way. That is done by *rescaling* the generator Q. In the listed problems, rescaling is discussed for many specific models. What follows is a brief general discussion.

When the Markov chain is irreducible, there exists a unique probability vector $\boldsymbol{\theta}$ satisfying

$$\boldsymbol{\theta} Q = 0, \qquad \boldsymbol{\theta} \mathbf{e} = 1,$$

where $\boldsymbol{\theta}$ is the *stationary probability vector,* or the *vector of steady-state probabilities* of Q. With $\boldsymbol{\theta}$ chosen as the initial probability vector, the *stationary rate* ρ at which transitions occur is

$$\rho = \sum_{i=1}^{m} \theta_i (-Q_{ii}).$$

The quantity ρ is also the expected number of transitions per unit of time in the stationary version of the Markov chain. In computing $\exp(Qt)$, it is helpful if ρ has a value significant to the problem at hand. In the absence of a compelling choice, we can set $\rho = 1$.

For any positive constant c, cQ is a generator with the same stationary probability vector $\boldsymbol{\theta}$ and with transition rate $c\rho$. Therefore, if we multiply Q by c, we can choose the transition rate. Computing $\exp(Qt)$ over $(0, t^*)$ is then equivalent to the evaluation of $\exp(cQt)$ over $(0, c^{-1}t^*)$. Multiplication of Q by c replaces the original time unit by c^{-1}. Such *rescaling* enables us to fix the transition rate ρ or to set the unit of time equal to some other interval significant to the problem at hand.

The organization of the computation of $\exp(Qt)$ is different for short and long intervals. These informal terms refer to short or long intervals for the rescaled chain. Without an appropriate choice of the time unit, that distinction is meaningless. The following algorithms to compute exponential matrices that are especially well-suited for Markov chains:

The Spectral Method: This method is mostly applicable for matrices Q

of rather low order m that are *diagonalizable*. That is, Q can be written as $Q = K^{-1}\Xi K$, where Ξ is a diagonal matrix with the eigenvalues of Q as diagonal elements. From the definition of the matrix exponential function, it then follows that

$$\exp(Qt) = K^{-1}\exp(\Xi t)K, \quad \text{for } t \geq 0.$$

The algorithm consists in diagonalizing Q and in implementing the stated formula. It requires complex arithmetic. If the spectral method is implemented, we suggest the use of library routines of proven high quality.

The Uniformization Method: This method arose from the observation that it is never advisable to compute $\exp(Qt)$ by forming a partial sum of the series

$$\exp(Qt) = \sum_{n=0}^{\infty} \frac{(Qt)^n}{n!}.$$

Rapid loss of significance results because of the negative diagonal elements of Q. For generators, however, there is a nice way of avoiding that difficulty. It also results in a useful algorithm.

Clearly, for sufficiently large positive λ, the matrix $\lambda I + Q$ is nonnegative. It suffices to choose λ such that $\lambda \geq \max\{-Q_{ii}\}$. The matrix defined by $P^\circ = \lambda^{-1}[\lambda I + Q]$ is then stochastic. Moreover,

$$\exp(Qt) = \exp(-\lambda t I + \lambda P^\circ t) = \sum_{n=0}^{\infty} e^{-\lambda t} \frac{(\lambda t)^n}{n!} P^{\circ n}, \qquad \text{for } t \geq 0.$$

We evaluate the Poisson density with parameter λt up to the index N for which the remaining probabilities are negligible. Next, we compute

$$E_N(t) = \sum_{n=0}^{N} e^{-\lambda t} \frac{(\lambda t)^n}{n!} P^{\circ n},$$

by Horner's algorithm in Section 3.2. The computed matrix $E_N(t)$ should be close to a stochastic matrix. When it is, we recommend making small numerical adjustments by renormalizing the row sums so that to machine accuracy, $E_N(t)$ is stochastic.

As described up to now, the uniformization method is suitable only if λt^* is moderately large. When that is not the case, we compute $\exp(Qt)$ by steps of size h on some interval $(0, t^*]$. Let r be an integer ≥ 2. For $1 \leq k < r$, we then evaluate $\exp(Qt)$ at values t satisfying $t^* k/r \leq t \leq t^*(k+1)/r$, by using the formula

$$\exp(Qt) = [\exp(Qr^{-1}t^*)]^k \exp[Q(t - kr^{-1}t^*)].$$

Runge - Kutta Methods: These are well-known numerical integration methods for ordinary differential equations. Here, they are essentially a variant of the uniformization method. For an appropriate step size h, we evaluate $\exp(Qh)$ to very high accuracy by the series in the uniformization method. By taking successive powers of $\exp(Qh)$ we compute $\exp(Qt)$ for $t = nh$, $n \geq 1$.

Taboo Probabilities

The conditional probabilities $P_{ij,H}(t)$, that the chain is in state j at time t and that, in moving from the initial state i to j, it has not visited the set H, are called *taboo probabilities.* As for discrete Markov chains many questions, such as first passage time distributions, are problems about taboo probabilities. In discussing these, we again partition the matrix Q as is illustrated in the following examples:

Example 5.1.1: Taboo Probabilities: We rearrange the states of the Markov chain in two sets, $\mathbf{1} = H^c$, the complement of H, and $\mathbf{2} = H$. The generator Q is now written in the form

$$Q = \begin{array}{c|cc|} \mathbf{1} & T(1,1) & T(1,2) \\ \mathbf{2} & T(2,1) & T(2,2) \end{array}.$$

We shall show that, for states i and j in $\mathbf{1}$, the taboo probability $P_{ij,H}(t)$ is the (i, j)-element of the matrix $V(t) = \exp[T(1, 1)t]$. The argument is the same as that leading to the Chapman - Kolmogorov equations. Considering all possible transitions in $(t, t + dt)$, we see that

$$V_{ij}(t + dt) = V_{ij}(t)[1 + Q_{jj}dt] + \sum_{\nu \neq i,\, \nu \in \mathbf{1}} V_{i\nu}(t) T_{\nu j}(1, 1)dt + o(dt).$$

Since $Q_{jj} = T_{jj}(1, 1)$, that leads to the differential equation

$$V'_{ij}(t) = \sum_{\nu \varepsilon 1} V_{i\nu}(t) T_{\nu j}(1, 1),$$

for $i \varepsilon 1$ and $j \varepsilon 1$. As $V(0) = I$, $V(t) = \exp[T(1, 1)t]$, for all $t \geq 0$.

The elementary conditional probability that the chain visits the set 2 for the first time in the interval $(t, t + dt)$ in the state $j \varepsilon 2$, given that it starts in $i \varepsilon 1$ at time zero, is the (i, j)-element of the matrix $\exp[T(1, 1)t]T(1, 2)\, dt$.

That is a useful matrix formalism with many applications. For $i \varepsilon 1$, and $j \varepsilon 2$, let ϕ_{ij} be the conditional probability that, starting in i, the chain eventually enters the set 2 in the state j. By the law of total probability, the matrix Φ with elements ϕ_{ij} is then given by

$$\Phi = \int_0^\infty \exp[T(1, 1)t]T(1, 2)dt = [-T(1, 1)]^{-1}T(1, 2),$$

provided the inverse exists. If Q is an irreducible generator, the Perron - Frobenius theorem in Appendix 1 implies that the eigenvalue with largest real part, $-\eta(1, 1)$, of $T(1, 1)$ is real and negative. That implies the existence of the inverse.

Example 5.1.2: Continuous Phase-Type Densities: Consider a finite Markov chain with a single absorbing state. We partition its transition probability matrix as

$$Q = \begin{array}{c} 1 \\ * \end{array} \left| \begin{array}{cc} T & \mathbf{T}^\circ \\ 0 & 0 \end{array} \right|,$$

where T is a matrix of order $m - 1$ and $\mathbf{T}^\circ$ is a column vector. The single absorbing state is denoted by $*$. The initial state of the Markov chain is chosen according to a probability vector $(\boldsymbol{\alpha}, \alpha_0)$. The probability $F(x)$ that absorption occurs no later than time x is given by

$$F(x) = 1 - \boldsymbol{\alpha} \exp(Tx)\mathbf{e}, \qquad \text{for } x \geq 0.$$

Clearly, $[\exp(Tx)\mathbf{e}]_i$ is the probability that, starting in state $i \neq *$, absorption has not occurred by time x. $F(\cdot)$ is a probability distribution with a jump of height α_0 at 0. On $(0,\infty)$, $F(\cdot)$ has a density $\phi(x)$, given by

$$\phi(x) = \boldsymbol{\alpha} \exp(Tx)\mathbf{T}^{\circ}, \qquad \text{for } x > 0.$$

$F(\cdot)$ is a *PH*-distribution. It is parameterized by the vector $\boldsymbol{\alpha}$ and the matrix T. The pair $(\boldsymbol{\alpha}, T)$ is called a *representation* of the *PH*-distribution. As discussed in Appendix 2, we always assume that the representation $(\boldsymbol{\alpha}, T)$ is irreducible. The mean μ'_1 of $F(\cdot)$ is calculated as follows:

$$\mu'_1 = \int_0^\infty [1 - F(x)]\, dx = \boldsymbol{\alpha} \int_0^\infty \exp(Tx)\, dx\, \mathbf{e} = \boldsymbol{\alpha}(-T)^{-1}\mathbf{e}.$$

The Laplace - Stieltjes transform of $F(\cdot)$ is given by

$$f(s) = \int_0^\infty e^{-sx}\, dF(x) = \alpha_0 + \boldsymbol{\alpha} \int_0^\infty \exp[-(sI - T)x]\, dx\, \mathbf{T}^{\circ} = \alpha_0 + \boldsymbol{\alpha}(sI - T)^{-1}\mathbf{T}^{\circ}.$$

By successive differentiations, we find that the kth noncentral moment of $F(\cdot)$ is given by

$$\mu'_k = k!\boldsymbol{\alpha}(-T)^{-k}\mathbf{e}, \qquad \text{for } k \geq 0.$$

Remark: Many problems in this chapter involve phase-type distributions. The reader is encouraged to study the more detailed discussion in Appendix 2 before attempting their solutions.

Steady-State Probabilities

For an irreducible generator Q there exists a *unique, positive* probability vector $\boldsymbol{\theta}$ that satisfies the equations $\boldsymbol{\theta}Q = 0$, and $\boldsymbol{\theta}\mathbf{e} = 1$, where $\boldsymbol{\theta}$ is the *stationary* probability vector or the vector of the *steady-state probabilities* $\{\theta_i\}$ of Q. Determining $\boldsymbol{\theta}$ is equivalent to computing the invariant probability vector of a stochastic matrix. From the definition of the matrix P°, we have

$$\mathbf{0} = \boldsymbol{\theta}Q = -\lambda\boldsymbol{\theta} + \lambda\boldsymbol{\theta}P^{\circ},$$

so that $\boldsymbol{\theta}$ is the invariant probability vector of P°. Therefore, every method to compute the steady-state probabilities of a stochastic matrix is easily adapted to the computation of $\boldsymbol{\theta}$.

Product Markov Chains and Related Constructions

It is often useful to combine two (or more) independent Markov chains into a single Markov chain. This construction leads to the Kronecker sum (see Appendix 1) of the generators. A similar construction is used to combine absorbing Markov chains. The following are illustrative examples.

Example 5.1.3: The Product Chain: Consider independent Markov chains respectively with m and n states and generators Q and R. The product Markov chain describes the simultaneous evolution of the two given chains. Its state space consists of the pairs (i, j), with $1 \le i \le m$, and $1 \le j \le n$, arranged in *lexicographic order,* that is the order $(1, 1), \cdots,$ $(1, n), (2, 1), \cdots, (2, n), \cdots, (m, n)$. If $\Phi_{ii', jj'}\, dt$ is the elementary probability of a transition $(i, j) \to (i', j')$ in $(t, t + dt)$, then

$$\Phi_{ii',jj'}\, dt = Q_{ii'}\, dt\, R_{jj'}\, dt = \mathrm{o}[(dt)^2], \quad \text{for } i \ne i',\ j \ne j',$$

$$\Phi_{ii,jj'}\, dt = [1 + Q_{ii}\, dt] R_{jj'}\, dt = R_{jj'}\, dt + \mathrm{o}[(dt)^2], \quad \text{for } j \ne j',$$

$$\Phi_{ii',jj}\, dt = Q_{ii'}\, dt [1 + R_{jj}\, dt] = Q_{ii'}\, dt + \mathrm{o}[(dt)^2], \quad \text{for } i \ne i'.$$

The diagonal elements of the generator Φ are therefore $\Phi_{ii,jj} = Q_{ii} + R_{jj}$. We see that the generator Φ of the product chain is the *Kronecker sum*

$$\Phi = Q \oplus R = Q \otimes I_n + I_m \otimes R$$

of the generators Q and R. When Q and R are irreducible, the stationary probability vector $\boldsymbol{\phi}$ of the product chain is the Kronecker product $\boldsymbol{\theta} \otimes \boldsymbol{\pi}$ of the stationary probability vectors $\boldsymbol{\theta}$ and $\boldsymbol{\pi}$ of Q and R. Clearly, $\boldsymbol{\theta} \otimes \boldsymbol{\pi}$ is just the mn-vector with components $\theta_i \pi_j$.

Example 5.1.4: Absorbing Markov Chains: We consider two independent Markov chains each with a single absorbing state as in Example 5.1.2. They correspond to PH-distributions with representations $(\boldsymbol{\alpha}, T)$ and $(\boldsymbol{\beta}, S)$ respectively. We now form the generator

$$Q^\circ = \begin{array}{c} 1 \\ * \end{array} \left| \begin{array}{cc} T \oplus S & \mathbf{T}^\circ \otimes \mathbf{e} + \mathbf{e} \otimes \mathbf{S}^\circ \\ 0 & 0 \end{array} \right| ,$$

with initial probability vector $(\boldsymbol{\alpha} \otimes \boldsymbol{\beta}, \alpha_0 + \beta_0 - \alpha_0\beta_0)$. In that Markov chain, absorption occurs when one of the given chains reaches its absorbing state. The time to absorption in the Markov chain Q° is the *minimum* of the absorption times in the orginal chains. Therefore, if X and Y are independent and have PH-distributions with the given representations, $Z = \min(X, Y)$ has a PH-distribution with representation $(\boldsymbol{\alpha} \otimes \boldsymbol{\beta}, T \oplus S)$. Similarly, by constructing the Markov chain with generator

$$Q^{\circ\circ} = \begin{array}{c|cccc|} \mathbf{1,2} & T \oplus S & I \otimes \mathbf{S}^{\circ} & \mathbf{T}^{\circ} \otimes I & \mathbf{0} \\ \mathbf{1} & 0 & T & 0 & \mathbf{T}^{\circ} \\ \mathbf{2} & 0 & 0 & S & \mathbf{S}^{\circ} \\ * & 0 & 0 & 0 & 0 \end{array} ,$$

with initial probability vector $(\boldsymbol{\alpha} \otimes \boldsymbol{\beta}, \beta_0\boldsymbol{\alpha}, \alpha_0\boldsymbol{\beta}, \alpha_0\beta_0)$, we see that the time to absorption in the Markov chain $Q^{\circ\circ}$ is the *maximum* of the absorption times in the original chains. If X and Y are independent random variables with PH-distributions, then $U = \max(X, Y)$ has a PH-distribution. Its representation requires $mn + m + n$ phases.

Remark: Similar constructions are very useful in setting up Markov chains for queues and reliability models. With practice, one soon can write generators for quite involved Markovian models in a transparent form. It is useful to think of appropriate subsets as *macro-states*. For example, in the generator $Q^{\circ\circ}$, the set denoted by $\mathbf{1, 2}$ corresponds to the case where neither of the original Markov chains have been absorbed. Conditionally, the two Markov chains then evolve independently and their transitions are governed by the matrix $T \oplus S$. Upon absorption of the first chain, we move to the macro-state $\mathbf{2}$ with rates described by the matrix $\mathbf{T}^{\circ} \otimes I$. The identity matrix indicates that, upon such a transition, there is no state change in the second chain. We recommend that readers who are not familiar with Kronecker products carry out the construction in detail for chains with two or three states. The conciseness of the Kronecker formalism is bound to be appreciated soon.

5.2 MAIN PROBLEM SET FOR CHAPTER 5

As for discrete-time Markov chains, a careful display of the generator Q of models arising in the problems is strongly recommended. Subsets of the state space relevant to taboo and absorption probabilities should be clearly identified. Numerical computations should be expedited by exploiting the special structure of the Markov chains. Many problems require knowledge of *PH*-distributions and of the Markovian arrival process. Prior study of the material in Appendices 2 and 3 is therefore recommended.

For some special infinite Markov chains, we can obtain analytic characterizations of the steady-state probabilities and some other features. Unless such material is classical, the reader will be asked to carry out the detailed derivations of such results. The statements of the related problems suggest the essential steps of such calculations.

Many problems call for qualitative interpretations of the numerical results. As Markov chains often describe models of practical interest, these interpretations should state what information the numerical answers convey about the physical behavior of the model.

EASIER PROBLEMS

Problem 5.2.1: For the two-state Markov chain with generator

$$Q = \begin{matrix} 1 \\ 2 \end{matrix} \begin{vmatrix} -\sigma_1 & \sigma_1 \\ \sigma_2 & -\sigma_2 \end{vmatrix},$$

obtain explicit formulas for the elements of the matrix $P(t) = \exp(Qt)$ in terms of the eigenvalues of Q. Fix the invariant probabilities π_1 and π_2 to be 1/3 and 2/3, respectively. That determines Q up to a constant k. How do the elements of $P(t)$ depend on k? Plot graphs of the elements $P_{11}(t)$ and $P_{22}(t)$ for several values of k.

The elementary conditional probability $\psi_1(t)\, dt$ that, given that the chain starts in state 1, a transition occurs in $(t, t + dt)$ is

$$\psi_1(t)\, dt = [P_{11}(t)\sigma_1 + P_{12}(t)\sigma_2]\, dt.$$

Therefore $\psi_1(t)$ is the *rate at which transitions occur* at time t when the chain is started in state 1. What is the limit ψ^* of $\psi_1(t)$ as $t \to \infty$? Find the value of k for which $\psi^* = 1$. For that k, plot $\psi_1(t)$ for $0 \leq t \leq 5$.

Problem 5.2.2: Adapt the methods for the computation of the vector $\boldsymbol{\pi}$ in Section 4.1 to the evaluation of the stationary probability vector $\boldsymbol{\theta}$ of an irreducible generator Q. Write general purpose subroutines to implement these methods and test them on a number of examples.

Problem 5.2.3: T is a principal submatrix of an irreducible generator Q. $\boldsymbol{\beta}$ is a given row vector of probabilities. Adapt the Gauss - Seidel method of Section 4.1 to compute the solution of the equation $\boldsymbol{\beta} = -\mathbf{x}T$. Use that method to solve Problem 5.2.4. Identify the matrix T and the vector $\boldsymbol{\beta}$. What initial solution would you choose for successive values of α?

Problem 5.2.4: The Markov chain with generator $Q(\alpha)$ is started in state 1. Let $\phi_{14}(\alpha)$ be the probability that, upon leaving the set $\mathbf{1} = \{1, 2, 3\}$, the chain enters state 4.

$$Q(\alpha) = \begin{array}{c|ccccc|} 1 & -10 & 9 & 1 & 0 & 0 \\ 2 & 2 & -6 & 1 & 1 & 2 \\ 3 & \alpha & 3 & -3-2\alpha & 0 & \alpha \\ 4 & 5 & 0 & 3 & -8 & 0 \\ 5 & 3 & 0 & 1 & 1 & -5 \end{array} .$$

Compute and plot the probability $\phi_{14}(\alpha)$ for $0 \le \alpha \le 5$.

Problem 5.2.5: For the Markov chain in Problem 5.2.4, compute and plot the steady-state probabilities $\theta_j(\alpha)$, $1 \le j \le 5$, for $0 \le \alpha \le 5$.

Problem 5.2.6: Rescaling: Consider an irreducible $m \times m$ generator Q and write $\sigma_i = -Q_{ii}$. $\boldsymbol{\theta}$ is the stationary probability vector of Q.

a. Explain why $\beta_i = \sigma_i^{-1}$ is the mean of a sojourn time in state i.

b. If $\sigma^* = \max\{\sigma_i\}$, verify that $P = I + \sigma^{*\,-1} Q$ is an irreducible stochastic matrix with invariant probability vector $\boldsymbol{\theta}$.

c. Write and test a computer code to compute $\boldsymbol{\theta}$ for a given generator Q.

d. Give a proof or an intuitive argument to show that the rate λ^* at which transitions occur in the steady-state version of a Markov chain with generator Q is given by $\lambda^* = \sum_{i=1}^{m} \theta_i \sigma_i$, and that the expected number of transitions in $(0, t)$ equals $\lambda^* t$.

e. *Rescaling:* Show how a positive constant a is to be chosen so that the stationary transition rate λ^* corresponding to the generator aQ has a given value. Explain why that corresponds to the choice of a time unit for the corresponding Markov chain. For your test examples in part c, print the rescaled generators for which $\lambda^* = 1$.

Problem 5.2.7: Time-Dependent Probabilities: The time-dependent transition probability matrix $P^*(t)$ of a Markov chain Q is given by

$$P^*(t) = \exp(Qt), \quad \text{for } t \geq 0.$$

We wish to examine $P(t)$ over an interval $(0, T)$ during which, on average, 50 transitions occur. How do you rescale the Markov chain and choose T? Next, implement the uniformization method. Defining σ^* and P as in Problem 5.2.6, write

$$\exp(Qt) = e^{-\sigma^* t}\exp(\sigma^* tP) = \sum_{k=0}^{\infty} e^{-\sigma^* t}\frac{(\sigma^* t)^k}{k!}P^k.$$

a. Let $C_n(t)$ be the matrix obtained by truncating the series at the index $k = n$. Obtain a bound on the vector $\mathbf{e} - C_n(t)\mathbf{e}$. Explain how that bound can be used as a stopping criterion in the computer code.

b. Discuss the merits of this (very good) procedure, but consider what happens when $\sigma^* t$ is large.

c. How would you implement the uniformization method to compute $P^*(rh)$, for $0 \leq r \leq N$, where $h > 0$ is a given *step* and N is a rather large positive integer? Think of h as a fraction, say 1/5, of the time unit of the chain and $N = 250$.

Write a code to implement this procedure and test it on a number of generators of dimension 10 and 20. Consider some cases where there is a difference of, say, two orders of magnitude between the smallest and the largest σ_i for the given generator. To avoid large output, print (or preferably, plot) only the components of the first row of $P^*(t)$. In computing $C_n(t)$, adapt Horner's algorithm of Section 3.2.

Problem 5.2.8: The Uniformization Method: The uniformization method performs even better in computing $\exp(Q^\circ t)$ for a principal submatrix Q° of a generator. The corresponding matrix P° is then substochastic and the series for $\exp(Q^\circ t)$ converges faster. Use the uniformization method to construct an algorithm to compute the distribution $F(\cdot)$ and the density $\phi(\cdot)$ of a PH-distribution with a given representation (α, T) of up to 50 phases.

First, rescale the distribution so that its mean $\mu_1' = \alpha(-T)^{-1}\mathbf{e} = 1$. Set the step $h = 0.05$ and use the uniformization method to compute the substochastic matrix $\exp(Th)$ very accurately. Every one of its elements should be evaluated to within 10^{-10}. Then compute and plot the values $F(kh)$ and $\phi(kh)$ for a sufficiently large number of k.

Problem 5.2.9: Time-Dependent Probabilities: When the matrix Q can be written in the form $Q = K^{-1}\Xi K$, where Ξ is a diagonal matrix with the eigenvalues ξ_j of Q as its diagonal elements, the matrix $\exp(Qt)$ is evaluated as $\exp(Qt) = K^{-1}\exp(\Xi t)K$, for $t \geq 0$. Here, $\exp(\Xi t)$ is a diagonal matrix with $\exp(\xi_j t)$ as diagonal elements. That is *spectral method*. It is a useful approach for working with generators of low dimension. We suggest using a library routine to diagonalize Q. Write a code to implement the spectral method for the same test examples as in Problem 5.2.7.

Problem 5.2.10: The generator Q of a three-state Markov chain is

$$Q = \begin{array}{c} 1 \\ 2 \\ 3 \end{array} \left| \begin{array}{ccc} -10 & 9 & 1 \\ 4 & -5 & 1 \\ 150 & 50 & -200 \end{array} \right| .$$

We construct the new 6×6 generator

$$Q_1 = \begin{array}{c} 1' \\ 1'' \\ 2' \\ 2'' \\ 3' \\ 3'' \end{array} \left| \begin{array}{cccccc} -20 & 20 & 0 & 0 & 0 & 0 \\ 0 & -20 & 18 & 0 & 2 & 0 \\ 0 & 0 & -10 & 10 & 0 & 0 \\ 8 & 0 & 0 & -10 & 2 & 0 \\ 0 & 0 & 0 & 0 & -400 & 400 \\ 300 & 0 & 100 & 0 & 0 & -400 \end{array} \right| ,$$

and think of the pair of states (i', i''), as a macro-state $\mathbf{i}$, for $i = 1, 2, 3$. Show that the Markov chain with generator Q_1 models the same behavior as that with the generator Q, *except* that the sojourns in states $\mathbf{i}$ now have Erlang distributions of order two rather than exponential distributions. The means of the sojourn times in the states $\mathbf{i}$ are the same as before, but their variability is smaller.

a. Compute the stationary probability vectors of both generators. For the second chain, set $\theta_i = \theta_{i'} + \theta_{i''}$ for $i = 1, 2, 3$. Compare these quantities to the steady-state probabilities of Q.

b. Use the uniformization method to compute the time-dependent transition probabilities of both Markov chains for $t = 0.1r$, $0 \leq r \leq 200$. For the second chain, set

$$P_{ij}^{\circ}(t) = P_{i'j'}(t) + P_{i'j''}(t), \quad \text{for } i, j = 1, 2, 3.$$

For each pair (i, j), plot these functions on the same graphs as those of the $P_{ij}(t)$ for the first chain. How are the time-dependent probabilities of the three macro-states affected by the reduced variability of the sojourn times? How should we modify Q when the sojourn times have Erlang distributions of order three, or, in general, of order k?

Problem 5.2.11: Solve Problem 5.2.10 by implementing the spectral method for the time-dependent probabilities.

Problem 5.2.12: Consider a sequence of intervals whose lengths are independent, exponentially distributed *alternatingly* with parameters θ and σ. There is an independent Poisson process of rate λ. Its events are registered only during the intervals with parameters θ. The process of registered events is called an *interrupted Poisson process.*

a. Show that the Laplace - Stieltjes transform $a(s)$ of the distribution $A(\cdot)$ of the time between two successive *registered* points is given by

$$a(s) = \frac{\lambda(s+\sigma)}{s^2 + (\lambda+\theta+\sigma)s + \lambda\sigma} = p\,\frac{\eta_1}{s+\eta_1} + (1-p)\,\frac{\eta_2}{s+\eta_2},$$

where $\eta_1 > \lambda > \eta_2 > 0$, and $p = (\lambda - \eta_2)(\eta_1 - \eta_2)^{-1}$. The quantities $-\eta_1$ and $-\eta_2$ are the zeros of the denominator. Note that $A(\cdot)$ is the *hyperexponential* distribution with parameters η_1, η_2, and p.

b. Show that, conversely, for any bona fide hyperexponential distribution $F(\cdot)$ with parameters η_1, η_2, and p, the quantities

$$\lambda = p\eta_1 + (1-p)\eta_2, \qquad \sigma = \eta_1\eta_2\,[p\eta_1 + (1-p)\eta_2]^{-1},$$

$$\theta = p(1-p)(\eta_1 - \eta_2)^2 \, [p\eta_1 + (1-p)\eta_2]^{-1}$$

are valid parameters for an interrupted Poisson process whose distribution $A(\cdot)$ agrees with $F(\cdot)$.

c. Explain why the successive events in an interrupted Poisson process form a renewal process. Note that the statements in parts *a* and *b* imply that such a process is ***equivalent*** to a hyperexponential renewal process. For a number of hyperexponential distributions with mean 1 and different coefficients of variation, compute the equivalent parameters θ, σ, and λ. Explain what these tell about the behavior of a hyperexponential renewal process.

Problem 5.2.13: $F(\cdot)$ is a continuous *PH*-distribution with representation (α, T). Rescale it to have mean one. Develop an efficient algorithm to determine the values x_1, x_2, and x_3 for which $F(x_1) = 0.9$, $F(x_2) = 0.95$, and $F(x_3) = 0.99$, respectively. Print x_1, x_2, and x_3 for the rescaled and for the original distributions. Solve the problem numerically for the *PH*-distribution with representation $\alpha = (1, 0, 0)$ and

$$T = \begin{vmatrix} -2 & 2 & 0 \\ 0 & -1.5 & 1.5 \\ 1 & 0 & -3 \end{vmatrix} .$$

Problem 5.2.14: Let τ be the absorption time in the Markov chain with initial probability vector $(\alpha, 0)$ and the generator

$$Q = \begin{vmatrix} T & \mathbf{T}^{\circ} \\ 0 & 0 \end{vmatrix} .$$

x_{95} is the 95th percentile of the *PH*-distribution of τ. What is the conditional probability ϕ_j, that the Markov process is in the transient state j, given that absorption has not occurred by time x_{95}? (or, informally, what are the probabilities of the phases of the paths that outlive the 95th percentile)? Write a computer code to evaluate the ϕ_j for the *PH*-distribution with representation (α, T). Implement your code for the Erlang distribution of various orders and interpret your numerical results.

Problem 5.2.15: Two machines A and B have exponential times-to-failure with parameters μ_1 and μ_2, respectively. Machine A is preferred; B is a backup unit. Whenever machine A is operational, it is used. B is in *cold standby*. That means that B cannot fail when it is not in use. When A fails, it starts an exponential repair time with parameter σ_1 and B is used. Should B also fail, its exponential repair time has parameter σ_2. If both machines require repair, a *single* repairman gives priority to repairing A. If only one machine is down, he works at its repair. After repair, a machine is returned good-as-new. Define the four states of the system as follows:

1: Machine A is in use, Machine B is OK.
2: Machine A is in use, Machine B is in repair.
3: Machine B is in use, Machine A is in repair.
4: Both machines are down, Machine A is in repair.

a. Set up a four-state Markov chain to describe the system. Write down its generator Q.

b. Calculate its steady-state probabilities $\{\pi(i)\}$ as fractions with simple numerators. Their denominator is a quadratic polynomial in σ_1.

c. Suppose that all the parameters other than σ_1 are fixed. Show that there is a unique value of σ_1 for which the fraction of time $\pi(4)$ that both machines are down has a given value, say 1/100.

d. For $\mu_1 = 0.01$, $\mu_2 = 0.05$, $\sigma_2 = 1$, find σ_1 such that $\pi(4) = 0.01$.

e. For the parameters found in part *d*, compute the time-dependent probabilities $P_{ij}(t)$ on the interval [0, 500].

Problem 5.2.16: Jobs arriving according to a Poisson process of rate λ are assigned to one of two exponential servers, I or II. Unit I has a service rate μ_1; Unit II, a service rate μ_2. Customers arriving when both servers are occupied are lost. If both are free, the arrival goes to either I or II with probabilities p or $q = 1 - p$. Any arrival that finds one server free is processed by that unit.

a. Set up a four-state Markov chain for that system and derive explicit formulas for its steady-state probabilities.

b. For $p = 0.5$ and given μ_1 and μ_2, find the value λ_{90} of λ for which 90%

of the arriving customers are admitted.

c. Setting the arrival rate equal to λ_{90} and for given μ_1 and μ_2, find the value of p for which the fraction of customers lost is smallest.

d. Determine the value of p for which the fraction of time that processor I is in use is largest.

Remark: With the more versatile PH-distributions for the service times this model is the subject of Problem 5.2.53. For the present problem explicit expressions are obtained for the various items of interest. The equations for λ_{90} and for p can be solved by elementary methods. The reader is encouraged to plot performance measures, such as the fraction of lost input, as functions of the various parameters.

Problem 5.2.17: Q is an irreducible generator of an m-state Markov chain. Explain why the element $M_{ij}(x)$ of the matrix

$$M(x) = \int_0^x \exp(Qu)\, du$$

is the conditional expected value of the time spent in state j in the interval $(0, x)$, given the chain starts in state i.

a. If $\boldsymbol{\theta}$ is the stationary probability vector of Q, show that the matrix $\mathbf{e}\boldsymbol{\theta} - Q$ is nonsingular.

b. Fully justify the derivation of the formula

$$M(x) = \mathbf{e}\boldsymbol{\theta}x + [I - \exp(Qx)](\mathbf{e}\boldsymbol{\theta} - Q)^{-1}$$

in Appendix 1. It is the analogue of the formula for discrete Markov chains in Problem 4.4.23.

AVERAGE PROBLEMS

Problem 5.2.18: $F(\cdot)$ is a probability distribution on $[0, \infty)$ with finite positive mean μ'_1.

a. If a random variable X has the probability distribution

$$F(x;\alpha) = \sum_{k=1}^{\infty} (1-\alpha)\alpha^{k-1} F^{(k)}(x),$$

with $0 < \alpha < 1$, show that, as $\alpha \to 1$, the limit distribution of $(1-\alpha)X$ is exponential with mean μ'_1.

b. Let $F(\cdot)$ be a *PH*-distribution with representation (β, S), where β is a probability vector. Show that $F(x;\alpha)$ is a *PH*-distribution with representation $(\beta, S + \alpha S^{\circ}\beta)$.

c. Develop an algorithm to find the ordinate x^* for which $F(x^*) = 0.99$. Next, find the value of α for which $F(x^*;\alpha) = 0.25$. Use the result in part *a* to obtain a starting solution in the search for α. Implement your code for Erlang distributions $F(\cdot)$ with mean one and of various orders.

Problem 5.2.19: On the interval (0, 1) a Poisson process of rate λ_1 generates $N_1(a)$ arrivals until time a. Between a and $2a$, a second Poisson process of rate λ_2 takes over and generates $N_2(a)$ arrivals. If $N_1(a) > N_2(a)$ the first process operates during the interval $(2a, 1]$ and vice versa. If $N_1(a) = N_2(a)$ one of the processes is chosen at random and operates during the remaining time. A total of K events occur in (0, 1].

a. For given λ_1 and λ_2, find the value $a_1^* \leq 1/2$ for which the expectation $E(K)$ is maximum.

b. $L(a)$ is the smallest integer for which $P\{K \geq L(a)\} \geq 0.75$. Determine the value $a_2^* \leq 1/2$ for which $L(a)$ is largest. Implement your algorithm for $\lambda_1 = 50$ and $\lambda_2 = 45$.

Remark: This is a continuous analogue of the model in Problems 2.3.37 and 2.3.38. Explain why it offers a good approximation to that model when N is very large and the probabilities $p(1)$ and $p(2)$ are small.

Problem 5.2.20: The generator Q of an irreducible Markov chain is partitioned as in Example 5.1.1. Time is rescaled so that in steady-state transitions occur at unit rate. Each time the Markov chain *enters* the set 1, a monitoring device is activated. That device has an exponential lifetime with parameter γ. Whenever the chain leaves set 1, the monitor is deactivated. Consider macro-states, 1^* and 1°, corresponding to the situations that the chain is in **1** with the monitor active, or with the monitor expired. Set **2** is as before.

a. Show that the monitored system is described by the Markov chain with generator

$$Q^* = \begin{matrix} 1^* \\ 1^\circ \\ 2 \end{matrix} \left| \begin{matrix} T(1,1) - \gamma I & \gamma I & T(1,2) \\ 0 & T(1,1) & T(1,2) \\ T(2,1) & 0 & T(2,2) \end{matrix} \right| .$$

b. Partition the stationary probability vector of Q^* into row vectors $\theta(1^*)$, $\theta(1^\circ)$, and $\theta(2)$. Why does $\theta(2)$ not depend on γ? Develop an efficient algorithm to compute the stationary transition probabilities of the Markov chain Q^* as functions of γ.

c. For $0.10 \le \gamma \le 10$, plot the quantity

$$K(\gamma^{-1}) = \frac{\theta(1^*)e}{\theta(1^*)e + \theta(1^\circ)e},$$

as a function of the mean lifetime γ^{-1} of the monitor. Implement your code for several Markov chains whose behavior is well understood. What does that graph tell you? Give a probabilistic interpretation for $K(\gamma^{-1})$. Specifically, let **1** and **2** respectively be the first two and the last two states in the chains with generators

$$Q_1 = \left| \begin{matrix} -5 & 4 & 1 & 0 \\ 6 & -8 & 1 & 1 \\ 0 & 5 & -10 & 5 \\ 1 & 0 & 3 & -4 \end{matrix} \right| , \quad Q_2 = \left| \begin{matrix} -5 & 1 & 4 & 0 \\ 6 & -8 & 1 & 1 \\ 0 & 5 & -10 & 5 \\ 1 & 0 & 3 & -4 \end{matrix} \right| .$$

Remarks: *a*. The graphs in part *c* provide a measure of how strongly the states in set **1** communicate with each other. Let us call this the *stickiness* of set **1**. $K(\gamma^{-1})$ is the steady-state fraction of sojourns in the set **1** while the monitor is active. In studying its graph for a single Markov chain, we see why it is important to rescale time to have a steady-state transition rate of 1. However, in comparing the graphs for two related Markov chains on the same state space, a common time scale should be used. What is the simple relation between $K_1(\gamma^{-1})$ and $K_2(\gamma^{-1})$ when a common time scale is chosen for both chains?

b. Once the steady-state probability vector has been computed for a value γ_0 it can easily be updated for other values by using an iterative method. How would you do that? It is now advisable to choose the initial γ_0 near the middle of the interval of interest.

Problem 5.2.21: For the partitioned Markov chain in Problem 5.2.20, let the lifetime of the monitor have a *PH*-distribution with representation $(\boldsymbol{\beta}, S)$. Every time the chain enters set 1, the monitor is activated according to the initial probability vector $\boldsymbol{\beta}$, independently of the past.

a. Show that, with appropriate definitions of the macro-states, the Markov chain now has the generator

$$Q^* = \begin{array}{c} 1^* \\ 1^\circ \\ 2 \end{array} \left| \begin{array}{ccc} T(1,1) \oplus S & I \otimes \mathbf{S}^\circ & T(1,2) \otimes \mathbf{e} \\ 0 & T(1,1) & T(1,2) \\ T(2,1) \otimes \boldsymbol{\beta} & 0 & T(2,2) \end{array} \right| .$$

b. What is the detailed form of Q^* when the monitor has an Erlang distributed lifetime of order r with scale parameter γ? Exploit the special structure of Q^* to develop an efficient algorithm to compute the invariant probability vector of that Markov chain. Define the analogue of the function K in Problem 5.2.20.

c. As in Problem 5.2.20, plot and interpret graphs of the function K for various orders of the Erlang distribution. Why is that generalization of interest?

Remark: Iterative methods, in combination with the use of the special structure of Q^*, are recommended for computing the stationary probability vectors for this model.

Problem 5.2.22: The Erlang Renewal Process: Consider the *stationary* renewal process with underlying Erlang density

$$\phi_m(u) = e^{-\lambda u} \frac{(\lambda u)^{m-1}}{(m-1)!} \lambda, \quad \text{for } u \geq 0.$$

Let $p(n; t)$ be the probability that n renewals occur in $(0, t)$.

a. Use properties of the Poisson process to show that for $t \geq 0$,

$$p(0;\,t) = \sum_{k=0}^{m-1} \left[1 - \frac{k}{m} \right] e^{-\lambda t} \frac{(\lambda t)^k}{k!},$$

and

$$p(n;\,t) = \frac{1}{m} \sum_{r=1}^{m} \sum_{j=1}^{m} e^{-\lambda t} \frac{(\lambda t)^{nm-m+r+j-1}}{(nm-m+r+j-1)!}, \quad \text{for } n \geq 1.$$

With careful attention to summation limits, rewrite the expression for $p(n;\,t)$ as a single sum.

b. Set $\lambda = m$. Why is the expected value of the density $\{p(n;\,t)\}$ now equal to t?

c. Compute and plot the densities $\{p(n;\,t)\}$ for $t = 5$, and for $1 \leq m \leq 10$. Discuss the qualitative implications of your numerical results.

d. In Appendix 2, a formula for the variance of the number of events in $(0,\,t)$ is given for a *PH*-renewal process. For the Erlang renewal process, work out that formula in scalar form. Use the resulting expression to calculate the standard deviations of the densities computed in part *c*.

Problem 5.2.23: Based on the treatment of the *PH*-renewal function $H(t)$ in Appendix 2, write an efficient code to compute $H(t)$ and the renewal density $H'(t)$. In computing the matrix $\exp(Q^* t)$, use the *uniformization method* with appropriate caution, so that your results remain accurate even for larger t. Why is it advisable to rescale $F(\cdot)$ so that $\mu'_1 = 1$?

Implement your code for particular distributions such as the Erlang with m phases and the hyperexponential distribution. Explore other distributions for which we can anticipate substantial (and nearly periodic) clustering of points in the renewal process. For example, let $F(\cdot)$ be a mixture of an exponential distribution with a small mean and an Erlang distribution of order 10 with a larger mean. Plot the renewal functions and densities. Which features of the renewal process are apparent from these graphs?

Problem 5.2.24: X has a *delayed PH*-distribution $F_a(\cdot)$, with delay parameter $a > 0$ if $X - a$ has a *PH*-distribution. Let that *PH*-distribution have the representation $(\boldsymbol{\alpha}, T)$ with $\boldsymbol{\alpha}\mathbf{e} = 1$.

a. Show how the probability density $\{b_k\}$ with

$$b_k = \int_0^\infty e^{-\lambda u} \frac{(\lambda u)^k}{k!} dF_a(u), \quad \text{for } k \geq 0$$

can be computed without numerical integrations. The sequence $\{b_k\}$ is the density of the number of arrivals in a Poisson process of rate λ during an interval of length X with distribution $F_a(\cdot)$.

b. Let $F^\circ(\cdot)$ be a mixture of the form $F^\circ(x) = \sum_{j=1}^{n} p_j F_{a_j}(x)$. The p_js are positive probabilities whose sum is one. The $F_{a_j}(\cdot)$s are delayed *PH*-distributions. Their representations and delay parameters may all be different.

c. Rescale $F^\circ(\cdot)$ to have mean one. Write a subroutine to compute and plot the rescaled distribution $F^\circ(\cdot)$ and its density.

d. Write a second subroutine to evaluate and plot the density $\{b_k\}$ corresponding to $F^\circ(\cdot)$ for several values of λ with $0 < \lambda \leq 1$.

Remark: Mixtures of delayed *PH*-distributions form a versatile class. They can serve to illustrate the effects of qualitatively different service time distributions on the steady-state queue length for the $M/G/1$ and related queueing models. See Section 4.2.3 and Problem 4.4.36.

Problem 5.2.25: A system consists of modules I and II, respectively, with m and n components all placed in parallel and with independent exponential lifetimes. Components in module I have a failure rate λ; those in II, a failure rate μ. In addition, any failure of a component in module I induces a shock to module II that causes any surviving components in II to fail according to independent Bernoulli trials with probability p, $0 \leq p < 1$. The system fails when all components in one of the modules have expired. In a type-I failure a component from module I is involved in the last failure. In a type-II failure, module II fails through the expiration of its last surviving component.

a. Show that the lifetime of the system has a *PH*-distribution and construct its representation for $m = 3$ and $n = 4$.

b. Prove that, if k and r components survive in the modules, the expected remaining lifetime $E(k, r; p)$ of the system satisfies the recurrence relation

$$E(k, r; p) = \frac{1}{\lambda k + \mu r} + \frac{\mu r}{\lambda k + \mu r} E(k, r-1; p)$$

$$+ \frac{\lambda k}{\lambda k + \mu r} \sum_{\nu=0}^{r} \binom{r}{\nu} p^{\nu}(1-p)^{r-\nu} E(k-1, r-\nu; p)$$

and specify the appropriate boundary conditions. Write a subroutine to evaluate the mean lifetime of the system for given parameters m, n, λ, μ, and p. For $\lambda = 1$, $\mu = 0.1$, $m = 4$, and $n = 6$, plot the ratio

$$R(m, n; p) = E(m, n; \frac{p)}{E(m,} n; 0),$$

for $0 \le p \le 0.2$ by increments of 0.01. What does that graph tell you?

c. Use the special structure of the representation of the *PH*-lifetime distribution to develop an efficient vector recurrence to compute the probabilities $\phi_1(r)$ and $\phi_2(k)$ that type-*I* or type-*II* failures occur and that respectively r or k components remain alive in the other module. Evaluate these probabilities for the parameter values in part *b*.

Problem 5.2.26: We wish to compute tables of the successive convolutions $F^{(j)}(x)$, $1 \le j \le N$, of a *PH*-distribution $F(\cdot)$ with representation (β, S). An efficient procedure will be described for $N = 3$, but it is applicable for general N values. It is used in solving Problem 5.2.60. For simplicity, β is a probability vector. The convolution $F^{(3)}(\cdot)$ is the absorption time distribution in the Markov chain

$$Q = \begin{vmatrix} S & S^{\circ}\beta & 0 & 0 \\ 0 & S & S^{\circ}\beta & 0 \\ 0 & 0 & S & S^{\circ} \\ 0 & 0 & 0 & 0 \end{vmatrix},$$

with initial probability vector $[\beta, 0, 0, 0] = [\beta_3, 0]$. Writing S_3 for the matrix in the *PH*-representation of $F^{(3)}(\cdot)$, we evaluate the vector

$$\mathbf{v}(x) = \beta_3 \exp(S_3 x),$$

which is partitioned into row vectors $\mathbf{v}_1(x)$, $\mathbf{v}_2(x)$, and $\mathbf{v}_3(x)$ of the same dimension as β. Explain why for $j = 1, 2, 3$, and $x \ge 0$,

$$F^{(j)}(x) = 1 - \mathbf{v}_j(x)\mathbf{e}.$$

Write an efficient code using the uniformization method to implement that procedure for a general $N \leq 20$. Scale the x-axis appropriately. Implement your algorithm for various PH-distributions and with $N = 5$.

Problem 5.2.27: Initially devices A_1, A_2, and A_3 are equipped respectively, with one component of type 1, type 2, and type 3. These have independent, exponential lifetimes, respectively with parameters μ_1, μ_2, and μ_3. The three types are interchangeable in their use. As it is not known which type is most reliable, the following replacement policy is proposed:

> If all three devices use components of different types and a failure occurs, replace the failed component with equal probability by a component of one of the other types.
> If two devices use type i and the third a different type j and the component of type j fails, replace it by a component of type i. However, if one of the type i components fails, replace it by one of the type that is not in use.
> When all devices use components of the same type, use that type in all future replacements.

We are interested in the waiting time until, for the first time, all units use the same type. It is an absorption time in a Markov chain with *ten* states, three of which are absorbing. List the states as (1, 2, 3), (1, 1, 2), (1, 1, 3), (2, 2, 1), (2, 2, 3), (3, 3, 1), (3, 3, 2), (1, 1, 1), (2, 2, 2), (3, 3, 3). For example, (1, 1, 3) signifies that two components of type 1 and one component of type 3 are in use. That state description does not keep track of the component type used by each unit. Construct the generator for the Markov chain. Write a subroutine to evaluate the absorption probabilities in the states (1, 1, 1), (2, 2, 2), and (3, 3, 3). Write further routines to compute

> *a.* The mean and the variance of the time until absorption
> *b.* The conditional means and variances of the times until absorption, given that absorption occurs in state (i, i, i), $i = 1, 2, 3$
> *c.* The mean numbers of each type of component used up to and including absorption.

Assume, without loss of essential generality, that $1 = \mu_1 \geq \mu_2 \geq \mu_3$. Implement your code for various choices of μ_1, μ_2, and μ_3. Interpret the numerical results for the various examples. Do you think this is a good policy for selecting the best type to be used?

Problem 5.2.28: Items used in successive replacements have independent, identically distributed lifetimes X_k with common probability density $\phi(\cdot)$ on $(0, \infty)$. At time zero a new item is put in use. The device is inspected at times ra, $a > 0$, $r \geq 1$. If it is working, no action is taken. If it is found to have failed, it is replaced by a new item. This process is continued indefinitely. The *downtime* of an item is the difference between the times of its replacement and failure.

a. Show that the expectation $M(a)$ of the time between replacements is

$$M(a) = \sum_{k=1}^{\infty} ka \cdot \int_{ka-a}^{ka} \phi(u)\, du.$$

The mean expected downtime $D(a)$ per item is

$$D(a) = \sum_{k=1}^{\infty} \int_{0}^{a} u\,\phi(ka - u)\, du.$$

b. Show that the mean $C(a)$ of the ratio of the downtime to the time between replacements is given by

$$C(a) = \sum_{k=1}^{\infty} \int_{0}^{a} \frac{u}{ka}\phi(ka - u)\, du.$$

c. If $\phi(\cdot)$ is the exponential density of mean 1, verify that

$$M(a) = \frac{a}{1 - e^{-a}}, \qquad D(a) = M(a) - 1, \quad \text{and}$$

$$C(a) = \left[1 - \frac{1 - e^{-a}}{a}\right] e^{a} \log(1 - e^{-a})^{-1}.$$

d. Compute and plot $D(a)/M(a)$ and $C(a)$ for $a = 0.025k$, $1 \leq k \leq 100$. Compare both graphs and discuss their qualitative differences.

e. Suppose you had to select a such that, over a long time period, the fraction downtime is 1/20. You could solve either of the equations

$$\frac{D(a)}{M(a)} = \frac{1}{20}, \qquad \text{or} \qquad C(a) = \frac{1}{20}.$$

You will find that the numerical values of a are quite different. Which of the two equations is appropriate and why?

Problem 5.2.29: For the model in Problem 5.2.28, the probability density $\psi(u; a)$ of the downtime is given by $\psi(u; a) = \sum_{k=1}^{\infty} \phi(ka - u)$, for $0 < u < a$. For several values of a, plot the densities $a\psi(ua; a)$ on (0, 1). In particular, set $a = 1$ (the mean lifetime) and $a = \log 2$ (the median lifetime). Interpret your results.

Problem 5.2.30: For the model in Problem 5.2.28, suppose that the inspections are done according to a Poisson process of rate $1/a$. Derive the joint distribution of the lifetime and the downtime of each item. For the exponential density $\phi(x) = e^{-x}$, obtain formulas for the same quantities as studied there and do the same numerical comparisons.

Problem 5.2.31: Write a computer program to perform the same tasks as in Problems 5.2.28 - 30 when $\phi(\cdot)$ is a PH-density of mean 1. Implement your code for four Erlang densities of various orders and for three hyperexponential densities. Discuss how the variability of the lifetimes affects the fraction of downtime. You need to do some numerical integrations involving PH-densities. Do these efficiently.

Problem 5.2.32: A device consists of Units I and II placed in series. Unit II has a lifetime distribution $F_2(\cdot)$ of phase type with representation (β, S). Unit I consists of m components in parallel with independent exponential lifetimes with parameter μ. Lifetimes of components in I are independent of those in Unit II.

a. The time-to-failure of Unit I has a probability distribution $F_1(\cdot)$ of phase type. What is its representation (α, T)?

b. Note that the lifetime of the device is $\min(T_1, T_2)$, where the lifetimes T_1 and T_2 of both units have PH-distributions. Develop an efficient algorithm for the mean lifetime E^* of the device. Explain why, without loss of generality, we can set $\mu = 1$. What is the time unit now? How should the representation (β, S) be modified?

c. Determine the smallest m for which $E^* \geq 4\beta(-S)^{-1}\mathbf{e} = 4E(T_2)$. Solve the problem numerically for $\mu = 2$ and $F_2(x) = 0.25E_1(\lambda_1; x) + 0.75E_5(\lambda_2; x)$, where $E_1(\lambda_1; x)$ is an exponential distribution of mean 2 and $E_5(\lambda_2; x)$ is an Erlang distribution of order 5 and mean 3.

Problem 5.2.33: For the device of Problem 5.2.32, determine the x_0 for which $F_2(x_0) = 0.9$. Next, find the smallest m for which the lifetime distribution $F(\,)$ of the device satisfies $F(10x_0) \leq 0.9$. Discuss a general approach to this problem, but obtain numerical results for the data in Problem 5.2.32. Describe an engineering scenario where that problem could arise.

HARDER PROBLEMS

Problem 5.2.34: Q is an irreducible generator with stationary probability vector $\boldsymbol{\theta}$. Prove the integration formula

$$I(t) = \int_0^t \int_0^u \exp(Qv)\, dv\, du$$

$$= \frac{1}{2}\mathbf{e}\boldsymbol{\theta}t^2 + [(\mathbf{e}\boldsymbol{\theta} - Q)^{-1} - \mathbf{e}\boldsymbol{\theta}]t - [I - \exp(Qt)](\mathbf{e}\boldsymbol{\theta} - Q)^{-2}.$$

Show that, as $t \to \infty$,

$$I(t) = \frac{1}{2}\mathbf{e}\boldsymbol{\theta}t^2 + [(\mathbf{e}\boldsymbol{\theta} - Q)^{-1} - \mathbf{e}\boldsymbol{\theta}]t - (\mathbf{e}\boldsymbol{\theta} - Q)^{-2}(I - \mathbf{e}\boldsymbol{\theta}) + o(1),$$

where o(1) stands for a matrix that tends to 0 as $t \to \infty$.

Problem 5.2.35: Establish the analogue of Property [PH-6] in Appendix 2 for continuous Markov chains. Assume that in the chain with generator

$$Q = \begin{vmatrix} T & \mathbf{T}^\circ \\ 0 & 0 \end{vmatrix}$$

every initial state leads to absorption. The generator is partitioned as

$$Q = \begin{array}{c|ccc|} 1 & T(1,1) & T(1,2) & \mathbf{T}^\circ(1) \\ 2 & T(2,1) & T(2,2) & \mathbf{T}^\circ(2) \\ * & 0 & 0 & 0 \end{array}.$$

The square submatrices $T(1, 1)$ and $T(2, 2)$ correspond to transitions *within* subsets 1 and 2. The submatrices $T(1, 2)$ and $T(2, 1)$ correspond to transitions from set 1 to 2 and vice versa. The initial probability vector is written in the correspondingly partitioned form $[\alpha(1), \alpha(2), \alpha_0]$.

a. Define τ as the time spent in the subset 1 prior to absorption in $*$. Prove that τ has a *PH*-distribution equal to that of the absorption time in the Markov chain with generator

$$Q^* = \begin{vmatrix} S & \mathbf{S}^\circ \\ 0 & 0 \end{vmatrix},$$

with initial probability vector $[\gamma, \gamma_0]$, where

$$S = T(1, 1) + T(1, 2)[-T(2, 2)]^{-1}T(2, 1), \qquad \mathbf{S}^\circ = \mathbf{T}^\circ(1) + T(1, 2)[-T(2, 2)]^{-1}\mathbf{T}^\circ(2),$$

$$\gamma = \alpha(1) + \alpha(2)[-T(2, 2)]^{-1}T(2, 1), \qquad \gamma_0 = \alpha_0 + \alpha(2)[-T(2, 2)]^{-1}\mathbf{T}^\circ(2).$$

Problem 5.2.36: Consider an $M/M/1$ of *finite capacity* m. T_i is the total time the queue spends *below* $i + 1$ during a busy period starting with one customer. Use the result in Problem 5.2.35 to compute the mean and standard deviation of T_i. Set the arrival rate $\lambda = 1$. Implement your code for $m = 20$ and for service rates μ slightly above or below 1. Tabulate these moments for $1 \leq i \leq m - 1$. What do your numerical results tell about the behavior of the queue during a busy period?

Problem 5.2.37: The model in Problem 5.2.27 can also be studied as a Markov chain with 27 states of which 3 are absorbing. Each state is a triplet (i_1, i_2, i_3) that keeps account of the component types in use by A_1, A_2, and A_3:

a. Write a subroutine which, for given values of μ_1, μ_2, and μ_3, constructs the generator of the Markov chain.

b. Compute the same quantities as in Problem 5.2.27, using this larger Markov chain.

c. By virtue of the result in Problem 5.2.27, the total time Y_{rj} that device A_r uses a component of type j *before* absorption in any one of the absorbing states has a *PH*-distribution. Write a code to compute, for all nine choices of r and j, the representations and the means and variances of the random variables Y_{rj}. Implement your code for various choices of μ_1, μ_2, and μ_3. Again, assume that $1 = \mu_1 \geq \mu_2 \geq \mu_3$, and interpret the numerical results for the various examples.

Problem 5.2.38: In a process described by an irreducible m-state Markov chain with generator Q, income is earned at a rate β_j whenever the chain is in state j. An independent Poisson process of rate λ generates times at which the chain is interrupted and immediately restarted. The chain is restarted, independently of the past, in a state chosen according to the probability vector $\boldsymbol{\alpha}$. $\boldsymbol{\beta}$ is the column vector with components β_j.

a. Show that the steady-state rate at which income accrues in that process is given by

$$\Psi(\lambda, \boldsymbol{\alpha}) = \frac{\boldsymbol{\alpha}(\lambda I - Q)^{-1}\boldsymbol{\beta}}{\boldsymbol{\alpha}(\lambda I - Q)^{-1}\mathbf{e}}.$$

b. For a given generator Q, rescale the original Markov chain so that, in steady state, transitions take place at unit rate. Next, for a given vector $\boldsymbol{\beta}$ and for $\lambda = 0.1k$, $1 \leq k \leq 40$, compute the vector $\boldsymbol{\alpha}(\lambda)$ for which $\Psi(\lambda, \boldsymbol{\alpha})$ is maximum. Illustrate the use of your code by generating numerical results for a number of well-chosen 10-state Markov chains.

Hint: Show that the function $\Psi(\lambda, \boldsymbol{\alpha})$ attains its maximum at a vertex of the simplex of vectors $\boldsymbol{\alpha}$.

Problem 5.2.39: The $M/PH/1$ Queue: Consider the $M/G/1$ queue with arrival rate λ and a service time distribution $H(\cdot)$ of phase type with representation $(\boldsymbol{\beta}, S)$ of order m. $\boldsymbol{\beta}$ is a probability vector. $H(\cdot)$ therefore does not have a jump at zero. The mean service time $\mu'_1 = \boldsymbol{\beta}(-S)^{-1}\mathbf{e}$. We assume that $\rho = \lambda\mu'_1 < 1$, so that the queue is stable.

Study that $M/PH/1$ queue as a Markov chain with states 0 and (i, j), $i \geq 1$, $1 \leq j \leq m$. State 0 corresponds to the empty queue; the state (i, j) signifies that there are i customers in the system and the service is in phase j. We write $\mathbf{i}$ for the set of states $\{(i, j)$: $1 \leq j \leq m$. Show that the generator Q of that chain is

$$Q = \begin{array}{c|cccccc|} 0 & -\lambda & \lambda\boldsymbol{\beta} & 0 & 0 & \cdots \\ 1 & \mathbf{S}^{\circ} & S-\lambda I & \lambda I & 0 & \cdots \\ 2 & 0 & \mathbf{S}^{\circ}\boldsymbol{\beta} & S-\lambda I & \lambda I & \cdots \\ 3 & 0 & 0 & \mathbf{S}^{\circ}\boldsymbol{\beta} & S-\lambda I & \cdots \\ 4 & 0 & 0 & 0 & \mathbf{S}^{\circ}\boldsymbol{\beta} & \cdots \\ 5 & 0 & 0 & 0 & 0 & \cdots \\ \cdot & \cdot & \cdot & \cdot & \cdot & \end{array} .$$

Denoting the steady-state probability vector of Q by $\mathbf{x} = [x_0, \mathbf{x}_1, \mathbf{x}_2, \cdots]$, write the steady-state equations

$$-\lambda x_0 + \mathbf{x}_1 \mathbf{S}^{\circ} = 0,$$

$$\lambda x_0 \boldsymbol{\beta} + \mathbf{x}_1 (S - \lambda I) + \mathbf{x}_2 \mathbf{S}^{\circ}\boldsymbol{\beta} = 0,$$

$$\lambda \mathbf{x}_{i-1} + \mathbf{x}_i (S - \lambda I) + \mathbf{x}_{i+1} \mathbf{S}^{\circ}\boldsymbol{\beta} = 0, \quad \text{for } i \geq 2.$$

Multiply each equation after the first, on the right by $\mathbf{e}$ to obtain that

$$\lambda \mathbf{x}_i \mathbf{e} = \mathbf{x}_{i+1} \mathbf{S}^{\circ}, \quad \text{for } i \geq 1.$$

For $i \geq 1$, replace $\mathbf{x}_{i+1}\mathbf{S}^{\circ}\boldsymbol{\beta}$ by $\lambda \mathbf{x}_i \mathbf{e}\boldsymbol{\beta}$ in the steady-state equations and deduce that for $i \geq 2$,

$$\mathbf{x}_i = \mathbf{x}_{i-1} R,$$

where $R = \lambda(\lambda I - \lambda \mathbf{e}\boldsymbol{\beta} - S)^{-1}$. Show further that $\mathbf{x}_1 = x_0 \boldsymbol{\beta} R$. It can be proved that R is a positive matrix with all eigenvalues inside the unit disk. Show by a direct calculation that the normalizing equation

$$1 = x_0 + \sum_{i=1}^{\infty} \mathbf{x}_i \mathbf{e} = x_0 + x_0 \boldsymbol{\beta} R (I - R)^{-1} \mathbf{e}$$

leads to $x_0 = 1 - \rho$. Conclude that

$$x_0 = 1 - \rho, \quad \text{and} \quad \mathbf{x}_i = (1 - \rho)\boldsymbol{\beta} R^i, \quad \text{for } i \geq 1.$$

Derive expressions in terms of R for the first two moments of the steady-state queue length density $\{x_0, \mathbf{x}_1\mathbf{e}, \mathbf{x}_2\mathbf{e}, \cdots\}$. Write a code to compute the matrix R, the vectors $\mathbf{x}_i$, and the steady-state queue length density and its mean and variance. Implement your code for various PH-densities $H(\cdot)$ but with the same value of ρ. What do your numerical results reveal about how the steady-state queue length density differs for various service time distributions?

Reference: Neuts, M. F., *"Matrix-Geometric Solutions in Stochastic Models: An Algorithmic Approach,"* Baltimore: The Johns Hopkins University Press, 1981, pp. 83 - 87.

Problem 5.2.40: The $M/PH/1$ Queue: Consider the $M/G/1$ queue with arrival rate λ and a service time distribution $H(\cdot)$ of phase type with representation $(\boldsymbol{\beta}, S)$ of order m. $H(\cdot)$ does not have a jump at zero. The mean service time $\mu'_1 = \boldsymbol{\beta}(-S)^{-1}\mathbf{e}$. Assume that $\rho = \lambda\mu'_1 < 1$, so that the queue is stable. For a stable $M/G/1$ queue, the steady-state waiting time distribution $W(\cdot)$ at an arrival is given by

$$W(x) = \sum_{k=0}^{\infty} (1-\rho)\rho^k H^{*(k)}(x), \qquad \text{for } x \geq 0,$$

where $H^{*(k)}(\cdot)$ is the k-fold convolution of the probability distribution

$$H^*(x) = \frac{1}{\mu'_1}\int_0^x [1 - H(u)]\, du, \qquad \text{for } x \geq 0.$$

a. Show that $H^*(\cdot)$ is a PH-distribution with representation $(\boldsymbol{\pi}, S)$, where $\boldsymbol{\pi} = \mu'^{-1}_1\boldsymbol{\beta}(-S)^{-1}$.

b. Show that $W(\cdot)$, a geometric mixture of the successive convolutions of $H^*(\cdot)$, is a PH-distribution with representation $\boldsymbol{\gamma} = \rho\boldsymbol{\pi}$, $K = S + \rho\mathbf{S}^\circ\boldsymbol{\pi}$. Notice that $\boldsymbol{\gamma}\mathbf{e} = \rho$, so that $W(0+) = 1 - \rho$. State formulas for the mean and variance of the waiting time distribution.

c. Write an efficient computer program to evaluate and plot the distributions $H(\cdot)$ and $W(\cdot)$ and their densities. Print their means and variances.

d. Use your code to determine the value of λ for which 5% of the customers will have to wait longer than $5\mu'_1$. Obtain and interpret numerical results for five different service time distributions $H(\cdot)$. Scale the service time distributions so that $\mu'_1 = 1$.

Problem 5.2.41: The $M/PH/1$ Queue with Finite Capacity: Consider the $M/PH/1$ queue with arrival rate λ and a service time distribution $H(\cdot)$ of phase type with representation $(\boldsymbol{\beta}, S)$ of order m. $H(\cdot)$ does not have a jump at 0 and is scaled so that its mean $\mu'_1 = \boldsymbol{\beta}(-S)^{-1}\mathbf{e} = 1$. There is a waiting room of size K. Customers arriving when there are $K+1$ jobs in the system are lost. There is a threshold L satisfying $1 \le L \le K$. With at most L jobs in the system, regular service is given. Whenever more than L jobs are present, service is dispensed at an accelerated rate $\theta \ge 1$.

Study that $M/PH/1$ queue as a Markov chain with states 0 and (i, j), $1 \le i \le K+1$, $1 \le j \le m$. State 0 corresponds to the empty queue; state (i, j) signifies that there are i customers in the system and the service is in phase j. The set of states $\{(i, j): 1 \le j \le m\}$ is **i**.

a. Show that the generator Q of that Markov chain is of the form displayed here for $L = 3$ and $K = 5$:

$$Q = \begin{array}{c|ccccccc|} 0 & -\lambda & \lambda\boldsymbol{\beta} & 0 & 0 & 0 & 0 & 0 \\ 1 & \mathbf{S}^\circ & S-\lambda I & \lambda I & 0 & 0 & 0 & 0 \\ 2 & 0 & \mathbf{S}^\circ\boldsymbol{\beta} & S-\lambda I & \lambda I & 0 & 0 & 0 \\ 3 & 0 & 0 & \mathbf{S}^\circ\boldsymbol{\beta} & S-\lambda I & \lambda I & 0 & 0 \\ 4 & 0 & 0 & 0 & \theta\mathbf{S}^\circ\boldsymbol{\beta} & \theta S-\lambda I & \lambda I & 0 \\ 5 & 0 & 0 & 0 & 0 & \theta\mathbf{S}^\circ\boldsymbol{\beta} & \theta S-\lambda I & \lambda I \\ 6 & 0 & 0 & 0 & 0 & 0 & \theta\mathbf{S}^\circ\boldsymbol{\beta} & \theta S \end{array} .$$

Partition the steady-state probability vector $\mathbf{x}$ of Q as $[x_0, \mathbf{x}_1, \mathbf{x}_2, \cdots, \mathbf{x}_{K+1}]$. Use the same argument as in Problem 5.2.40 to show that

$$\mathbf{x}_i = x_0\, \boldsymbol{\beta} R^i, \quad \text{for } 1 \le i \le L,$$

$$\mathbf{x}_i = x_0\, \boldsymbol{\beta} R^L [R(\theta)]^{i-L}, \quad \text{for } L \le i \le K,$$

and $\mathbf{x}_{K+1} = \lambda\theta^{-1}\mathbf{x}_k(-S)^{-1}$, where the matrices R and $R(\theta)$ are defined by

$$R = \lambda(\lambda I - \lambda\mathbf{e}\boldsymbol{\beta} - S)^{-1}, \quad \text{and} \quad R(\theta) = \lambda(\lambda I - \lambda\mathbf{e}\boldsymbol{\beta} - \theta S)^{-1}.$$

Show that x_0 is given by

$$x_0 = \left\{ 1 + \sum_{i=1}^{L} \beta R^i \mathbf{e} + \sum_{i=1}^{K-L} \beta R^L [R(\theta)]^i \mathbf{e} + \lambda\theta^{-1} \beta R^L [R(\theta)]^{K-L} (-S)^{-1} \mathbf{e} \right\}^{-1}.$$

c. Explain why the steady-state fraction of lost arrivals is $\phi(\theta) = \mathbf{x}_{K+1}\mathbf{e}$. Note that for $\theta = 1$, we obtain the ordinary $M/PH/1$ queue of finite capacity. For $K = 10$ and $\lambda = 0.70 + 0.05k$, $1 \le k \le 10$, compute $\phi(1)$.

d. Next, set $L = 5$ and subsequently $L = 8$. For the same λ as in part *c*, compute the values of θ for which $\phi(\theta) = 0.1\phi(1)$. Implement your code for an Erlang service distribution of order 3 and for two hyperexponential distributions with different coefficients of variation. Remember to normalize the service time distribution so that $\mu'_1 = 1$. Discuss the sensitivity of θ to the variability of the service time distribution and to the threshold L.

Problem 5.2.42: The $PH/M/1$ Queue: Arrivals to a single server with exponential services of rate μ occur according to a PH-renewal process. The representation of the PH-distribution $F(\cdot)$ of the interarrival times is (α, T), where α is a probability vector. The mean interarrival time is $\lambda'_1 = \alpha(-T)^{-1}\mathbf{e}$. In numerical computations, we rescale so that $\lambda'_1 = 1$. The $PH/M/1$ queue may be studied as the Markov chain with generator

$$Q = \begin{array}{c|ccccc|} 0 & T & \mathbf{T}^\circ\alpha & 0 & 0 & \cdots \\ 1 & \mu I & T-\mu I & \mathbf{T}^\circ\alpha & 0 & \cdots \\ 2 & 0 & \mu I & T-\mu I & \mathbf{T}^\circ\alpha & \cdots \\ 3 & 0 & 0 & \mu I & T-\mu I & \cdots \\ 4 & 0 & 0 & 0 & \mu I & \cdots \\ 5 & 0 & 0 & 0 & 0 & \cdots \\ \cdot & \cdot & \cdot & \cdot & \cdot & \end{array}.$$

The state (i, j) signifies that there are i customers in the system and that the PH-arrival process is in its phase j. Suppose that the stationary probability vector $\mathbf{x}$ of Q exists and is partitioned as $\mathbf{x}_0, \mathbf{x}_1, \mathbf{x}_2, \cdots$. Similarly to what was done in Problem 5.2.40, we try to construct a *matrix-geometric* solution of the form $\mathbf{x}_i = \mathbf{x}_0 R^i$, for $i \ge 0$. Suppose that there is a nonnegative matrix R that satisfies the equation

$$\mu R^2 + R(T - \mu I) + \mathbf{T}^\circ\alpha = 0. \qquad (1)$$

If we substitute vectors of the form $\mathbf{x}_i = \mathbf{x}_0 R^i$ in the steady-state equations, we see that a valid (unique) stationary vector $\mathbf{x}$ is obtained if there is a positive vector $\mathbf{x}_0$ for which

$$\mathbf{x}_0(T + \mu R) = 0, \tag{2}$$

and in addition the vector $\sum_{i=0}^{\infty} \mathbf{x}_0 R^i = \mathbf{x}_0(I - R)^{-1}$ is finite. If that last requirement is satisfied, $\mathbf{x}_0(I - R)^{-1}\mathbf{e} = 1$ since the steady-state probabilities must sum to one. All eigenvalues of the matrix R should therefore lie inside the unit disk.

a. Try finding a nonnegative solution to equation (1) by successive substitutions. Rewrite (1) as

$$R = \mu R^2(\mu I - T)^{-1} + \mathbf{T}^{\circ}\boldsymbol{\alpha}(\mu I - T)^{-1},$$

and perform successive substitutions starting with $R(0) = 0$. Verify that every iterate is of the form $R(n) = \mathbf{T}^{\circ}\mathbf{u}(n)$, where the vectors $\mathbf{u}(n)$ are positive and coordinate-wise increasing in n. This suggests that we try a solution of the form $R = \mathbf{T}^{\circ}\mathbf{u}$, where $\mathbf{u}$ is a positive row vector. We see that $\mathbf{u}R = (\mathbf{u}\mathbf{T}^{\circ})\mathbf{u}$, so that $\eta = \mathbf{u}\mathbf{T}^{\circ}$ is an eigenvalue of R with a corresponding positive eigenvector. (From the theory of nonnegative matrices, this implies that η is the eigenvalue of R with largest modulus.) To have a matrix-geometric solution of the desired form, we must find a value of $\eta < 1$. Substituting that trial solution into (1) we obtain

$$\mu\eta\mathbf{T}^{\circ}\mathbf{u} + \mathbf{T}^{\circ}\mathbf{u}(T - \mu I) + \mathbf{T}^{\circ}\boldsymbol{\alpha} = 0.$$

Multiply that equation on the left by the positive vector $\boldsymbol{\alpha}(-T)^{-1}$ to obtain an equation implying that $\mathbf{u} = \boldsymbol{\alpha}[\mu(1 - \eta)I - T]^{-1}$, and that $\eta = \mathbf{u}\mathbf{T}^{\circ}$ should satisfy

$$f(\mu - \mu\eta) = \eta, \tag{3}$$

where $f(s) = \boldsymbol{\alpha}(sI - T)^{-1}\mathbf{T}^{\circ}$ is the Laplace - Stieltjes transform of the interarrival time distribution. By examining the graph of $f(\mu - \mu z)$ on $[0, 1]$, show that there is a unique $\eta < 1$ for which (3) holds if and only if $\mu\lambda'_1 > 1$, i.e., if the service rate μ exceeds the arrival rate $(\lambda'_1)^{-1}$ of customers. In that case, the *PH/M/1* queue is said to be *stable*. Verify that the matrix R is explicitly given by

$$R = \mathbf{T}^{\circ}\alpha[\mu(1-\eta)I - T]^{-1},$$

and that for $i \geq 1$, $R^i = \eta^{i-1}R$. Show further that $R\mathbf{e} = \mu^{-1}\mathbf{T}^{\circ}$ and

$$(I - R)^{-1} = I + (1-\eta)^{-1}R.$$

b. Show that the vector $\mathbf{x}_0$ is given by $\mathbf{x}_0 = c\alpha(-T)^{-1}[\mu(1-\eta)I - T]^{-1}$, where c is a constant to be determined. Use the normalizing equation to show that

$$1 = c\alpha(-T)^{-1}[\mu(1-\eta)I - T]^{-1}[\mathbf{e} + \mu^{-1}(1-\eta)^{-1}\mathbf{T}^{\circ}].$$

From that equation deduce that $c = \mu(\lambda'_1)^{-1}(1-\eta)$. For that last derivation recall that the Laplace - Stieltjes transform of

$$F^*(x) = \frac{1}{\lambda'_1}\int_0^x [1 - F(u)]\, du,$$

is given by

$$f^*(s) = \frac{1}{\lambda'_1}\alpha(-T)^{-1}(sI - T)^{-1}\mathbf{T}^{\circ},$$

and that

$$\frac{1}{\lambda'_1}\alpha(-T)^{-1}(sI - T)^{-1}\mathbf{e} = \frac{1 - f^*(s)}{s}.$$

Verify that, in final form, the stationary probability vector $\mathbf{x}$ is given by

$$\mathbf{x}_0 = \mu(1-\eta)(\lambda'_1)^{-1}\alpha(-T)^{-1}[\mu(1-\eta)I - T]^{-1},$$

$$\mathbf{x}_i = (1-\eta)\eta^{i-1}(\lambda'_1)^{-1}\alpha[\mu(1-\eta)I - T]^{-1}, \quad \text{for } i \geq 1,$$

since $\mathbf{x}_0\mathbf{T}^{\circ} = (1-\eta)(\lambda'_1)^{-1}$.

c. Set $\rho = (\lambda'_1\mu)^{-1} < 1$; then ρ is the *traffic intensity* of the $PH/M/1$ queue. Now verify that the marginal density $y_i = \mathbf{x}_i\mathbf{e}$ of the number of customers in the system is given by

$$y_0 = 1 - \rho, \quad \text{and} \quad y_i = \rho(1-\eta)\eta^{i-1}, \quad \text{for } i \geq 1.$$

Explain why the steady-state probability ϕ_i that there are i customers in the system just before an arrival is given by

$$\phi_i = \frac{\mathbf{x}_i\mathbf{T}^\circ}{\sum_{v=0}^{\infty} \mathbf{x}_v\mathbf{T}^\circ},$$

and verify that $\phi_i = (1-\eta)\eta^i$, for $i \geq 0$.

d. Write a computer code to evaluate the steady-state probability vector $\mathbf{x}$ for a given interarrival time distribution $F(\cdot)$. Choose μ such that successively $\rho = 0.05k$ for $1 \leq k \leq 19$. Print out only the essential items to characterize the vector $\mathbf{x}$. Avoid printing a long column of vectors $\mathbf{x}_i$!

Remark: This, together with the $M/PH/1$ queue, is one of few examples for which a matrix-geometric solution can be written explicitly. See, Neuts, M. F., "*Matrix-Geometric Solutions in Stochastic Models: An Algorithmic Approach,*" Baltimore: The Johns Hopkins University Press, 1981, Chapter 3. In Chapter 6, this model serves to explore some aspects of the steady-state behavior through visualization.

Problem 5.2.43: The $H_2/M/1$ Queue: This is the particular $PH/M/1$ queue for which the interarrival time distribution is *hyperexponential* with two exponential components. We write the probability distribution $F(\cdot)$ as

$$F(x) = 1 - pe^{-\lambda x} - qe^{-\lambda\theta x}, \quad \text{for } x \geq 0.$$

The parameters satisfy $0 < p = 1 - q < 1$, $\lambda > 0$, and $0 < \theta < 1$.

a. Verify that $\lambda'_1 = 1$ for $\lambda = p + q\theta^{-1}$. From now on, let λ be so chosen. Derive a formula for the standard deviation $\sigma(p, \theta)$, which is also the coefficient of variation of $F(\cdot)$. For a few selected values of p, and of θ, plot a graph of $\sigma(p, \theta)$ as a function of the other variable.

b. Based on the results in Problem 5.2.42, derive explicit analytic expressions for η and for the components of **u** for the stable $H_2/M/1$ queue.

c. Write a computer code to evaluate the steady-state densities of the queue length at an arbitrary time and just prior to an arbitrary arrival. Select five pairs of parameters p and θ for which the values of $\sigma(p, \theta)$ are quite different. For each pair print out the queue length densities for μ chosen such that $\rho = 0.05k$, $10 \le k \le 19$. You will find that the results are highly sensitive to the value of $\sigma(p, \theta)$. Think about the behavior of the queue that explains those results.

Problem 5.2.44: A closed queueing system consists of two servers, I and II, and n jobs circulating forever between them. Each job is processed by a unit and, upon completion of its service, immediately moves to the other unit. Each server processes one job at a time. Jobs arriving while a server is occupied queue up. All service times are assumed to be independent. Those dispensed by Units I and II have service time distributions $H_1(\cdot)$ and $H_2(\cdot)$, respectively.

a. When $H_1(\cdot)$ and $H_2(\cdot)$ are exponential with parameters μ_1 and μ_2, show that the system is described by an $(n+1)$-state Markov chain. Write down its generator $Q(n)$ and derive its steady-state probabilities π_i, $0 \le i \le n$. What are the π_is when $\mu_1 = \mu_2$?

b. $H_1(\cdot)$ is exponential with rate μ_1 and $H_2(\cdot)$ is a PH-distribution with representation $(\boldsymbol{\beta}, S)$ of order m. Adapt the result for the $M/PH/1$ queue of finite capacity in Problem 5.2.41 to obtain matrix formulas for the steady-state probabilities of the Markov chain describing the model.

c. Set up the Markov chain when both service time distributions are of phase type. Discuss an algorithm to compute its steady-state probabilities. Write a program to implement your algorithm for $n = 10$ with PH-distributions with two phases each. Obtain numerical results for five well-chosen pairs of *hyperexponential* distributions with the same means but different coefficients of variation. What do your numerical results suggest about the effect of variability in the service times on the behavior of the queues in front of each server?

Reference: For a detailed discussion of the model with PH-service times see Latouche, G. and Neuts, M. F., "The superposition of two PH-renewal processes," in *"Semi-Markov Models: Theory and Applications,"* (J. Janssen, ed.) New York: Plenum Press, Proc. Symp. on Semi-Markov

Processes, Brussels, Belgium, June 1984, 131 - 77, 1986.

Problem 5.2.45: For an irreducible m-state Markov chain in steady state, we wish to select a subset of $m_1 < m$ states with the highest probability of being visited at least once during an interval of time $[0, t]$. For large m that combinatorial optimization problem is very difficult. We ask you to set up the problem and to solve it in some particular cases. For m_1 selected states we partition the generator as

$$Q = \begin{matrix} 1 \\ \\ 2 \end{matrix} \begin{vmatrix} T(1, 1) & T(1, 2) \\ & \\ T(2, 1) & T(2, 2) \end{vmatrix},$$

where 1 consists of the selected states and 2 of the others. The steady-state probability vector $\boldsymbol{\theta} = [\boldsymbol{\theta}(1), \boldsymbol{\theta}(2)]$ has been computed and the time scale is such that the rate of transitions equals 1.

a. $\Psi(t)$ is the probability that set 1 is visited at least once during $[0, t]$. Explain why $\Psi(t) = 1 - \boldsymbol{\theta}(2)\exp[T(2, 2)t]\mathbf{e}$. For a given t, we need to determine the set of m_1 states for which $\Psi(t)$ is maximum. In general, that can be done only by complete enumeration of all the subsets of cardinality m_1. Clearly, that is a formidable task.

b. By explicit calculation, verify that the Markov chain with generator

$$Q = \frac{1}{5} \begin{vmatrix} -4 & 1 & 3 \\ 2 & -5 & 3 \\ 2 & 4 & -6 \end{vmatrix}$$

has stationary transition rate 1. For that chain, solve the problem for $m_1 = 1$ and $t = 2$.

c. If you had to pick $m_1 = 2$ states so as to maximize $\Psi(t)$, which would you choose? Based on insights gained from this example, how can we solve the problem in part *a* in general for $m_1 = m - 1$?

c. Consider a birth and death process on the integers $0 \le i \le 9$. Denote the (positive) transition rates $i \to i + 1$ by λ_i for $0 \le i \le 8$, and the transition rates $i \to i - 1$ by μ_i for $1 \le i \le 9$. For given λ_i and μ_i, compute the steady-state probabilities θ_i. Set the stationary transition rate equal to 1.

Develop an algorithm to find the set of four *consecutive* states with the highest probability of being visited at least once in $[0, t]$. Implement your code for a number of examples of your choice, but select these to obtain qualitatively interesting results. For instance, for given λ_i and μ_i, see how the optimal set changes when you successively set $t = 0.5k$, $1 \le k \le 10$. $\Psi(t)$ has to be computed for six different coefficient matrices $T(1, 1)$ each of order 6. Write an efficient subroutine to do that. When the λ_i and μ_i do not depend on i, the chain describes a finite $M/M/1$ queue. For that case, compare examples where λ is slightly larger, equal to, or slightly smaller than μ.

Quasi-Stationary Densities: The following five problems deal with the quasi-stationary density of a finite subset of a Markov chain. That notion involves a nice application of the Perron - Frobenius theory of irreducible nonnegative matrices summarized in Appendix 1.

Problem 5.2.46: In Problem 4.4.60, a method is described to compute the Perron - Frobenius eigenvalue η of an irreducible nonnegative matrix A. An irreducible *semi-stable* matrix A has negative diagonal elements, nonnegative off-diagonal elements, and $A\mathbf{e} \le 0$.

a. Translate all mathematical statements in Problem 4.4.60 and adapt Elsner's procedure to compute the eigenvalue $-\eta^*$ of maximum real part of an irreducible semi-stable A and the corresponding eigenvector $\mathbf{u}$. Show also how to compute the normalized right eigenvector $\mathbf{v}$.

b. Modify the code for Problem 4.4.60 to evaluate $-\eta^*$, $\mathbf{u}$, and $\mathbf{v}$. Test your code for a number of semi-stable matrices. Specifically, let A be a generator of an m-state Markov chain for which $\eta^* = 0$, $\mathbf{u} = \boldsymbol{\theta}$, and $\mathbf{v} = \mathbf{e}$.

Problem 5.2.47: Quasi-Stationary Densities: For an irreducible m-state Markov chain, Q is partitioned as

$$Q = \begin{array}{c} 1 \\ \\ 2 \end{array} \left| \begin{array}{cc} T(1, 1) & T(1, 2) \\ & \\ T(2, 1) & T(2, 2) \end{array} \right| .$$

The $m_1 \times m_1$ matrix $T(1, 1)$ is irreducible. Its Perron - Frobenius eigenvalue and the corresponding left and right eigenvectors are denoted by $-\eta^*$, $\mathbf{u}$, and $\mathbf{v}$, respectively. The eigenvectors are normalized as in Appendix 1. The probability vector $\mathbf{u}$ is called the *quasi-stationary density* of set 1.

a. As for discrete-time Markov chains, (see Problem 4.4.61), explain the probabilistic significance of **u** and η^*.

b. The quasi-stationary density depends only on the matrix $T(1, 1)$. The set **1** may therefore also be a finite subset of an infinite Markov chain. Consider the $M/M/1$ queue with service rate $\mu = 1$. The vector $\mathbf{u}(k)$ is the quasi-stationary density of the set $\mathbf{1} = \{0, \cdots, k\}$, $k \geq 1$. The quantity $-\eta^*(k)$ is the eigenvalue of maximum real part of the corresponding matrix $T(1, 1)$. What is the significance of $\eta^*(k)$ and of $\mathbf{u}(k)$ for the $M/M/1$ queue?

c. Adapt Elsner's algorithm to the simple special structure of $T(1, 1)$ to compute $\eta^*(k)$ and $\mathbf{u}(k)$. For a given value of λ, plot $\eta^*(k)$ for $1 \leq k \leq 20$, as well as graphs of the quasi-stationary densities $\mathbf{u}(k)$ for $10 \leq k \leq 20$. Implement your code for λ equal to 0.40, 0.75, 0.90, 1, 1.25, and 2. Discuss the qualitative significance of your numerical results.

Hint: In implementing Elsner's algorithm for k, use the vector $\mathbf{u}(k-1)$ to construct a convenient starting solution.

Problem 5.2.48: Quasi-Stationary Densities for the $M/M/1$ Queue: For the $M/M/1$ queue in Problem 5.2.47, set $k = 10$. Use Elsner's algorithm to compute and plot η^* as a function of the arrival rate λ on the interval [0.25, 1.50] by steps of 0.05. Also, for 10 selected values of λ, plot the quasi-stationary densities $\mathbf{u}(\lambda)$. Discuss the significance of these graphs.

Hint: Start with $\lambda = 1$. For that value, the probability vector $\mathbf{u}(0)$ with all equal components is an excellent initial vector for Elsner's iteration. In implementing Elsner's algorithm for the next larger or smaller λ, use the preceding vector $\mathbf{u}(\lambda)$ as a convenient starting solution.

Problem 5.2.49: Quasi-Stationary Densities for the $M/M/1$ Queue: Explain why, for the $M/M/1$ queue in Problem 5.2.48, all sets of states of the form $\{i, i+1, \cdots, i+k\}$ with $i \geq 1$ have the same quasi-stationary densities. The eigenvalue η^* depends only on k and λ. Modify the programs for Problems 5.2.47 and 5.2.48 to perform the same numerical tasks for the set of states $\{1, \cdots, k+1\}$. As always, the numerical results should tell something of interest about the behavior of the queue.

Problem 5.2.50: Quasi-Stationary Densities for the $M/M/1$ Queue: Consider the $M/M/1$ queue of Problem 5.2.48 with $\lambda < 1$ in steady-state.

a. $\phi(j; i, k)$ is the conditional probability that at time zero, there are $i + j$

customers present, given that there are between i and $i + k$ customers in the system. Show that, for $i \geq 0$ and $k \geq 1$,

$$\phi(j; i, k) = \frac{(1 - \lambda)\lambda^j}{1 - \lambda^{k+1}}, \quad \text{for } 0 \leq j \leq k.$$

We see that the density $\{\phi(j; i, k)\}$ is the same for all values of i. Consider the set of states $\mathbf{1} = \{1, \cdots, C + 1\}$. Let $\tau(k)$ be the first time that an $M/M/1$ queue with initial probability density $\alpha_j = \phi(j; 1, k)$, for $1 \leq j \leq k$, and $\alpha_j = 0$ for other j, leaves set $\mathbf{1}$.

b. Develop a code to compute the mean, the variance, the density, and the distribution of $\tau(k)$. For $k = 10$ and for $0.55 \leq \lambda \leq 0.95$ by steps of 0.1, compare the distribution of $\tau(k)$ to the quasi-stationary density of the set $\mathbf{1}$. For each value of λ, also print the density $\{\phi(j; 1, k)\}$ and the vector $\mathbf{u}(k)$. What do these numerical results tell us?

Problem 5.2.51: For the two-machine model of Problem 5.2.15 suppose that the time-to-failure distributions are PH-distributions, respectively, with representations $[\alpha(1), T(1)]$ and $[\alpha(2), T(2)]$ with m_1 and m_2 phases. If machine A fails, it enters an exponential repair time with parameter σ_1 and B is used. Should machine B also fail, its exponential repair time has parameter σ_2. As in Problem 5.2.15, Machine A is used whenever possible. If both machines are down, a single repairman gives priority to the repair of machine A. Following any repairs, a machine is returned good-as-new.

a. Set up a Markov chain to describe the system. How many states are needed? Write down its generator Q.

b. Partition the steady-state probability vector of Q in a natural manner. Obtain concise matrix formulas for it in terms of the given parameters.

c. Suppose that all parameters other than σ_1 are fixed. Show that there is a unique value of σ_1 for which the fraction of time $\pi(4)$ that both machines are down has a given value, say 1/100.

d. Set $\sigma_2 = 1$. Choose various PH-distributions for the times-to-failure. Normalize their representations so that their means agree with those of the exponential distributions in Problem 5.2.15. For each example, find the σ_1 for which $\pi(4) = 0.01$. Discuss the effect of random variability in the times-to-failure.

Problem 5.2.52: Choosing Between Two Servers: Jobs arrive according to a Poisson process of rate λ to a service system with two processors. Every admitted job is processed by one of the two servers. The service times are independent. For jobs served by processor I, service times have a PH-distribution with representation $[\beta(1), S(1)]$. Those handled by II have a PH–distribution with representation $[\beta(2), S(2)]$. Jobs arriving when both processors are busy are lost. If only one processor is idle, an arriving job goes there.

a. Suppose that, when both are free, an arriving job chooses processor I or II with probability p or $q = 1 - p$ according to independent Bernoulli trials. Explain why the system is described by the Markov chain with generator

$$Q = \begin{array}{c} 0 \\ 1 \\ 2 \\ 1,2 \end{array} \left| \begin{array}{cccc} -\lambda & \lambda p\, \beta(1) & \lambda q\, \beta(2) & 0 \\ \mathbf{S}^{\circ}(1) & S(1) - \lambda I & 0 & \lambda I \otimes \beta(2) \\ \mathbf{S}^{\circ}(2) & 0 & S(2) - \lambda I & \beta(1) \otimes \lambda I \\ 0 & I \otimes \mathbf{S}^{\circ}(2) & \mathbf{S}^{\circ}(1) \otimes I & S(1) \oplus S(2) \end{array} \right| .$$

State 0 signifies that both servers are idle. The symbols 1, 2, and 1, 2 denote macro-states indicating which servers are busy and the phases of their service.

b. Construct an efficient algorithm to compute the stationary probabilities of Q. For $p = 0.5$ and for given PH service time distributions, determine the arrival rate λ_{90} for which 90% of all customers will be admitted. In what follows, fix the arrival rate at that value.

c. In steady state, what fraction of admitted customers is served by processor I? What fraction of admitted customers find one of the servers busy? For what fraction of time is I (resp. II), the only busy server?

d. Develop an algorithm to find the value of p for which the fraction of lost customers is smallest.

e. Determine the value of p for which the fraction of time that processor I is in use is maximum.

f. Implement your algorithm for selected number of pairs of service time distributions with interesting qualitative features. For example, let both distributions have the same mean but quite different variances. Discuss the qualitative implications of your numerical results.

Remark: Problem 5.2.16 deals with the case of exponential service times. For the corresponding four-state Markov chain, several of the quantities of interest can be written explicitly. That simple case provides accuracy checks for the code for general *PH*-distributions.

Problem 5.2.53: In the system described in Problem 5.2.16, a customer arriving when both servers are idle goes to the faster of the two. Suppose that we have a total available service rate $\mu = \mu_1 + \mu_2$.

a. Without loss of generality, assume that $\mu_1 \geq \mu_2$. For Poisson arrivals, find the value of μ_1 for which the fraction of lost customers is minimum.

b. Next, let the customers arrive according to a *PH*-renewal process. Its interarrival time distribution has the representation $(\boldsymbol{\alpha}, T)$ and the arrival rate is $\lambda = [\alpha(-T)^{-1}\mathbf{e}]^{-1}$.

Show that this two-server loss system is described by the Markov chain with generator

$$Q = \begin{array}{c|cccc|} 0 & T & \mathbf{T}^{\circ}\alpha & 0 & 0 \\ 1 & \mu_1 I & T - \mu_1 I & 0 & \mathbf{T}^{\circ}\alpha \\ 2 & \mu_2 I & 0 & T - \mu_2 I & \mathbf{T}^{\circ}\alpha \\ 12 & 0 & \mu_2 I & \mu_1 I & T + \mathbf{T}^{\circ}\alpha - \mu I \end{array} .$$

Develop an efficient algorithm for the stationary probability vector $\boldsymbol{\pi}$ of Q. Explain why the steady-state fraction of lost customers is

$$\phi(\mu_1, \mu_2) = \frac{1}{\lambda}\boldsymbol{\pi}_{12}\mathbf{T}^{\circ},$$

where $\boldsymbol{\pi}_{12}$ is the vector of steady-state probabilities corresponding to the set of states 12. Write a computer program to determine the value of μ_1 for which $\phi(\mu_1, \mu_2)$ is minimum. Select a number of *PH*-distributions with the same mean. Also choose various values of μ. Use your code to explore the optimal allocation of service rate. Draw qualitative conclusions.

Reference: For Poisson arrivals and c servers with a given total service rate μ, the rates $\mu_1 > \cdots > \mu_c$ for which the loss probability is smallest have been determined in A. Tahara and T. Nishida, "Optimal allocation of service rates for multi-server Markovian queue." *Journal of the Operations Research Society of Japan,* 18, 90 - 96, 1975.

Problem 5.2.54: A Simple Epidemic Model: Initially, a population consists of n healthy individuals and one infectious individual. All individuals remain alive throughout the epidemic. $J(t)$ is the number of healthy individuals at time t. In a Markovian model for the epidemic it is assumed that, if $J(t) = j$, then independently of the past, a healthy person becomes infected in $(t, t + dt)$ with elementary probability $\beta j(n - j + 1)\, dt$. The rate of infection is therefore proportional to the numbers of healthy and infected individuals. It is customary to normalize the time scale by setting $\beta = 1$.

a. Describe the epidemic as a Markov chain with states $0, 1, \cdots, n$ and with $J(0) = n$. What is the distribution of the time T_j to reach the state j, $0 \le j \le n - 1$? State expressions for its mean and variance. Write a subroutine to compute these moments.

b. Compute the probabilities $p_j(t) = P\{J(t) = j \mid J(0) = n\}$ for t in $(0, t^*)$. Use an informative step size. Also compute and plot the conditional mean $\mu(t)$ of $J(t)$. Implement your code for $n = 10, 20, 30$. What are the qualitative implications of your results?

Reference: Bailey, Norman T. J., "*The Elements of Stochastic Processes with Applications to the Natural Sciences,*" New York: John Wiley & Sons, 1964, pp. 164 - 169.

CHALLENGING PROBLEMS

Problem 5.2.55: A *stationary PH*-renewal process with underlying continuous distribution $F(\cdot)$ of representation (α, T) and mean μ'_1 describes the arrival of shocks to a system. The system fails when M shocks have occurred. M is a random variable with a discrete *PH*-density with representation (β, S). Both α and β are probability vectors. Therefore, M is positive and $F(\cdot)$ does not have a jump at 0.

a. Show that the lifetime τ of the system has a *PH*-distribution $G(\cdot)$ with representation (γ, L) given by

$$\gamma = \theta^* \otimes \beta, \qquad L = T \otimes I + T^{\circ}\alpha \otimes S,$$

where $\theta^* = (\mu'_1)^{-1}\alpha(-T)^{-1}$.

b. Code an algorithm to compute the mean, the variance, the distribution, and the density of $G(\cdot)$.

c. Choose three different *PH*-distributions $F(\cdot)$ with mean 1 and four different discrete *PH*-densities for M with mean 3. Print the items specified in part *b* for all 12 pairs of data and plot the densities and distributions functions of τ.

Reference: Neuts, M. F. and Bhattacharjee, M. C., "Shock models with phase type survival and shock resistance," *Naval Research Logistics Quarterly,* 28, 213 - 219, 1981.

Problem 5.2.56: The $PH/M/c$ **Queue:** Consider the same arrival process as in Problem 5.2.42, but assume that customers are served by one of $c > 1$ servers, each with exponential processing times of rate μ. This model, the $PH/M/c$ queue, may be studied as a Markov chain whose generator Q is similar to that for the $PH/M/1$ queue. Its structure is block tridiagonal. The first two blocks in the 0-row are T and $\mathbf{T}^{\circ}\alpha$, followed by zero matrices. In the rows $i = 1, \cdots, c-1$, the three nonzero blocks are

$$i\mu I, \quad T - i\mu I, \quad \mathbf{T}^{\circ}\alpha.$$

For all rows with $i \geq c$, they are

$$c\mu I, \quad T - c\mu I, \quad \mathbf{T}^{\circ}\alpha.$$

If the stationary probability vector $\mathbf{x}$ of Q exists and is partitioned as $\mathbf{x}_0, \mathbf{x}_1, \mathbf{x}_2, \cdots$, a solution of the form $\mathbf{x}_0, \mathbf{x}_1, \cdots \mathbf{x}_{c-1}, \mathbf{x}_{c-1}R, \mathbf{x}_{c-1}R^2, \cdots$ can be constructed.

a. Proceeding as in Problem 5.2.42, show that, if and only if $c\mu\lambda'_1 > 1$, there exists a matrix R of the form $R = \mathbf{T}^{\circ}\alpha[c\mu(1 - \eta)I - T]^{-1}$, which satisfies

$$c\mu R^2 + R(T - c\mu I) + \mathbf{T}^{\circ}\alpha = 0,$$

and whose largest eigenvalue $\eta < 1$. That eigenvalue is the solution in $(0, 1)$ of the equation $f(c\mu - c\mu\eta) = \eta$.

b. Substitute the proposed form of the vector $\mathbf{x}$ into the steady-state equations. It remains to determine positive vectors $\mathbf{x}_0, \mathbf{x}_1, \cdots \mathbf{x}_{c-1}$, satisfying

$$\mathbf{x}_0 T + \mu \mathbf{x}_1 = \mathbf{0},$$

$$\mathbf{x}_{i-1} \mathbf{T}^\circ \boldsymbol{\alpha} + \mathbf{x}_i (T - i\mu I) + (i+1)\mu \mathbf{x}_{i+1} = \mathbf{0},$$

for $1 \le i \le c-2$, and

$$\mathbf{x}_{c-2} \mathbf{T}^\circ \boldsymbol{\alpha} + \mathbf{x}_{c-1}[T - (c-1)\mu I + c\mu R] = \mathbf{0},$$

as well as the normalizing equation $\sum_{i=0}^{c-2} \mathbf{x}_i \mathbf{e} + \mathbf{x}_{c-1}(I - R)^{-1}\mathbf{e} = 1$.

c. Multiply each of the first c of these equations by $\mathbf{e}$, to verify that

$$\mathbf{x}_{i-1} \mathbf{T}^\circ = i\mu \mathbf{x}_i \mathbf{e}, \quad \text{for } 1 \le i \le c-1.$$

For each such i, replace $\mathbf{x}_{i-1} \mathbf{T}^\circ \boldsymbol{\alpha}$ by $i\mu \mathbf{x}_i \mathbf{e} \boldsymbol{\alpha}$ in the appropriate equation to obtain that

$$\mathbf{x}_0 = \mu \mathbf{x}_1 (-T)^{-1}, \quad \mathbf{x}_i = (i+1)\mu \mathbf{x}_{i+1} (i\mu I - i\mu \mathbf{e}\boldsymbol{\alpha} - T)^{-1},$$

for $1 \le i \le c-2$ and

$$\mathbf{x}_{c-1}[(c-1)\mu I - (c-1)\mu \mathbf{e}\boldsymbol{\alpha} - T - c\mu R] = 0.$$

It can be shown that all the inverses in these equations exist and are positive matrices. The last equation determines $\mathbf{x}_{c-1}$ up to a multiplicative constant. The vectors $\mathbf{x}_i$, $0 \le i \le c-2$, are so determined up to a common constant which is obtained from the normalizing equation.

d. Write a code to compute the vector $\mathbf{x}$ for the stable $PH/M/c$ queue. Test your code on a number of examples. As in Problem 5.2.42, report only essential information.

e. Derive expressions for the marginal densities of the queue length at an arbitrary time and just prior to an arbitrary arrival. As in Problem 5.2.42, you should obtain simple expressions for these densities, especially for the terms with indices $\geq c$. Add a subroutine to your code to report these densities, their means, and variances.

f. Customers are served in order of arrival. The waiting time W_A is the difference between the start of a customer's service and his arrival. W_t, the virtual waiting time at t is the time a hypothetical customer joining the queue at that time would have to wait. Derive simple expressions for the steady-state distributions of W_A and W_t. Add a subroutine to your code to print the parameters and first two central moments of these distributions.

Problem 5.2.57: Removing Servers in a *PH/M/c* Queue: The code developed in Problem 5.2.53 can be used to solve the following interesting problem. First show that the fraction of time $\psi(c)$ that all servers are busy is $\psi(c) = \mathbf{x}_{c-1} R (I - R)^{-1} \mathbf{e}$, and that the smallest number of servers of rate μ for which the queue is stable is

$$c^* = \min \{i : i\mu > (\lambda'_1)^{-1} = 1\}.$$

Start with some $c > c^*$ for which $\psi(c) \leq 0.10$. Successively reduce the number of servers until either inequality is violated. Use the code of Problem 5.2.53 to examine the impact of the successive removals of servers on the steady-state distributions of queue lengths and waiting times.

Remark: The removal of servers is particularly significant if the variance of $F(\cdot)$ is large. Arrivals then tend to occur in clusters separated by long intervals. For some c the probability $\psi(c)$ may be very small because all servers are occupied only after a cluster of arrivals. See what happens to the descriptors of the queue if we remove servers. Choose distributions $F(\cdot)$ that reflect a variety of input streams. Based on your numerical results, discuss how the queue behaves.

Problem 5.2.58: The $H_2/M/c$ Queue: With the hyperexponential distribution $F(\cdot)$, as in Problem 5.2.43, we can get explicit formulas for the vectors $\mathbf{x}_i$ and for the other descriptors of the queue. That is possible because only 2×2 matrices need to be inverted and there is an explicit formula for η. Derive these explicit formulas. Write them in simple modular forms. Use them in a code to compute the queue descriptors and to examine the removal of servers as in Problem 5.2.57.

Remark: This problem serves two purposes. It shows how complicated formulas become when matrix notation is abandoned in favor of *scalar* expressions. Also, the $H_2/M/c$ model is one of the simplest queues with non-Poisson arrivals. The program for this problem is useful in showing the effect of non-Poisson arrivals to novices to queueing theory.

Problem 5.2.59: In order for a certain task to be completed, a certain device must be open for the entire duration T of the task. T is a random variable independent of whether the device is open or closed. Once the task is started it cannot be interrupted. However, the task fails if the device closes while it is being performed. The state of the device is described by an m-state Markov chain with an irreducible generator Q. If the chain is in one of its states $1, \cdots, r$ with $r < m$, the device is open; in the remaining states, it is closed. The matrix Q is partitioned as

$$Q = \begin{array}{c} 1 \\ 2 \end{array} \left| \begin{array}{cc} Q(1,1) & Q(1,2) \\ Q(2,1) & Q(2,2) \end{array} \right|,$$

according to sets **1** of open and **2** of closed states. The stationary probability vector $\boldsymbol{\theta}$ of Q is similarly partitioned as $[\boldsymbol{\theta}(1), \boldsymbol{\theta}(2)]$.

Assume that, at time zero, a task arrives and that the Markov chain is in steady state. The task is lost unless the device is open. It will be *completed* if and only if the chain remains open for its entire duration. If T has a PH-distribution with n phases and representation $(\boldsymbol{\gamma}, S)$, derive a matrix formula for the probability p that the task will be completed. Write a program to evaluate p. Implement your code for some interesting examples of your choice.

Hint: Using Kronecker products and the formalism of PH-distributions to show that

$$p = [\boldsymbol{\theta}(1)\mathbf{e}]^{-1}[\boldsymbol{\theta}(1) \otimes \boldsymbol{\gamma}][-Q(1,1) \otimes I - I \otimes S]^{-1}[\mathbf{e} \otimes \mathbf{S}^{\circ}].$$

Calculate p explicitly when Q is a 2×2 matrix and T has an exponential distribution. Use that special case to test your general code.

Problem 5.2.60: Standby Redundancy: A system consists of m components in series. The component i is backed up by $n_i - 1$ identical components in cold standby. Together these n_i components make up *module* i. The lifetimes of all components are independent. Those in module i have a lifetime distribution $F_i(\cdot)$ with mean μ'_i. In what follows, let $F_j(\cdot)$ be a

PH-distribution with representation $[\boldsymbol{\beta}_j, S_j]$ and without jump at 0. Upon failure, any component is immediately replaced by another component in the module. The system fails when all components in one of its modules have expired. The m-tuple $\mathbf{n} = (n_1, \cdots, n_m)$ of the numbers of components in the various modules is called a *configuration*. Its reliability $R(x; \mathbf{n}) = R(x; n_1, \cdots, n_m)$ at $x \geq 0$ is the probability $P\{T > x\}$, where T is the lifetime of the system.

a. Show that

$$R(x; \mathbf{n}) = \prod_{j=1}^{m} [1 - F_j^{(n_j)}(x)],$$

where $F_j^{(n_j)}(\cdot)$ is the n_j-fold convolution of $F_j(\cdot)$.

b. For a given probability α, say $\alpha = 0.80$, let x^* be the unique solution of the equation $R(x^*; 1, 1, \cdots, 1) = \alpha$. Write a subroutine to find x^* for m given PH-distributions $F_j(\cdot)$.

We think of $(0, x^*]$ as the interval during which *early system failures* are likely. We shall add components to modules to increase the reliability at x^* from α to θ, where $\alpha < \theta < 1$. Because of technological constraints there can be at most N_j components in module j for $1 \leq j \leq m$. The N_j are given positive integers. If $n_j < N_j$, we say that module j is *unsaturated*. We propose three strategies to search for a configuration with reliability at least θ. For a current configuration $\mathbf{n}$ with $R(x^*; \mathbf{n}) < \theta$ and $n_j \leq N_j$, for $1 \leq j \leq m$, a component will be added to one of the unsaturated modules.

Strategy 1: Add the component to the unsaturated module i with the smallest *mean time-to-failure* $n_i \mu'_i$.

Strategy 2: Add the component to the module i for which

$$F_i^{(n_i)}(x^*) = \max_j \{F_j^{(n_j)}(x^*)\},$$

where j ranges over the indices for which $n_j < N_j$. Module i has the highest failure probability in $(0, x^*]$ among the unsaturated modules.

Strategy 3: Specify a probability $0 < \gamma < 1$. For the unsaturated modules

j, compute the unique x_j^* for which

$$F_j^{(n_j)}(x_j^*) = \gamma.$$

Add the component to the module i for which x_i^* is smallest. In that strategy we strengthen the module whose reliability decreases most rapidly over time to a specified γ.

In all three search strategies, we stop as soon as the desired reliability θ is achieved. We report the designs obtained by each strategy and their corresponding reliabilities. If in a search all modules become saturated before we attain the reliability θ, a message is printed and the reliability of the saturated configuration is reported.

c. Develop an algorithm to implement all three search strategies. Apply your code to a system with three modules and with $N_1 = 5$, $N_2 = 4$, $N_3 = 3$. The PH-distributions $F_j(\cdot)$ have the representations

$$\beta_1 = (0.8, 0.2), \qquad S_1 = \begin{bmatrix} -0.5 & 0 \\ 0 & -0.2 \end{bmatrix},$$

$$\beta_2 = (0.6, 0.4), \qquad S_2 = \begin{bmatrix} -0.1 & 0 \\ 0 & -0.2 \end{bmatrix},$$

$$\beta_3 = (0.8, 0.1, 0.1), \qquad S_3 = \begin{bmatrix} -0.6 & 0 & 0 \\ 0 & -0.3 & 0 \\ 0 & 0 & -0.2 \end{bmatrix}.$$

Determine x^* for $\alpha = 0.8$ and let the increased reliability θ be 0.98. For Strategy 3, use $\gamma = 0.2$. In numerical computations, set the time unit equal to the largest mean of the given PH-distributions. In the final printout report the results in terms of the original time unit. Implement your algorithm for other data of your choice. Interpret the numerical results. Give reasons for preferring one or the other search strategy in practice.

Hint: In computing the convolutions needed in this problem, use the algorithm of Problem 5.2.26.

Reference: Neuts, M. F. and Robinson, D. G., "An algorithmic approach to increased reliability through standby redundancy," *IEEE Transactions in Reliability*, 38, 430 - 435,1989.

Problem 5.2.61: A Distribution of Claims: A time span X has distribution $F(\cdot)$ on $(0, \infty)$. Insurance claims arrive according to an independent Poisson process of rate λ. The sizes of the successive claims are independent with common distribution $H(\cdot)$ on $(0, \infty)$. Y is the total of all claims submitted in $(0, X)$.

a. Show that the Laplace - Stieltjes transform $\phi(s)$ of Y is given by $\phi(s) = f[\lambda - \lambda h(s)]$. Let $f(s)$ and $g(s)$ be the Laplace - Stieltjes transforms of $F(\cdot)$ and $G(\cdot)$. Express the mean and variance of Y in terms of the moments of $F(\cdot)$ and $G(\cdot)$.

b. If F and G are continuous PH-distributions with representations (α, T) and (β, S), show that the distribution $\Phi(\cdot)$ of Y is of phase type and construct its representation.

c. Rescale the distribution H so that its mean is one. Write an efficient code to compute the distribution $\Phi(\cdot)$ for various values of λ under the assumptions of part *b*. Implement your code for the case where $F(\cdot)$ is an Erlang distribution of order 5 and mean 10 and $H(\cdot)$ a hyperexponential distribution with two components. Choose several hyperexponential distributions with different coefficients of variation.

d. Under PH-assumptions, the distribution Φ has an exponential asymptote. Compute the parameters of that asymptote and for each of your examples examine numerically how fast the graph of Φ approaches its asymptote.

Hint: Use properties [PH-4] and [PH-5] in Appendix 2.

Problem 5.2.62: The $M/PH/1$ Queue: This problem deals with the distribution of the largest queue length M attained during a busy period of an $M/PH/1$ queue. We construct the generator Q as in Problem 5.2.39. All elements in the first row are set to zero, so that zero is an absorbing state. To start from the start of the service of a customer arriving to an empty queue, the initial state is chosen in set 1 according to vector $\boldsymbol{\beta}$. There will be at most $k \geq 1$ customers present during a busy period if and only if the Markov chain reaches the absorbing state 0 without reaching the set $\boldsymbol{k} + 1$. Clearly, $P\{M = 1\} = \boldsymbol{\beta}(\lambda I - S)^{-1}\mathbf{S}^{\circ}$. To compute $P\{M \leq k\}$ for $k > 1$, define the vectors $\mathbf{y}_i(k)$, where $y_{ij}(k)$ is the probability that, starting in state (i, j)

with $1 \le i \le k$, $1 \le j \le m$, absorption in 0 occurs without visiting the set $k+1$.

a. Show that the vectors $\{\mathbf{y}_i(k)\}$ satisfy the equations

$$\mathbf{y}_1(k) = (\lambda I - S)^{-1}\mathbf{S}^\circ + \lambda(\lambda I - S)^{-1}\mathbf{y}_2(k),$$

$$\mathbf{y}_i(k) = (\lambda I - S)^{-1}\mathbf{S}^\circ\boldsymbol{\beta}\mathbf{y}_{i-1}(k) + \lambda(\lambda I - S)^{-1}\mathbf{y}_{i+1}(k),$$

for $1 \le i < k$, and

$$\mathbf{y}_k(k) = (\lambda I - S)^{-1}\mathbf{S}^\circ\boldsymbol{\beta}\mathbf{y}_{k-1}(k).$$

Explain why $P\{M \le k\} = \boldsymbol{\beta}\mathbf{y}_1(k)$, for $k \ge 2$.

b. For successive values of k, solve that system of equations by block Gauss - Seidel iteration. That is, in going from $k-1$ to k, use the solution computed for $k-1$ as the starting solution and implement the iteration

$$\mathbf{y}_1^{(n+1)}(k) = (\lambda I - S)^{-1}\mathbf{S}^\circ + \lambda(\lambda I - S)^{-1}\mathbf{y}_2^{(n)}(k),$$

$$\mathbf{y}_i^{(n+1)}(k) = (\lambda I - S)^{-1}\mathbf{S}^\circ\boldsymbol{\beta}\mathbf{y}_{i-1}^{(n+1)}(k) + \lambda(\lambda I - S)^{-1}\mathbf{y}_{i+1}^{(n)}(k)$$

for $1 \le i < k$, and

$$\mathbf{y}_k^{(n+1)}(k) = (\lambda I - S)^{-1}\mathbf{S}^\circ\boldsymbol{\beta}\mathbf{y}_{k-1}^{(n+1)}(k),$$

for successive values of n. Write an efficient code and define an appropriate stopping criterion.

c. The random variable M has a proper probability distribution if and only if the traffic intensity $\rho = \lambda\mu'_1 \le 1$. In writing a code to handle all cases, compute $P\{M \le k\}$ for successive k values until that probability exceeds 0.99 or until k reaches 25, whichever comes first.

Implement your code for λ successively equal to 0.75, 0.95, 1, and 1.25. Use various service time distributions scaled so that $\mu'_1 = 1$. Specifically, use the Erlang distribution of order 3 and two hyperexponential distributions with different coefficients of variation. Discuss the qualitative implications of your numerical results.

Problem 5.2.63: The $M/PH/1$ Queue: How should the procedure in Problem 5.2.62 be modified if the busy period is started at the beginning of a service and there are c customers initially? Make that minor modification to the code written for Problem 5.2.62 and examine for the same examples how the distribution of M changes for $c = 2, 3, 4$.

Problem 5.2.64: The Growth of the $M/PH/1$ Queue: Consider the queueing model in Problem 5.2.62 with the same initial conditions. The random variable $\tau(k)$ is the first time that there are k customers present in the queue. State 0 is now not absorbing; that first passage time may span several busy periods. For the same examples as suggested in Problem 5.2.62, compute the mean $E[\tau(k)]$ for $2 \le k \le 10$. Note that $\tau(k)$ has a PH-distribution and exploit the special structure of its representation. As in Problem 5.2.62, solve an appropriate system of linear equations by block Gauss - Seidel iteration.

Problem 5.2.65 : The Growth of the $M/PH/1$ Queue: For the queue in Problem 5.2.62, the random variable $\tau'(k)$ equals $j > 1$ if the number of customers in the system reaches k for the first time during the jth service. For the same examples as suggested in Problem 5.2.62, compute the mean $E[\tau'(k)]$ for $2 \le k \le 10$.

Problem 5.2.66: Customers requiring service from a single processor arrive to a pool of active customers according to a Poisson process of rate λ. While in the pool, each customer (independently) issues requests for access to the server. Each customer's requests form a Poisson process of rate σ. While the server is busy, such requests go unanswered. However, upon becoming free, the server accepts the customer who makes the first request thereafter. Even customers arriving to an empty system are not served immediately and must issue requests for access. The service times of successive customers are independent with the common PH-distribution $H(\cdot)$ of representation (β, S), m phases and mean μ'_1.

a. Define a continuous-time Markov chain with states $0, 1, \cdots$. State i signifies that the server is free and there are i customers in the pool. In addition, for $i \ge 1$, there are macro-states i consisting of m pairs (i, j). These signify that there are i customers in the system and the job in service is in its phase $j = 1, \cdots, m$. With the generator Q given by

$$Q = \begin{array}{c|cccccccc}
0 & -\lambda & 0 & \lambda & 0 & 0 & 0 & 0 & \cdots \\
1 & \mathbf{S}^\circ & S - \lambda I & 0 & \lambda I & 0 & 0 & 0 & \cdots \\
1 & 0 & \sigma\boldsymbol{\beta} & -\lambda-\sigma & 0 & \lambda & 0 & 0 & \cdots \\
2 & 0 & 0 & \mathbf{S}^\circ & S - \lambda I & 0 & \lambda I & 0 & \cdots \\
2 & 0 & 0 & 0 & 2\sigma\boldsymbol{\beta} & -\lambda-2\sigma & 0 & \lambda & \cdots \\
3 & 0 & 0 & 0 & 0 & \mathbf{S}^\circ & S - \lambda I & 0 & \cdots \\
3 & & \cdots & & \cdots & & \cdots & &
\end{array}$$

explain why this Markov chain describes the evolution of the number of customers in the system and of the service phase.

b. We write the invariant probability vector $\boldsymbol{\theta}$ of Q, if it exists, in the partitioned form $x_0, \mathbf{y}_1, x_1, \mathbf{y}_2, x_2, \mathbf{y}_3, x_3, \cdots$, where x_i and $\mathbf{y}_i$ correspond to state i and macro-state i. We introduce the generating functions $X(z) = \sum_{i=0}^{\infty} x_i z^i$, and $\mathbf{Y}(z) = \sum_{i=1}^{\infty} \mathbf{y}_i z^{i-1}$. Using the steady-state equations, show that $X(z)$ and $\mathbf{Y}(z)$ satisfy

$$\mathbf{Y}(z) = \sigma X'(z)\boldsymbol{\beta}[\lambda(1 - z) - S]^{-1}, \tag{1}$$

$$\sigma z X'(z) + \lambda(1 - z)X(z) = \mathbf{Y}(z)\mathbf{S}^\circ. \tag{2}$$

Recalling that the Laplace - Stieltjes transform of the *PH*-distribution with representation $(\boldsymbol{\beta}, S)$ is given by $\phi(s) = \boldsymbol{\beta}(sI - S)^{-1}\mathbf{S}^\circ$, eliminate $\mathbf{Y}(z)$ from (1) and (2) to obtain the differential equation

$$\frac{X'(z)}{X(z)} = \frac{\lambda}{\sigma} \frac{1 - z}{\phi[\lambda(1 - z)] - z}. \tag{3}$$

Deduce the equations

$$X(z) = X(1) \exp\left[-\frac{\lambda}{\sigma}\int_z^1 \frac{1 - u}{\phi[\lambda(1 - u)] - u}\, du\right], \tag{4}$$

$$\mathbf{Y}(z) = \lambda \frac{1 - z}{\phi[\lambda(1 - z)] - z} X(z)\boldsymbol{\beta}[\lambda(1 - z) - S]^{-1}. \tag{5}$$

Show that $X(z)$ is a valid generating function if and only if the equation $z = \phi[\lambda(1 - z)]$ has no root in the interval $(0, 1)$ and only a single root at $z = 1$. As for the $M/G/1$ queue, that is the case if and only if $\rho = \lambda\mu'_1 < 1$.

From $X(1) + \mathbf{Y}(1)\mathbf{e} = 1$, deduce that $X(1) = 1 - \rho$.

c. Define $\phi^*(z)$ by $\phi^*(z) = [\rho(1-z)]^{-1}\ [1 - \phi[\lambda(1-z)]]$, and set $\Phi(z) = (1-\rho)[1 - \rho\phi^*(z)]^{-1}$. Adapt the result stated in Problem 3.6.15 to obtain a stable recurrence relation for the computation of the terms of the density $\{\phi_i\}$ with generating function $\Phi(z)$. Show that $X(z)$ may be rewritten as

$$X(z) = (1-\rho)\prod_{i=1}^{\infty} \exp\left[-\frac{\lambda\phi_{i-1}}{(1-\rho)i\sigma}(1-u^i)\right]. \tag{6}$$

Explain why the product is a generating function of the type discussed in Problem 3.6.26. Rewrite equation (5) as

$$\mathbf{Y}(z) = \lambda z\,\mathbf{Y}(z)(\lambda I - S)^{-1} + \frac{\lambda}{1-\rho}\Phi(z)X(z)\boldsymbol{\beta}(\lambda I - S)^{-1}.$$

Expand both sides in power series to obtain a stable recurrence for the vectors $\mathbf{y}_i$, for $i \geq 1$.

Derive expressions for $X'(1)$ and $\mathbf{Y}'(1)$ and use these as accuracy checks in the computation of the x_i and the vectors $\mathbf{y}_1$. Encode your algorithms in a carefully written and well-structured program. Implement your program for the cases where $H(\cdot)$ is the Erlang distribution of orders 1 and 5, and the hyperexponential distribution

$$H(x) = 1 - \frac{1}{2}e^{-x} - \frac{1}{2}e^{-x/9}.$$

Choose the value of λ such that $\rho = 0.8$. Vary the calling rate σ and discuss its effect on the fraction of idle time $X(1)$ of the server and on the conditional queue length densities, given that the server is idle or busy.

Reference: Neuts, M. F. and Ramalhoto, M. F. "A service model in which the server is required to search for customers," *J. Appl. Prob.*, 21, 157 - 166, 1984.

Problem 5.2.67: The Number of Transitions in a Markov Chain: We first consider the joint generating function of the numbers $N_{hk}(t)$ of transitions from state h to state k during $(0, t]$ in an m-state Markov chain with generator Q. We make the convention that $N_{hh}(t) = 0$ for $1 \leq h \leq m$. Let Z be a matrix with elements z_{hk} where $z_{hh} = 1$, for $1 \le h \le m$. The *Schur product* $A \circ B$ of $m \times m$ matrices A and B is the matrix with elements $A_{hk}B_{hk}$. $J(t)$ is the state of the Markov chain at time t. We write $\Phi(Z; t)$ for the $m \times m$ matrix with elements

$$\Phi_{ij}(Z; t) = E[\prod_{h,k} z_{hk}^{N_{hk}(t)} I\{J(t) = j\} \mid J(0) = i].$$

a. Show that for $1 \le i, j \le m$, and $t \ge 0$,

$$\Phi_{ij}(Z; t) = \delta_{ij}\exp(Q_{ii}t) + \sum_{r \ne i} \int_0^t \exp[Q_{ii}(t-u)]\, Q_{ir} z_{ir} \Phi_{rj}(Z; u).$$

Establish the equivalent system of differential equations

$$\frac{d}{dt}\Phi_{ij}(Z; t) = \sum_{r=1}^{m} Q_{ir} z_{ir} \Phi_{rj}(Z; u),$$

with $\Phi_{ij}(Z; 0) = \delta_{ij}$. Conclude that $\Phi(Z; t) = \exp[Q \circ Zt]$, for $t \ge 0$.

In what follows, Q is irreducible and we set $Z^\circ = z\,\mathbf{e}\,\mathbf{e}^T + (1-z)I$, where $\mathbf{e}$ is a column vector with all components equal to one. Clearly, Z° is an $m \times m$ matrix with all diagonal elements equal to one and all off-diagonal elements equal to z. For that special choice of Z, we obtain generating functions of the *number* $N^\circ(t)$ *of transitions* in the interval $(0, t]$. Check that the first derivative with respect to z of $Q \circ Z^\circ$ is $Q + \Delta(\mathbf{q})$, where $\Delta(\mathbf{q})$ is a diagonal matrix with the quantities $-Q_{ii}$ as its diagonal elements. All higher order derivatives of Z° are zero matrices.

b. Write $\Phi(Z^\circ; t)$ as

$$\Phi(Z^\circ; t) = \sum_{n=0}^{\infty} \frac{t^n}{n!}(Q \circ Z^\circ)^n,$$

and verify that

$$M_1^\circ(t) = \left[\frac{\partial}{\partial z}\Phi(Z^\circ; t)\right]_{z=1} = \sum_{n=1}^{\infty} \frac{t^n}{n!} \sum_{r=0}^{n-1} Q^r[Q + \Delta(\mathbf{q})]Q^{n-r-1},$$

and

$$M_2^\circ(t) = \left[\frac{\partial^2}{\partial z^2}\Phi(Z^\circ; t)\right]_{z=1}$$

$$= 2\sum_{n=2}^{\infty}\frac{t^n}{n!}\sum_{r=0}^{n-2}\sum_{v=0}^{r} Q^v[Q + \Delta(\mathbf{q})]Q^{r-v}[Q + \Delta(\mathbf{q})]Q^{n-r-2}.$$

Deduce from the formula for $M_1^\circ(t)$ that $\boldsymbol{\theta} M_1^\circ(t)\mathbf{e} = \boldsymbol{\theta}\mathbf{q}t$, so that, as stated in the text, the rate ρ of transitions in the stationary version of the Markov chain is $\rho = \boldsymbol{\theta}\mathbf{q}$. Deduce from the formula for $M_2^\circ(t)$ that

$$\boldsymbol{\theta} M_2^\circ(t)\mathbf{e} = 2\boldsymbol{\theta}\Delta(\mathbf{q})\sum_{n=2}^{\infty}\frac{t^n}{n!}Q^{n-2}\mathbf{q} = 2\boldsymbol{\theta}\Delta(\mathbf{q})\int_0^t\int_0^u \exp(Qv)\,dv\,du\,\mathbf{q}.$$

Use the integration formula in Problem 5.2.34 to obtain an explicit matrix formula for the variance $\mathrm{Var}^\circ(t)$ of the number of transitions in an interval of length t in the stationary version of the Markov chain. Show that as $t \to \infty$, $\mathrm{Var}^\circ(t) = A_0 + A_1 t + o(1)$, and obtain explicit formulas for the coefficients A_0 and A_1 of the linear asymptote.

Write a program to evaluate and plot $\mathrm{Var}^\circ(t)$ and the *dispersion function* $[\mathrm{Var}^\circ(t)]^{1/2}/(\boldsymbol{\theta}\mathbf{q}t)$ for a given generator Q. Implement your code for several choices of Q and discuss the qualitative information that is apparent from the graphs. Choose some matrices Q with significant differences in the sizes of the diagonal elements.

Problem 5.2.68: A Markov Chain with a Reward Function: Consider an m-state irreducible Markov chain $\{J(t)\}$ with generator Q. For every unit of time spent in state j, a reward a_j accrues. Equivalently, for a piecewise constant random function $a_{J(t)}$ with value a_j when $J(t) = j$, the total reward $R(t)$ over an interval $(0, t)$ is given by

$$R(t) = \int_0^t a_{J(t)}\,dt, \qquad \text{for } t \geq 0.$$

Let $V(x; t)$ be the matrix with elements

$$V_{ij}(x;\, t) = P\{R(t) \leq x,\, J(t) = j \mid J(0) = i\}.$$

$F(x;\, t) = P\{R(t) \leq x\} = \boldsymbol{\theta} V(x;\, t)\mathbf{e}$ is then the probability distribution of the reward during an interval of length t in the stationary Markov chain.

a. Show that

$$V_{ij}(x;\, t) = \delta_{ij} e^{Q_{ii}t} U(x - a_i t) + \sum_{r \neq i} \int_0^t e^{Q_{ii}u} Q_{ir} V_{rj}(x - a_i u;\, t - u)\, du$$

for $x \geq 0$ and $t \geq 0$. $U(v) = 1$, for $v \geq 0$, and $U(v) = 0$, for $v < 0$.

Deduce from that equation that the Laplace - Stieltjes transforms

$$V^*_{ij}(s;\, t) = \int_0^\infty e^{-sx}\, dV_{ij}(x;\, t)$$

satisfy the equations

$$\frac{\partial}{\partial t} V^*_{ij}(s;\, t) = -a_i s V^*_{ij}(s;\, t) + \sum_{r=1}^{m} Q_{ir} V^*_{rj}(s;\, t),$$

with $V^*_{ij}(s;\, 0) = \delta_{ij}$. $V^*(s;\, t) = \{V^*_{ij}(s;\, t)\}$ may therefore be written as

$$V^*(s;\, t) = \exp[(Q - \Delta(\mathbf{a})s)t],$$

where $\Delta(\mathbf{a})$ is a diagonal matrix with the diagonal elements a_i.

b. Establish that the mean $\mu'_1(t)$ of $F(x;\, t)$ equals $\mu'_1(t) = \boldsymbol{\theta}\mathbf{a}t$, for $t \geq 0$.

c. Show that the second moment $\mu'_2(t)$ of $F(x;\, t)$ is given by

$$\mu'_2(t) = 2\sum_{n=2}^{\infty} \frac{t^n}{n!} \boldsymbol{\theta}\Delta(\mathbf{a})Q^{n-2}\Delta(\mathbf{a})\mathbf{e} = 2\boldsymbol{\theta}\Delta(\mathbf{a})\int_0^t \int_0^u \exp(Qv)\, dv\, du\, \mathbf{a}.$$

Use the integration formula in Problem 5.2.34 to obtain an explicit matrix formula for the variance Var(t) of $F(x;t)$. Show that as $t \to \infty$, Var(t) = $A_0 + A_1 t + o(1)$, and obtain explicit formulas for the coefficients A_0 and A_1 of the linear asymptote.

c. Write a program to evaluate and plot Var(t) and the *dispersion function* $[\mathrm{Var}(t)]^{1/2}/\mu_1(t)$ for a given generator Q and for several choices of the parameters a_j. What do these graphs tell you about the reward processes of your various examples?

Hint: To obtain the differential equations for the Laplace - Stieltjes transforms we take the Laplace transform of both sides of the first equation. Setting

$$V_{ij}^{\circ}(s;t) = \int_0^{\infty} e^{sx} V_{ij}(x;t)\,dx,$$

this yields the equation

$$V_{ij}^{\circ}(s;t) = \delta_{ij} s^{-1} \exp[(Q_{ii} - a_i)t] + \sum_{r \neq i} \int_0^t e^{Q_{ii}u} Q_{ir}\,du \int_{a_i u}^{\infty} e^{-sx} V_{rj}(x - a_i u;\, t - u)\,dx.$$

Now first make the change of variable $x - a_i u = y$ in the rightmost integral and multiply both sides of the equation by $\exp[-(Q_{ii} - a_i)t]$. That yields

$$\exp[-(Q_{ii} - a_i)t] V_{ij}^{\circ}(s;t) = \frac{1}{s} + \sum_{r \neq i} \int_0^t \exp[-(Q_{ii} - a_i)(t-u)] Q_{ir} V_{rj}^{\circ}(s;\, t-u).$$

Set $t - u = v$ in that equation and differentiate both sides with respect to t. Simplifying, we obtain that

$$\frac{\partial}{\partial t} V_{ij}^{\circ}(s;t) = -a_i s V_{ij}^{\circ}(s;t) + \sum_{r=1}^{m} Q_{ir} V_{rj}^{\circ}(s;t).$$

However, $V_{ij}^{*}(s;t) = s V_{ij}^{\circ}(s;t)$, and the equation for the Laplace - Stieltjes transform follows.

Problem 5.2.69: An Epidemic Model: Initially, a population consists of m infectious and n healthy individuals. Through encounters between sick and healthy individuals, some of the latter become infectious. Infectious persons eventually either become immune or die. Healthy individuals are not removed from the population. We ignore deaths from other causes, and healthy individuals cannot become immune without contracting the disease first. The epidemic is described by a Markov chain with states (i, j), the numbers of infected and of healthy individuals. The state indices i and j satisfy $0 \leq j \leq n$ and $0 \leq i + j \leq m + n$. The remaining $m + n - i - j$ individuals have been removed by immunity or death.

In an interval $(t, t + dt)$, a state change $(i, j) \to (i + 1, j - 1)$ (an infection) can occur with elementary probability $\beta i^{\alpha} j \, dt$, or a state change $(i, j) \to (i - 1, j)$ (the removal of an infective) with elementary probability $\rho i \, dt$. The parameters β and ρ are the infection and removal rates, and α, $0 < \alpha \leq 1$, is a parameter used to model the degree of interaction between infectious and healthy individuals; with smaller α there is less exposure of healthy individuals to those who are infectious.

a. Write the generator Q of the Markov chain and specify its initial conditions for $m = 2$ and $n = 4$. List the states for successive decreasing values of j and for increasing i values for each j. For general m and n, how many states are there? Which states are absorbing, how many such states are there, and what do they signify? If we partition Q into blocks corresponding to the successive j values, what is its general structure and how are the blocks defined?

b. Write a code to compute the probabilities of absorption in the various absorbing states. Also compute the conditional mean absorption times, given that the epidemic terminates in each absorbing state.

c. Write a carefully planned program to compute the time-dependent probabilities for this Markov chain. Your code should be able to handle values of n up to 20 and of m up to 5.

d. After testing your code for small m and n, use it to examine the effect of the parameter α on the behavior of the epidemic for $n = 20$ and $m = 2$. Notice that the time unit may be chosen such that $\beta = 1$. Discuss how the epidemic behaves for various values of ρ and α.

Reference: The model with $\alpha = 1$ is known as the *general stochastic epidemic*. It is possible to derive somewhat involved recurrence relations for the probability generating functions of associated random variables. See J. Gani, "On the general stochastic epidemic," *Proc. Fifth Berkeley Symposium on Math. Statist. and Prob.* Berkeley: Univ. of California Press, 1966, pp. 271 - 279.

Problem 5.2.70: This problem deals with a model for the insertion of tasks during periods of availability generated by a partitioned Markov process. We consider an irreducible Markov process with $m_1 + m_2$ states and with a generator Q partitioned in the form

$$Q = \begin{array}{c} 1 \\ \\ 2 \end{array} \left| \begin{array}{cc} Q(1,1) & Q(1,2) \\ & \\ Q(2,1) & Q(2,2) \end{array} \right| .$$

Tasks can be initiated only while the Markov process is in set 1 of the first m_1 states and must be entirely completed while the process is in set 1. If a task is ongoing and the process switches to a state in set 2 of the other m_2 states, the task is aborted and terminates. Arrivals of tasks during sojourns in set 2 are noted, but such tasks are said to be terminated by nonacceptance.

The durations of successive accepted tasks are mutually independent, and also independent of the Markov chain Q. They have a PH-distribution with irreducible representation (α, T). Counting from the termination of the preceding task, whether by completion, abortion or nonacceptance, the time until the next arrival has a (common) PH-distribution with irreducible representation (β, S). The successive interarrival times are conditionally independent of the Markov chain Q and of the durations of all preceding tasks.

Consider macro-states corresponding to the qualitative statements: "The Markov process is in set 1 and there is no active task," "The Markov process is in set 1 and a task is current," and "The Markov process is in set 2." Show that this system is represented by the Markov process with generator

$$Q^* = \left| \begin{array}{ccc} Q(1,1) \oplus T & I \otimes \mathbf{T}^{\circ}\boldsymbol{\beta} & Q(1,2) \otimes I \\ I \otimes \mathbf{S}^{\circ}\boldsymbol{\alpha} & Q(1,1) \oplus S & Q(1,2) \otimes \mathbf{e}\boldsymbol{\alpha} \\ Q(2,1) \otimes I & 0 & Q(2,2) \oplus (T + \mathbf{T}^{\circ}\boldsymbol{\alpha}) \end{array} \right| .$$

Write a general code to compute the stationary probability vector $\boldsymbol{\pi}$ of the Markov process Q^*, partitioned according the three macro-states. From it, determine the fraction of arriving tasks that are not accepted, accepted but aborted, and completed. Also determine the fraction of time during which useful work is done, that is time spent on tasks that are eventually completed. Further determine the fraction of time wasted on tasks that are eventually aborted.

For the simplest case where $Q(1, 1) = -Q(1, 2) = -\sigma_1$, and $Q(2, 2) = -Q(2, 1) = -\sigma_2$, and the durations of the tasks and the arrival times are exponential with parameters μ and λ, respectively, obtain explicit scalar expressions for all the required quantities. Test the accuracy of your general algorithm by constructing matrix parameters for the Markov chain Q that make it stochastically equivalent to that simple special case.

When you are convinced of the correctness of your general code use it to examine the following question: Let the Markov chain Q model an alternating renewal process with Erlang sojourn times in the "up" state 1 and in the "down" state 2. Similarly, let the phase-type distributions of the task durations and of the waiting times until the next arrivals of tasks be Erlang distributions of varying orders. Across a number of examples, keep the means of these Erlang distributions the same as in the simple example, i.e., respectively equal to σ_1^{-1}, σ_2^{-1}, λ^{-1}, and μ^{-1}. Can you draw qualitative conclusions about the effect of the variability of the various periods on the computed quantities of interest? Specifically, use the following values for the means:

$$\sigma_1^{-1} = 10, \quad \sigma_2^{-1} = 1, \quad \mu^{-1} = 1, \quad \lambda^{-1} = 5.$$

Remark: This model is related to that in Problem 3.2.60, but is considerably more complex.

Problem 5.2.71: A Loss System with Exponential Servers: This problem deals with a queueing system with m independent exponential servers who process customers at rates $\mu_1, \mu_2, \cdots, \mu_m$. Job arrivals form a renewal process with interarrival time distribution $F(\cdot)$. The Laplace-Stieltjes transform of $F(\cdot)$ is $f(s)$ and its mean is λ_1'. Customers who find all m servers busy are lost. Those who find servers free are served by the idle processor with the lowest number. The successive parts of this problem are steps toward proving some important properties of such a loss system.

a. Let there be only one server of rate μ. Show that the arrival times of the successive customers who find that server busy form a *renewal process* with underlying distribution $F^{\circ}(\cdot)$ that satisfies the equation

$$F^{\circ}(x) = \int_{0}^{x} [e^{-\mu u} + (1 - e^{-\mu u})F^{\circ}(x - u)] \, dF(u),$$

for $x \geq 0$. By taking the Laplace transform of both sides of that equation, show that the Laplace - Stieltjes transform of $F^{\circ}(\cdot)$ is

$$f^{\circ}(s) = \frac{f(s + \mu)}{1 - f(s) + f(s + \mu)},$$

and calculate the mean λ_1° of that distribution. Explain why the steady-state fraction of lost customers is $\lambda_1'/\lambda_1^{\circ}$.

b. If $F(\cdot)$ is a PH-distribution with representation (α, T), show the $F^{\circ}(\cdot)$ is the PH-distribution with representation $\gamma = (\alpha, 0)$ and

$$L = \begin{vmatrix} T - \mu I & \mu I \\ \mathbf{T}^{\circ}\alpha & T \end{vmatrix}.$$

c. When there $m \geq 2$ servers, explain why the arrivals who find servers $1, \cdots, i$, $i < m$ occupied form a renewal process with underlying distribution $F_i^{\circ}(\cdot)$. Write a recurrence relation for the probability distributions $F_i^{\circ}(\cdot)$ and for their means $\lambda_1^{\circ}(i)$.

d. Construct the PH-representation for the distribution $F_3^{\circ}(\cdot)$ and develop an algorithm to compute its mean and variance and the steady-state loss probability for a system with three servers. Implement your code for various hyperexponential distributions of the form

$$F(x) = 1 - pe^{-\theta_1 x} - (1 - p)e^{-\theta_2 x},$$

and for various choices of μ_1, μ_2, and μ_3.

Reference: For a discussion of this loss system, see J. Riordan, *"Stochastic Service Systems,"* New York: John Wiley & Sons, pp. 36 - 40, 1962.

CHAPTER 6

EXPERIMENTATION AND VISUALIZATION

6.1. INTRODUCTION

Simulating chance experiments is an important current use of computers. Techniques for generating variates with specified probability distributions are basic to such experiments. There is a vast literature on random variate generation and on the design and analysis of simulation experiments. This chapter does not deal extensively with these matters. We assume that the reader is acquainted with the fundamentals of simulation methodology. Algorithmic thinking, so necessary for the problems in the preceding chapters, is also essential to the proper design of computer experiments. In both, the abstract structure of the model needs to be identified and a proper sequence of modules must be written to compute or simulate that structure correctly. There is no inherent difference between algorithms for numerical computation and those to simulate the structure of experiments.

With graphical units now commonly available, more and more complex random phenomena can be visualized. The combination of computer experiments with a visualization of their results is a subject of major current research interest. Some problems in this chapter deal with models to aid the student in gaining access to that challenging area.

We assume familiarity with the notion of a *pseudo random number generator* and with the use of such generators in simulations. A reliable subroutine to generate uniform pseudo random variates should be available. The value returned by that subroutine is called a *uniform variate.* Unless stated otherwise, the letter U will denote a generic uniform variate. Using uniform variates, simulated values from other univariate or multivariate distributions can be generated. An excellent reference to the procedures for generating variates with prescribed distributions is

Devroye, L., *"Non-Uniform Random Variate Generation,"* New York: Springer-Verlag, 1986.

Library subroutines are available to generate variates from the classical probability distributions. As with special functions, it is usually safer to use a high-quality library routine than to write such subroutines oneself. The algorithms for variate generation are often based on pretty probability problems. We list a few of these in the problem section. The ideas underlying this chapter are well illustrated by these problems. Starting with the simple ingredient of uniform random variates, the structural relations between certain random variables are used to construct an experiment. The outcome of that experiment is a variate with a desired distribution. The same applies to the simulation of more complex structures. To generate a Markov chain, a queue, or a more involved stochastic process, we must relate its structure ultimately to uniform variates. The logical process that justifies such a procedure is entirely similar to the sequence of arguments to validate a numerical algorithm.

The Purpose of Experimentation: Many practically important probability distributions are intractable by analytic or algorithmic means. They are studied therefore by computer experiments. If the application warrants it, the experimental results are summarized in empirical formulas for future use. Some distributions useful in statistics can be obtained only in this manner. That is similar to the practice in fields such as structural engineering, aerodynamics, or materials science of using empirical formulas derived from laboratory experiments. The following is an example of a probability distribution studied experimentally. We use it to illustrate the empirical approach to be followed in many of the problems in this chapter.

Example 6.1.1: The elements U_{11}, U_{12}, U_{21}, U_{22} of a 2×2 matrix U are independent uniform random numbers in $[0, 1]$. The largest eigenvalue η of U is a random variable with values in $[0, 2]$. Its distribution is analytically so involved as to be intractable. This example and several related problems deal with an empirical study of the distribution of η.

This question arose during tests of a computer code for Elsner's algorithm for the Perron - Frobenius eigenvalue η of a nonnegative matrix. To that end, we generated $m \times m$ matrices with independent elements uniformly distributed in $[0, 1]$. For such matrices, η takes values in $[0, m]$. (Why is that?) We performed a number of experiments using many replications and with different orders m. In each case, we noticed that the sample mean $\overline{\eta}$ was very close to $m/2$. Moreover, histograms of the empirical results suggested that η has a nearly symmetric density about $m/2$.

It would have been an astonishing mathematical result had the density of η indeed been symmetric. We spent some time in vain attempts to prove this. Following the wise mathematical practice of checking a conjecture first for the simplest case, we focused on the case $m = 2$. After much calculation, we found that, even in that simple case, the analytic expression for the distribution of η is forbiddingly complicated. An experimental exploration of a conjecture should also deal with the simplest case first. We therefore did extensive simulations for 2×2 matrices. With η now given by

$$\eta = \frac{1}{2}\{U_{11} + U_{22} + [(U_{11} - U_{22})^2 + 4U_{12}U_{21}]^{1/2}\},$$

it is easily computed for many randomly generated matrices U. With modest computation times, we could do experiments with large sample sizes.

Many replications of our experiment, each with many matrices U, convinced us that the conjectured symmetry was not valid. In all cases, the sample mean $\bar{\eta}$ was found to be very close to but always less than 1. That is very unlikely for a symmetric density. The near-symmetry of the density is visually evident from histograms of the empirical values of η. However, whenever possible, experimental results should be subject to rigorous statistical tests. There are several tests for the symmetry of empirical distributions. The following one is based on the Kolmogorov - Smirnov (K–S) test, familiar from introductory statistics texts.

Let η_i be an empirical value of η. We define new data by setting $x_i = \eta_i$ if $0 \le \eta_i \le 1$ and $x'_i = 2 - \eta_i$ if $1 < \eta_i \le 2$. Under the null-hypothesis that η has a symmetric density, the $\{x_i\}$ and $\{x'_i\}$ are observations from the same density. The two-sample Kolmogorov - Smirnov test can now be applied to these sets of data. Accepting the null-hypothesis does not mean that η *has* a symmetric density, but only that the deviations from symmetry are sufficiently small that our data do not offer persuasive evidence against it. All this means here is that our eye does not deceive us and that the empirical distribution is indeed close to symmetric.

If the practical interest of the problem would warrant it, we could now fit a parametric distribution, say, a beta distribution symmetric on [0, 2], to the empirical data. If such a fit is found to be satisfactory and the estimated parameters of that distribution are substantially the same over various replications of the experiment, we can state that, *empirically*, the (intractable) density of η is closely approximated by a beta density with such and such parameters. In reporting such a finding, all information necessary for others to replicate our experiment should be given. That should include the type of computer and the random number generator used in the experiment, the sample size and the number of times each experiment was replicated, specifics on statistical tests and fitting procedures that were implemented, and so on. Even from this simple example, it is clear that proper computer experimentation is a demanding, time-consuming task. While the tools of simulation are well-developed, the science of computer experimentation and the standards for reporting its finding are still being developed. This chapter is only an introduction to an important and promising methodology.

References: Example 6.1.1 and a general discussion of computer experiments may be found in Neuts, M. F., "Computer experimentation in applied probability," in *"A Celebration of Applied Probability,"* A special volume (25A) of the Journal of Applied Probability, J. Gani, Editor, 31 - 43, 1988. Many more examples are discussed in Grenander, U. *"Mathematical Experiments on the Computer,"* New York: Academic Press, 1982. There is also a large literature on matrices with random elements. For the model in the example, it is known that, as $m \to \infty$, $m^{-1}\eta \to 1/2$, with probability 1. Recently, J. Silverstein also proved that the distribution of η rapidly converges to a normal limit law. Proofs of these theorems require advanced methods.

Displaying Experimental Results: The informative graphical display of statistical data is an art that is learned only through extensive practice. It requires a good critical sense, so that the graphs display exactly what they should convey. They should not be misleading and there should be no unwanted selection of the information. We can offer only the briefest introduction to this subject. We refer the reader to the literature on graphical statistics, exploratory data analysis, and to the manuals of major statistical software packages. The available software removes much of drudgery of graphics, yet the selection of proper display parameters is learned through practice.

Many problems ask for *histograms* to compare the densities of certain random variables, as in Example 6.1.1. The choice of the number and the width of class intervals is important. In many cases, these are dictated by the data that are generated. For comparisons, all histograms should be constructed on the same scale and with the same class widths.

To illustrate this, suppose that we obtain data values in [0, 1] (or some other specified interval). We divide [0, 1] in 100 (or 200 if a finer resolution is needed) intervals of equal length. Upon generating variates from the distribution of interest, we easily identify the subinterval to which each variate belongs. With a sample size n, say, of 100,000 we so fill an array containing the counts n_i of the observations in the various intervals $[0.01i$,

0.01(i + 1)). The empirical frequencies $p_i = n_i/n$ are calculated at the end. Next comes the display. The histogram should look like the unknown density with the apparent area under the curve equal to 1. We therefore plot a graph with the ordinates $100p_i$ at the midpoints of the class intervals. We can now compare histograms for different parameters of the model or with different samples sizes.

Two-dimensional data sets can be compared by three-dimensional histograms or by scatter plots. Both modes of display can be conveniently generated by library routines, but the choices of scale and classes remain up to the user. Scatter plots are useful in seeing where bivariate data tend to cluster. They also bring out differences between two or more sets of bivariate data. Three dimensional histograms are more informative and the point from which they are viewed can be varied. For comparisons the same viewpoint should obviously be used.

For stochastic processes, such as queues, there still are few commonly used visualizations. Specialized software systems for the dynamic visualization of simulated behavior of complex systems such as manufacturing plants or communications networks have been developed. Their discussion is beyond our scope. In Section 6.3, we stress the importance of displaying a *plume* of realizations (trajectories, sample paths) of the process. Because of the inherent dependence in the process, a single sample path, even when viewed over a long time interval, can be misleading. By simulating sets of path functions and by gathering summary statistics, the reader should be able to visualize the physical behavior of the simple stochastic models treated in the problems. Graphical displays should give a deeper understanding of the standard analytic descriptors of these models.

6.2. THE ALIAS METHOD

One method for variate generation is so important in the study of Markov chains that it merits a detailed discussion here. Suppose we wish to generate variates from a discrete probability density $\{p_1, p_2, \cdots, p_k\}$ on the values $\{x_1, x_2, \cdots, x_k\}$. Without loss of generality we can set $x_i = i$, for $1 \leq i \leq k$. The *alias method* is based on the following property:

The probability density $\{p_1, p_2, \cdots, p_k\}$ can always be written as

$$p_i = \frac{1}{k}\sum_{j=1}^{k} [q_j I\{r_j = i\} + (1 - q_j) I\{s_j = i\}], \quad 1 \le i \le k,$$

for k pairs of integers (r_j, s_j) satisfying $1 \le r_j \le k$, and $1 \le s_j \le k$, and for quantities q_j that are probabilities.

Before we sketch the proof of that property, we discuss its implications. Generating a variate with the density $\{p_1, p_2, \cdots, p_k\}$ can be done by an equivalent experiment in which we randomly select one of k experiments each having only two possible outcomes. If the experiment j is chosen, we select the outcomes r_j or s_j with conditional probabilities q_j and $1 - q_j$, respectively. Implementation of the procedure requires lists of the integers r_j, of their *aliases* s_j, and of the conditional probabilities q_j.

If we know these, generating a variate is easy. Starting with a first uniform variate U, we set $j = [kU] + 1$, to choose the experiment j. Using a second uniform variate U_1, we return the value r_j if $U_1 \le q_j$ and s_j otherwise. The variate with the density $\{p_i\}$ is therefore generated at the fixed cost of two uniform variates, one multiplication and one comparison. Some library routines further use the fact that $[kU]$ and $U_1 = kU - [kU]$ are independent random variables to save the separate generation of U_1. For some pseudo random number generators that is not advisable when k is large.

The proof of the property proceeds by induction. It is obviously valid for $k = 2$. Assume its validity for $k < n$ and consider the case $k = n$. Let r_1 and s_1, respectively, be indices for which p_i is smallest and largest. Set $q_1 = kp_{r_1}$, then clearly $0 \le q_1 \le 1$. Also

$$p'_{s_1} = p_{s_1} - \frac{1 - q_1}{k} \ge 0,$$

because $p_{s_1} \ge k^{-1}$. In the original list of probabilities $\{p_i\}$, delete p_{r_1} and replace p_{s_1} by p'_{s_1}. The remaining probabilities sum to $1 - k^{-1}$. By the induction hypothesis, the density $\{p'_i\}$ which concentrates on $n - 1$ outcomes can be written as an equiprobable mixture of $n - 1$ experiments each with two alternatives. The algorithm to compute the (r_j, s_j) and the corresponding q_j proceeds along the lines of the proof. A library routine to compute the so-called *alias tables* $\{(r_j, s_j); q_j\}$ is commonly available. In substantial simulations of Markov chains, the storage and the setup computation required by the alias method are worthwhile because of the modest and fixed computational cost of generating each subsequent state.

The alias method can also be adapted to special probability densities with infinite support. For the given density $\{p_i\}$, determine an index k^* for which the probability $p_1 + \cdots p_{k^*} = 1 - \varepsilon$, where ε is small. Using a first uniform variate U_0, check whether $U_0 \leq 1 - \varepsilon$. If so, generate the variate from the conditional density $\{(1-\varepsilon)^{-1}p_i\}$ on $1, \cdots, k^*$ by the alias method. If the complementary event occurs, determine the smallest index $i > k^*$ for which $\sum_{j=k^*+1}^{i} p_i > 1 - U_0$, and return the value i for the variate. It is implicitly assumed that terms in the tail of the density $\{p_i\}$ can be accurately computed.

Simulating Finite Markov Chains: The utility of the alias method is obvious in simulating the paths of finite discrete-time Markov chains. Each row of the transition probability matrix P of the Markov chain defines a multinomial trial by which the next state is chosen. It therefore suffices to compute the alias tables for each row of P. If the current state is i, the next state is chosen by generating a variate using the corresponding alias table. In computing the alias table for a row i only the positive elements in that row are used. Moreover, any special features of the transition probabilities can be exploited to reduce the number of alias tables that need to be computed or to expedite the simulation. Such common sense techniques are explored in some of the problems.

6.3. UNDERSTANDING STEADY-STATE BEHAVIOR

One of the most important notions in the theory of stochastic processes is that of a system *in steady state*. The following is an informal discussion of that subject for Markov chains. In some of the problems, the reader is asked to visualize the behavior of Markovian models to gain a better intuitive understanding of their steady-state regime. In our discussion, we focus on the simple case of an m-state irreducible aperiodic Markov chain $\{J_n\}$ with transition probability matrix P. As $n \to \infty$, the n–step transition probabilities P_{ij}^n tend to the components π_j of the invariant probability vector π. For large n, knowledge of the initial state J_0 is unimportant to the probability of the event $\{J_n = j\}$. A recurrent Markov chain forgets its initial conditions. A consequence of that limit is that, as $n \to \infty$, the *joint* conditional probability

$$P\{J_n = j, J_{n+1} = j_1, J_{n+2} = j_2, \cdots, J_{n+k} = j_k | J_0 = i\} \to \pi_j P_{jj_1} P_{j_1 j_2} \cdots P_{j_{k-1} j_k}.$$

For large n, the initial state becomes unimportant to the entire future of the Markov chain. Moreover, if the initial state is chosen according to the probabilities $\pi_1, \cdots, \pi_m$, we have that

$$P\{J_n = j, J_{n+1} = j_1, J_{n+2} = j_2, \cdots, J_{n+k} = j_k\} = \pi_j P_{jj_1} P_{j_1 j_2} \cdots P_{j_{k-1} j_k}.$$

The expression on the right-hand side does not depend on n. The joint probabilities of any $k + 1$ successive states are now *translation invariant,* so that any arbitrary epoch can be chosen as the time origin. With these initial conditions, we obtain the stationary (or steady-state) version of the process.

The preceding results can be visualized by simulating a fairly large set of paths of the Markov chain and simultaneously displaying the results on a computer screen. We shall call such a set a *plume* of paths. If we start all paths from the same state i, there is usually a discernible, more or less common, initial behavior to all paths. That initial transient behavior depends on state i. A similar behavior is apparent if the initial state is chosen according to a probability vector $\mathbf{p}(0)$ other that π.

In a simulation of a large number of paths with each initial state chosen according to π, the behavior of the plume of paths is substantially the same over any time segment $n, n + 1, \cdots, n + k$. Each individual path has, of course, a specific initial state so that it does not make much sense to talk of *one* realization as reaching steady-state.

Clearly, only a finite set of paths can be simulated. The mathematical notion of the stationary version of the Markov chain deals with the idealized, infinite set of all possible paths of the process. That set cannot be visualized. Let us refer to the set of all possible paths with given initial conditions as the *ensemble* of paths with those initial conditions. The ensemble of paths for the stationary version of the chain will be called the *stationary ensemble.* The convergence theorem for positive recurrent Markov chains says that, probabilistically, the ensemble of paths starting from any initial conditions eventually looks like the stationary ensemble. Let us view the stationary ensemble at an arbitrary time zero. Then π_j is the probability of the sub-ensemble of paths that happen to be in state j at time zero. That is a useful way to think about the steady-state probabilities. We illustrate this by the following example:

As in Example 4.1.1, consider a irreducible Markov chain with its states partitioned into sets, 1 and 2. The transition probability matrix P is partitioned as

$$P = \begin{array}{c|cc|} 1 & T(1,1) & T(1,2) \\ 2 & T(2,1) & T(2,2) \end{array}.$$

The invariant probability vector $\boldsymbol{\pi}$ of P is accordingly partitioned as $[\boldsymbol{\pi}(1), \boldsymbol{\pi}(2)]$. The quantities $\boldsymbol{\pi}(1)\mathbf{e}$ and $\boldsymbol{\pi}(2)\mathbf{e}$ may be interpreted as the probabilities of the sets of paths in the stationary ensemble that are in set 1, respectively 2, at time zero.

The quantity $[\boldsymbol{\pi}(1)\mathbf{e}]^{-1}\pi_j$, with $j \in 1$, is the conditional probability of set of paths that are in the state j at time zero. Choosing the initial probability vector $[\boldsymbol{\pi}(1)\mathbf{e}]^{-1}\boldsymbol{\pi}(1)$ therefore corresponds to viewing the subset of paths in the stationary ensemble that happen to be in set 1 at time zero. Similarly, the initial probability vector $[\boldsymbol{\pi}(2)T(2,1)\mathbf{e}]^{-1}\boldsymbol{\pi}(2)T(2,1)$ corresponds to considering the subset of paths in steady-state that have *entered* set 1 at time zero. Such choices of initial conditions are useful in simulation experiments for Markov chains and in visualizing their path behavior. For instance, for a queueing model we may choose 1 to correspond to an unusually long queue. By choosing the initial probabilities to correspond to an entry into 1, a display of many paths with those initial conditions will show how the queue typically recovers from a heavy load condition.

Periodic Markov Chains: For discrete Markov chains, it is interesting to think through, in the light of the preceding discussion, what happens in an irreducible, *periodic* Markov chain. Suppose that all states are of period $d > 1$. By relabeling the states, the transition probability matrix P can then be rewritten as

$$P = \begin{array}{c|cccccc|} 1 & 0 & P(1) & 0 & 0 & \cdots & 0 \\ 2 & 0 & 0 & P(2) & 0 & \cdots & 0 \\ 3 & 0 & 0 & 0 & P(3) & \cdots & 0 \\ & \cdots & & & \cdots & & \\ d & P(d) & 0 & 0 & 0 & \cdots & 0 \end{array},$$

where $P(1), \cdots, P(d)$ are stochastic matrices of order $m^* = m/d$. The period d is clearly a divisor of the number m of states. If we start the chain at any state in the set 1, it moves next to a state in 2, next to a state in 3, and so on, to return to some state in 1 at time d. The limit theorem for the n-step transition probabilities P_{ij}^n must be modified to say that the subsequence

$$P_{ij}^{\nu_{ij} + dn} \to \pi_j,$$

where ν_{ij} is the smallest number of steps needed to reach the set to which j belongs from the set that contains the state i. A periodic Markov chain "remembers" the set in the partitioned Markov chain to which its initial state belongs. To what does the stationary ensemble of such a chain correspond? First we calculate the invariant vector $\boldsymbol{\pi}$ which is partitioned as $\boldsymbol{\pi}(1)$, $\boldsymbol{\pi}(2)$, $\cdots$, $\boldsymbol{\pi}(d)$. We readily see that $\boldsymbol{\pi}(1) = d^{-1}\boldsymbol{\pi}^*$, $\boldsymbol{\pi}(2) = d^{-1}\boldsymbol{\pi}^* P(1)$, $\boldsymbol{\pi}(3) = d^{-1}\boldsymbol{\pi}^* P(1)P(2)$, $\cdots$, $\boldsymbol{\pi}(d) = d^{-1}\boldsymbol{\pi}^* P(1)P(2) \cdots P(d-1)$, where $\boldsymbol{\pi}^*$ is the invariant probability vector of the irreducible, aperiodic stochastic matrix $P(1)P(2) \cdots P(d)$. It is clear that $\boldsymbol{\pi}(r)\mathbf{e} = 1/d$ for $1 \le r \le d$.

Consider a Markov chain with $m^* = m/d$ states, but with nonstationary transition probabilities. The successive transitions are performed in a periodic manner with the stochastic matrices $P(1)$, $P(2)$, $\cdots$,$P(d)$. Also keep track of where that process is in its cycle by labeling the states accordingly. We so obtain the original periodic Markov chain. The stationary version of that Markov chain is obtained by choosing the place r in the cycle at random and then selecting the state in r according to $d\boldsymbol{\pi}(r)$.

Continuous-Time Markov Chains: The steady-state ensemble of an irreducible Markov chain with generator Q is obtained by choosing the initial state according to the stationary probability vector $\boldsymbol{\theta}$. We can again construct various initial probability vectors that correspond to finding the Markov chain in certain subsets at the arbitrary time origin zero. Choices that correspond to conditioning on the occurrence of specific transitions at time zero require some explanation. While those choices involve some technical mathematical issues beyond the scope of this discussion, they can be easily understood by using elementary probabilities. The elementary probability that, in steady state, a transition occurs in the interval $(0, dt)$ is given by

$$\rho\, dt = \sum_{j=1}^{m} \sum_{i \ne j} \theta_i Q_{ij}\, dt = \sum_{j=1}^{m} \theta_i [-Q_{ii}]\, dt.$$

The elementary probability of a transition from i to j in $(0, dt)$ is $\theta_i Q_{ij}\, dt$, for $i \neq j$. The ratio

$$\phi(j; i) = \frac{\theta_i Q_{ij}}{\rho}$$

can therefore be interpreted as the conditional probability of a state change $i \to j$, given that a transition occurs at time zero. We shall agree that $\phi(i; i) = 0$. The initial probability vector with components $\sum_{i=1}^{m} \phi(j; i)$ therefore corresponds to choosing time zero at an arbitrary visit to state j.

The corresponding constructions for partitioned generators are particularly useful. Write Q as

$$Q = \begin{array}{c} 1 \\ \\ 2 \end{array} \left| \begin{array}{cc} T(1, 1) & T(1, 2) \\ & \\ T(2, 1) & T(2, 2) \end{array} \right| ,$$

and $\boldsymbol{\theta}$ as $[\boldsymbol{\theta}(1), \boldsymbol{\theta}(2)]$. To place the time origin at an arbitrary entry into the set **1**, we start the chain with the initial probability vector $[\boldsymbol{\theta}(2)T(2, 1)\mathbf{e}]^{-1}$ $[\boldsymbol{\theta}(2)T(2, 1), \mathbf{0}]$.

Visualizing the Behavior of Infinite Markov Chains: With some care, the behavior of some infinite Markov chains can be visualized on a computer screen. Obviously, we can follow the path only while it is in the finite portion of the state space we choose to represent on the screen. A common error is to restrict the simulation to that finite portion, but then we are not visualizing the correct Markov chain. We must allow the path to go "off the screen" and by scrolling, allow it to return to it eventually.

To make this concrete, suppose we plot paths of a symmetric random walk on the integers, or equivalently the difference between the numbers of heads and tails in coin tossing. The current state is depicted by an asterisk placed on a horizontal line on the screen. The initial position 0 is in the middle of the screen and we track the walk between $-K$ and K. The successive positions of the walk are marked on successive lines. By scrolling, we can so follow the path for as long as we wish. Alternatively, we can plot segments of the path using a horizontal time axis. On a high-resolution screen several segments can be depicted on the same screen or we can refresh the screen for each successive segment.

If we restrict the walk to $[-K, K]$, we will depict a positive recurrent

Markov chain. A typical path will spend roughly half the time on each side of 0 and, depending on the size of K, crosses quite often between the positive to the negative sides. To depict the unbounded random walk properly, we do not plot its position when it is outside of the interval $[-K, K]$. Our simulation program keeps track of the position and resumes plotting when the path returns to that interval. We so see the very different behavior typical of a null-recurrent Markov chain. Over a long simulation, the path now typically spends most of the time on one side of the axis. That is an important point. It is very interesting to visualize the behavior of a critical $M/G/1$ queue in the same manner to understand what happens when the arrival and service rates are exactly equal. Some problems in this chapter deal with that.

6.4. MAIN PROBLEM SET FOR CHAPTER 6

EASIER PROBLEMS

Problem 6.4.1: When asked to generate a random string of m zeros and n ones, a student successively picked independent locations in a string of length $m + n$. A symbol 1 was entered into each location until n *different* locations had been filled. Whenever a location came up that had been picked earlier, that drawing was discarded. Explain why this is valid but inefficient. Describe an efficient method that requires a fixed number of uniform variates. Compare the expected number of uniform random variates for the first method to the *fixed* number for the efficient procedure.

Problem 6.4.2: Write a subroutine to generate a random string of m zeros and n ones. X is the number in $(0, 1)$ whose $m + n$ first binary digits consist of m zeros and n ones in random order. Use that subroutine to obtain an estimate based on 100 replications, each of 10,000 variates, of the mean $E(X)$ and construct a confidence interval for your estimate. Implement your code for five choices of m and n.

Problem 6.4.3: Normal Variates: The Box - Muller Method: *a.* If U_1 and U_2 are independent uniform random variates show that

$$X_1 = (-2 \log U_1)^{1/2} \cos 2\pi U_2, \quad \text{and} \quad X_2 = (-2 \log U_1)^{1/2} \sin 2\pi U_2,$$

are independent normal with mean zero and variance one. Write a subroutine to generate pairs of independent normal variates by using that property.

b. The pairs $V_1 = 2U_1 - 1$ and $V_2 = 2U_2 - 1$ are the coordinates of a point chosen at random in the square with vertices (1, 1), (1, −1), (−1, 1), (−1, −1). Set $R = V_1^2 + V_2^2$ and generate pairs (V_1, V_2) until $R \leq 1$. When such a pair is found, set

$$X_1 = V_1\left[-2\,\frac{\log R}{R}\right]^{-1/2}, \quad \text{and} \quad X_2 = V_2\left[-2\,\frac{\log R}{R}\right]^{-1/2}.$$

Show that X_1 and X_2 are independent $N(0, 1)$ variates.

c. Compare the time required by the central processor to generate 10,000 normal variates by each of the methods in parts *a* and *b*.

Remark: The variant of the Box - Muller method in part *b* is more efficient as it avoids calls to compute trigonometric functions. However, it may require several pairs (V_1, V_2) before a point inside the circle is obtained. The problem asks you to compare both methods on your computer.

Problem 6.4.4: If $\{U_k\}$ is a sequence of independent, uniform random numbers and a a positive constant, show that the random variable

$$N = \min\,\{n : \prod_{k=1}^{n} U_k < e^{-a}\} - 1$$

has a Poisson distribution with mean a. That is the basis for a classical method to generate Poisson variates.

a. Next, consider a two-dimensional analogue of this model. Define the random variables $Z_1(n)$ and $Z_2(n)$ by

$$Z_1(n) = \prod_{k=1}^{n} U_{1,k}, \quad \text{and} \quad Z_2(n) = \prod_{k=1}^{n} U_{2,k},$$

where $\{U_{1,k}\}$ and $\{U_{2,k}\}$ are independent sequences of independent random numbers. Set $Z_1(0) = Z_2(0) = 1$. If k steps are required for the point with coordinates $[Z_1(n), Z_2(n)]$ to reach the set of (x_1, x_2) satisfying

$x_1^2 + x_2^2 < b^2$, $x_1 \geq 0$, $x_2 \geq 0$, set $N = k - 1$.

b. Obtain an analytic expression for the probability $P\{N > k\}$ in terms of a rather involved integral (which you are not being asked to compute numerically).

c. Consider the squares $0 \leq x_1 \leq b$, $0 \leq x_2 \leq b$, and $0 \leq x_1 \leq b/\sqrt{2}$, $0 \leq x_2 \leq b/\sqrt{2}$. If k_1 and k_2 trials are required to reach the larger, respectively the smaller square, set $N_1 = k_1 - 1$ and $N_2 = k_2 - 1$. Show that the identical marginal densities of N_1 and N_2 are the same as the density of the maximum of two independent Poisson random variables. Compute the marginal densities of N_1 and N_2 for a given value of b.

d. Study the distribution of the random variable N in part *b* by simulation. Specifically, do 1,000 replications of the first passage time and use these to obtain an empirical density for N. Implement your code for $b = 0.1$. Examine how well the empirical density of N can be approximated by a *mixture* of the densities of N_1 and N_2. How you fit a mixture of these two densities to the empirical density is left up to you, but you should justify your method.

Problem 6.4.5: We compare two ways of randomly partitioning the interval [0, 1] into three pieces. In Experiment *I*, two points are chosen independently and at random in [0, 1]. These determine three sub-intervals of [0, 1]. X_1 and X_2 are the lengths of the two leftmost intervals. Their joint probability density $\phi(u_1, u_2)$ is positive on the triangle C where $u_1 \geq 0$, $u_2 \geq 0$, and $u_1 + u_2 \leq 1$, and only there.

a. Verify that on C, $\phi(u_1, u_2) = 2$.

In Experiment *II*, we first choose point a at random in the unit interval. Then a second point is chosen at random in the *longest* of the intervals $[0, a]$ and $(a, 1]$. The lengths of the two leftmost intervals are Y_1 and Y_2 and we write $\psi(u_1, u_2)$ for their joint probability density, which is also positive on the triangle C only.

b. Verify that on C, $\psi(u_1, u_2)$ is given by

$$\psi(u_1, u_2) = \frac{1}{1 - u_1}, \qquad \text{for } u_1 + u_2 \leq 1/2,$$

$$= \frac{1}{1 - u_1} + \frac{1}{u_1 + u_2}, \qquad \text{for } u_1 \leq 1/2 \leq u_1 + u_2,$$

$$= \frac{1}{u_1 + u_2}, \qquad \text{for } 1/2 \le u_1 \le u_1 + u_2.$$

c. Calculate the marginal densities of $\phi(u_1, u_2)$ and $\psi(u_1, u_2)$ and check that ψ is a proper joint probability density.

d. Clearly, the probability $P\{X_1 \le X_2\} = 1/2$. What is $P\{Y_1 \le Y_2\}$?

e. A person lacking the background to follow the calculations for parts *c* and *d* believes that both ways of dividing [0, 1] in three random intervals are equivalent. How would you use a simulation experiment to show that they are not? What graphical displays would you use?

Problem 6.4.6: In yet another way of partitioning the interval (0, 1) in three intervals, the first point is chosen at random. Then one of the two intervals is chosen with probability 1/2 and a point is selected at random in it. Define the random variables Y_1 and Y_2 as in Problem 6.4.5 and find their joint density. Answer the same questions as in Problem 6.4.5.

Problem 6.4.7: The elements of a 3×3 matrix U are independent uniform random numbers. Write the canonical equation for its eigenvalues in the form $x^3 + A_2x^2 + A_1x + A_0 = 0$, where the random variables A_0, A_1, and A_2 are functions of the elements of U. The Perron eigenvalue η of U is a simple, positive root of that equation. The other two roots are either real of complex conjugate. Let p^* be the probability that all three eigenvalues are real.

For a simulation experiment, calculate the sample size n needed to construct a confidence interval $(\bar{p}^* - 0.005, \bar{p}^* + 0.005)$ that covers the true p^* with probability at least 0.95. With that value of n, perform 10 replications of your experiment and print the 10 estimates $\bar{p}^*$ of p^*.

Hint: The cubic equation has three real roots if and only if $q^3 + r^2 \le 0$, where

$$q = \frac{1}{3}A_1 - \frac{1}{9}A_2^2, \quad \text{and} \quad r = \frac{1}{6}(A_1A_2 - 3A_0) - \frac{1}{27}A_2^3.$$

The desired quantity p^* is the probability of the corresponding event which is a complicated subset of the nine-dimensional unit cube. To determine n, use classical formulas for a confidence interval for the parameter p of binomial trials.

Problem 6.4.8: If the elements of the 3×3 matrix in Problem 6.4.7 are uniform in the interval $[0, T]$, $T > 0$, show that the probability p^* does not depend on T. In general, show that if the elements have a common distribution $F(\cdot)$, with a *scale parameter* λ, the probability $p^*(F)$ that all three eigenvalues are real does not depend on λ. As in Problem 6.4.7, determine $p^*(F)$ experimentally when $F(x) = 1 - \exp(-\lambda x)$, for $x \geq 0$.

Problem 6.4.9: For the 3×3 matrix U in Problem 6.4.7, we wish to determine experimentally the probability p° that the other two eigenvalues are complex with negative real parts.

a. Let η be the Perron eigenvalue of U. We know that the other two roots are complex if and only if $q^3 + r^2 > 0$, where q and r are as defined in Problem 6.4.7. Show that, if that condition holds, the complex roots have (the same) negative real part if and only if

$$U_{11} + U_{22} + U_{33} - \eta < 0. \tag{1}$$

b. Organize your code as follows: For any randomly generated matrix U, test whether the two eigenvalues other than η are complex. If that is the case, use Elsner's algorithm to determine whether condition (1) is satisfied. Notice that there is no need to calculate η precisely, so that you should organize that portion of your code to halt as soon as the validity of (1) can be decided.

c. Can you construct a different solution of this problem by using explicit analytic conditions for the cubic equation to have two complex roots with negative real part?

Problem 6.4.10: Consider a random permutation $\boldsymbol{\pi} = [\pi(1), \cdots, \pi(n)]$ of the integers $1, \cdots, n$ and define the *total variation* TVar $(\boldsymbol{\pi})$ of that permutation by TVar $(\boldsymbol{\pi}) = \sum_{i=1}^{n-1} | \pi(i+1) - \pi(i) |$.

a. Determine the set of possible values of the function TVar $(\boldsymbol{\pi})$.

b. Use the algorithm developed in Problem 1.3.33 to compute the exact probability density of TVar $(\boldsymbol{\pi})$ for $n = 5, \cdots, 10$.

c. Do five replications, each of 50,000 trials, in which independent random permutations of the integers $1, \cdots, 10$ are generated and the empirical probability density of TVar $(\boldsymbol{\pi})$ is determined. Perform χ^2 goodness of fit tests for each replication to get an idea of how well the empirical and the computed frequencies agree.

A very nice subject for visualization, and an excellent teaching aid, is the behavior of sums of independent random variables as predicted by the great theorems of probability theory. The well-known *quincunx* of Francis Galton which is found in some museums is an early visualization of the convergence of the binomial to the normal distribution. With modest graphics capabilities, many other limit laws can be visualized. The following three problems deal with such experiments.

Problem 6.4.11: Sums of Independent Random Variables: $\{X_i, i \geq 1\}$ is a sequence of independent, identically distributed random variables with common mean μ. By the *law of large numbers,* the random variables $Y_n = n^{-1}(X_1 + \cdots + X_n)$ converge to μ with probability one. If the mean does not exist, the sequence $\{Y_n\}$ does not converge. Various theorems imply that its behavior is then decidely erratic.

a. If the X_i are normal with mean zero and variance one, what is the distribution of Y_n? Use the result in Problem 6.4.3, to generate 10 sequences of 1,000 $N(0, 1)$ random variables. For each, compute and plot the corresponding sequence $\{Y_n\}$.

b. Next, concatenate the normal variates into five sequences $\{Z_j\}$ of length 2,000. Form the ratios $C_j = Z_{2j-1}/Z_{2j}$. What is the (common) distribution of the random variables C_j? As in part *a*, compute and plot the sequences $\{Y'_n\}$ for the C_j-sequences? Why do the $\{Y_n\}$ and the $\{Y'_n\}$ behave so totally differently?

Problem 6.4.12: Sums of Independent Random Variables: If U is a uniform variate, then $Y = U^{-1/\alpha}$ with $\alpha > 0$ has a Pareto distribution $F_P(\cdot)$. What is the probability density of Y? For which α does $F_P(\cdot)$ have a finite variance? A finite mean?

a. Define X by $X = Y$ with probability 1/2 and $X = -Y$ with probability 1/2. X has a symmetric density $\phi(u)$ that is positive only for $|u| > 1$. Its mean or variance is finite if and only if the corresponding moment of the Pareto distribution exists.

Generate and store 20,000 uniform variates. Use these to generate three sequences of variates $\{X_i\}$ with the same distribution as that of X. For the first sequence, choose α so that the variance is finite; for the second, choose α such that the mean exists but not the variance. For the third, let also the mean fail to exist. For each sequence, compute and plot the successive averages $n^{-1}(X_1 + \cdots + X_n)$ and discuss their behavior.

Repeat that experiment several times. Implement it with different triples of values of α that meet the required conditions. For example, choose the first α a little larger than the critical value for the variance to exist, and the second a little larger than that for the mean to exist. Then, keeping the first α, choose the second just a little smaller than the critical value for the variance to exist. What do your experimental results suggest about the speed of convergence in the law of large numbers?

Remark: We use the Pareto distribution only because it is easy to generate variates with that distribution. Also, the existence of its moments depends on α only. The perceived effect of the failure of the mean or variance to exist is quite typical. It can be similarly demonstrated also for other distributions.

Problem 6.4.13: Sums of Independent Random Variables: For the symmetric random variables X constructed in Problem 6.4.12, choose α such that the variance σ^2 is finite. Since their mean is zero, the *central limit theorem* guarantees that the normalized sums $S_n^\circ = [\sigma n^{1/2}]^{-1}(X_1 + \cdots + X_n)$ have approximately a $N(0, 1)$ distribution as $n \to \infty$. Generate 1,000 samples, each of $n = 100$ variates X_j. For each sample, calculate the corresponding S_n° and plot a histogram of the 1,000 values so obtained. Repeat this five times for three choices of the parameter α. Based on your plots, do you consider the normal approximation adequate for $n = 100$? To remove subjectivity from your decision, apply a χ^2 goodness of fit test for each of the 15 replications. What are your conclusions and how does the value of α enter into the picture?

AVERAGE PROBLEMS

Problem 6.4.14: A Parlor Game: In a group of m persons, one person is designated banker. The banker deals five cards from a standard deck to each player. Next five cards from the remaining stock are turned over. For the ith card, each person holding a card with the same face value must pay i tokens to the pot. Each person starts with a large number of tokens so that all players can make all required payments. Thereupon, five more cards are turned over, but now the banker must pay i tokens to each player holding a card with the same face value as the ith card. When these payouts cannot be covered from the money in the pot, the role of banker passes to the next person. Otherwise the current banker gets what is left in the pot and continues holding the bank.

Before each game, the deck is thoroughly shuffled. We assume that all drawings are random. Since $5(m + 2)$ cards are used, the game can be played by up to eight persons. It can be shown that each game is fair in that the expected number of tokens gained by each player in each game is 0. What makes the game interesting is the large random variability in the returns, particularly those of the banker.

Do a simulation study of this game for $m = 5$. For samples of 1,000 games each, study such features as the number of games for which the same person holds the bank and the probability density of the returns to the various players. State qualitative conclusions from your experiment. Next, repeat the experiment for $m = 4$ and $m = 6$. Discuss how the game appears to depend on the number of players.

Problem 6.4.15: For the experiment in Example 6.1.1, apply the Kolmogorov - Smirnov two-sample test for the null-hypothesis that the distribution is symmetric. Apply the test for the sample sizes 10,000, 100,000, and 1,000,000. Clearly, these samples are sufficiently large that the asymptotic test may be applied. Note that you need to store the data η_i in two arrays to be sorted. Strive for efficiency in computation and storage.

Problem 6.4.16: For the experiment in Example 6.1.1, it is interesting to see how the nearly symmetric density comes about. We do that by varying the distribution of the elements U_{ij} of the 2×2 matrix. As an extreme case, let each element of U be 0 or 1 with probability 1/2. As efficiently as possible (exploit symmetries), verify that for the 16 possible matrices, η takes the values 0, 1, $(1 + \sqrt{5})/2$, and 2, with probabilities 3/16, 10/16, 2/16, and 1/16, respectively.

Next, choose the elements at random from the values i/k, $0 \le i \le k$, where $k = 10, 20, 50, 100$. For each k, generate 10,000 matrices and plot a histogram of the values of η. Compare the empirical densities for these k to the empirical density obtained when the U_{ij} are uniformly distributed on (0, 1).

Hint: To obtain plots for easy comparisons and to avoid unnecessary storage, divide the interval (0, 2) into 100 equal segments and have an array of counters for each segment. Find the interval to which each computed η_i belongs and increment the corresponding counter by 1. Then, divide all counts by 10,000 and plot the resulting frequencies.

Problem 6.4.17: Do the same experiment as in Example 6.1.1, but sample the U_{ij} from the density

$$\phi(u) = (\alpha + 1) \mid 1 - 2u \mid^{\alpha}, \quad \text{for } 0 < u < 1,$$

where $\alpha > -1$. A generic element of the matrix is generated by

$$\frac{1}{2} [1 - |1 - 2U^{\circ}|^{\frac{1}{\alpha+1}}],$$

where U° is a uniform random variate. Try $\alpha = -0.75$ and also a few large positive values of α. Explain what happens in the latter cases.

Also use the density $\phi(u) = (2\alpha)^{-1}$, for $0 < u < \alpha$, or $1 - \alpha < u < 1$, and 0 elsewhere. Set α successively equal to 0.5, 0.45, 0.35, 0.25, 0.15, and 0.05. What happens to the nearly symmetric density which is still found for $\alpha = 0.5$ as we move probability toward the endpoints?

Problem 6.4.18: The elements U_{11}, U_{12}, U_{21}, U_{22} of a 2×2 matrix U are independent and uniformly distributed on [0, 1]. Show that both eigenvalues of U are real and that Z_1, the larger of the two, lies in (0, 2). Show that the smaller eigenvalue Z_2 is positive with probability 1/2. Generate 10,000 such random matrices U and examine the joint empirical distribution of their eigenvalues. Plot a scatter diagram and histograms of the observed Z_1 and Z_2. As noted in Example 6.1.1, the density of Z_1 is nearly symmetric on [0, 2]. Identify other empirical properties of the eigenvalues and confirm or dismiss them by additional experiments, using different sample sizes and possibly different random number generators.

Problem 6.4.19: Do the experiment in Example 6.1.1 for 10,000 matrices of order m. For each m, plot a frequency chart on the interval (0, 1) of the values of $y_i = m^{-1}\eta_i$. As in Problem 6.4.16, divide (0, 1) into 100 equal classes and determine the interval to which y_i belongs. Notice that this requires only modest accuracy in the computation of η by Elsner's algorithm. The iteration will therefore converge rapidly. To avoid long processing times, set the stopping criterion accordingly. Do the experiment for

$m = 5$, 10, 15, and 20, and compare the successive empirical frequency charts. What does your experiment suggest happens as m becomes large?

Remark: It is known that with probability 1, $m^{-1}\eta \to 1/2$, as $m \to \infty$, but the proof relies on advanced methods. J. Silverstein recently proved that the distribution of η rapidly converges to a normal limit law.

Problem 6.4.20: For the problems dealing with the near-symmetry of the distribution of η where we group the data, we can no longer apply the Kolmogorov - Smirnov test. How would you adapt the classical χ^2-test to test for symmetry in these cases? Implement the procedure for the experiment in Example 6.1.1.

Problem 6.4.21: Billiards Revisited: If U is uniform on (0, 1) and $0 < p < 1$, show that

$$N = \left[\frac{\log U}{\log (1 - p)}\right],$$

has the geometric density $p_k = p^k(1 - p)$, on $k \geq 0$.

Now consider a random walk on the pairs (i, j) of nonnegative integers, which proceeds as follows: The walk starts at (0, 0). It *alternatingly* moves horizontally to the right or vertically upwards for geometrically distributed numbers of steps. Horizontal motions have a geometric density $\{p_1^k(1 - p_1)\}$ on $k \geq 0$. The corresponding density for the vertical motions has parameter p_2. With probability 1/2, the first motion is either horizontal or vertical. Assume that $0 < p_2 < p_1 < 1$. For a given positive integer n_1, let M be the point at which the walk reaches (or crosses) the line $i = n_1$.

For $n_1 = 100$, simulate the behavior of that random walk. Display its path on a computer screen. Now perform 200 replications of the experiment, each time selecting the direction of the first motion independently with probability 1/2. Record the values M_r for $1 \leq r \leq 200$ and let M^* be their median. This is a simulation of the game of Billiards discussed in Example 2.2.1. The median M^* should be a reasonable estimate of the quantity n_2 discussed there.

If you have written a code to find n_2 as in Problem 2.3.30, compare M^* to the value of n_2 obtained by the exact algorithm. Do so for several choices of p_1 and p_2. What do you think about this alternative method to set the handicap for the game of Billiards?

Problem 6.4.22: Consider the variant of Billiards in Problem 2.3.34. As in Problem 6.4.21, simulate and visualize the path traced by the scores during the game. Setting $p_2 = 0.40$, and $n_1 = n_2 = 25$, first simulate 1,000 replications of the game with $p_1 = 0.40$. Obtain a (rough) estimate of the probability the Player I wins. As the advantage given by the extra turn is substantial, it should be significantly smaller than 1/2. Now successively increase p_1 by steps of 0.05, each time repeating the simulation and estimating the probability that I wins. If you earlier solved Problem 2.3.34, compare its numerical results to those of the simulation.

Remark: The simulation is obviously not intended as an alternative to the exact algorithmic solution. The interest is in visualizing the path behavior and in appreciating the influence of the value of p_1.

Problem 6.4.23: Uniform Spacings: To divide the interval (0, 1) randomly in $m + 1$ parts we can generate m independent uniform variates and rank them as $Z_1 < \cdots < Z_m$ from smallest to largest. The intervals $(0, Z_1]$, $(Z_1, Z_2]$, $\cdots$,$(Z_{m-1}, Z_m]$, $(Z_m, 1)$ form a uniform partition of (0, 1). For a given m, the lengths Y_j, $1 \le j \le m + 1$ of these intervals are called the *uniform spacings*.

As an alternative, we can generate $m + 1$ independent exponential variates with common parameter λ. Their successive partial sums are $S_1, \cdots,$ S_{m+1}. For $1 \le j \le m$, the random variables Z_j are now defined by $Z_j = S_j(S_{m+1})^{-1}$. Show that both ways of generating the Z_j are equivalent. For $m = 19$, generate 10,000 sets $\{Z_1, \cdots, Z_m\}$ by each method. Record the processing time for both tasks. Which method is most efficient on your computer?

References: An introduction to the properties of spacings is found in Karlin, S. and Taylor, H. M., "*A First Course in Stochastic Processes,*" 2d. ed. New York: Academic Press, 1975. An extensive survey is Pyke, R.,"Spacings," *Journal of the Royal Statistical Society,* Series B, 27, 395 - 449, 1965.

Problem 6.4.24: Uniform Spacings: Use the second procedure described in Problem 6.4.23. Apply the Kolmogorov - Smirnov test for uniformity with $\alpha = 0.01$ to each of the 10,000 19-tuples $\{Z_j\}$ generated in Problem 6.4.23. Report the fraction of samples for which the null-hypothesis of uniformity is accepted.

a. Using these same variates, form a histogram of the density of the random variable

$$H(m) = \frac{H(Y_1, \cdots, Y_{m+1})}{\log(m+1)},$$

where $H(Y_1, \cdots, Y_{m+1}) = -\sum_{j=1}^{m+1} Y_j \log Y_j$, is the *entropy* of the probability density $Y_1, \cdots, Y_{m+1}$. It is largest when $Y_j = (m+1)^{-1}$, for $1 \le j \le m+1$.

b. Explain why all Y_j have the same marginal density $\phi(u) = m(1-u)^{m-1}$, for $0 \le u \le 1$.

c. Show that the expected value

$$E[-Y_1 \log Y_1] = -\int_0^1 u \log u \; m(1-u)^{m-1} \, du = \frac{1}{m+1} \sum_{i=2}^{m+1} \frac{1}{i}.$$

and deduce that

$$EH(m) = 1 - \frac{1 - \gamma + o(1)}{\log(m+1)},$$

where γ is the Euler-Mascheroni constant and o(1) denotes a quantity that tends to 0 as $m \to \infty$. For selected other values of m, do an empirical study of the distribution of $H(m)$.

Problem 6.4.25: Uniform Spacings: For the uniform spacings in Problem 6.4.23, consider the random variables

$$L_1 = \sum_{j=1}^{m+1} \left| Y_j - \frac{1}{m+1} \right|, \quad \text{and} \quad L_2 = \sum_{j=1}^{m+1} \left[Y_j - \frac{1}{m+1} \right]^2.$$

a. Show that $0 \le L_1 \le 2$ and $0 \le L_2 \le 2$. Derive formulas for the means of L_1 and L_2.

b. For selected values of m, perform an experimental study of the joint and the marginal distributions of L_1 and L_2. Does your experiment suggest that both random variables have asymptotically normal distributions?

Reference: For a detailed study of $L_1/2$ see B. Sherman, "A random variable related to the spacings of sample values,'' *Annals of Mathematical Statistics*, 21, 339 - 361, 1950.

Problem 6.4.26: Non-Uniform Spacings: Generate $m + 1$ independent variates with common continuous distribution $F(\cdot)$. Denote their successive partial sums by $S_1, \cdots, S_{m+1}$. For $1 \le j \le m$, consider the random variables $Z_j = S_j (S_{m+1})^{-1}$. Do the same tasks as in Problem 6.4.24 for the Z_j so generated. Report your empirical results for $m = 19$. First choose the $m + 1$ variates using the uniform distribution on (0, 1). Next, use two hyperexponential distributions with different coefficients of variation. Are the empirical acceptance probabilities for the K–S test and the empirical distributions of $H(m)$ very different from those found in Problem 6.4.24? Explain your results.

Problem 6.4.27: More Spacings: For a given $m > 1$, we partition the interval [0, 1] into $m + 1$ intervals as follows: A first point is chosen at random. After k, $1 \le k < m$, points have been picked, the $(k + 1)$st point is chosen at random in the *longest* of the $k + 1$ intervals determined so far. The case $m = 2$ is the subject of Problem 6.4.5. For larger values of m, the joint distribution of the resulting intervals is intractable. This problem deals with an experiment to explore properties of this random partition.

a. Develop and program an *efficient* procedure to generate the locations of the m points.

b. By $Y_1, \cdots, Y_{m+1}$, we denote the lengths of the intervals in the partition and by $Y_1^* \le \cdots \le Y_{m+1}^*$, the ordered values of these random variables. The sums S_j and S_j^* are defined by $S_j = \sum_{r=1}^{j} Y_r$, and $S_j^* = \sum_{r=1}^{j} Y_r^*$ for $1 \le j \le m$. $S_0 = S_0^* = 0$. Generate informative plots of the pairs $\{j/m, S_j\}$ and $\{j/m, S_j^*\}$, for $0 \le j \le m$. What do these tell you?

c. First for $m = 9$ and next for $m = 19$, generate 1,000 replications of the experiment. For each sample, evaluate the Kolmogorov - Smirnov statistic

$$D_m = \max_j \{ | S_j - j/m | \}.$$

In the Kolmogorov - Smirnov test with $\alpha = 0.05$ of the null-hypothesis H_0 that the m points *appear* to be uniformly and independently placed, we reject H_0 if $D_m > 1.36/\sqrt{m}$. Had the partition been generated as specified by the null-hypothesis, approximately 5% of the samples would lead to rejection of H_0. What fraction of the 1,000 samples $\{Y_j\}$ are rejected?

d. Form the sample means

$$\bar{S}_j = \frac{1}{1{,}000} \sum_{\nu=1}^{1{,}000} S_j(\nu), \quad \text{and} \quad \bar{S}_j^* = \frac{1}{1{,}000} \sum_{\nu=1}^{1{,}000} S_j^*(\nu),$$

and plot the pairs $\{j/m, \bar{S}_j\}$ and $\{j/m, \bar{S}_j^*\}$, for $0 \le j \le m$. Do these graphs suggest any qualitative conclusions? Which of the two plots do you find most suggestive?

Remark: One would expect the partition to have fewer long intervals than random placement. Do you experimental results show that? Replications of the experiment and also trials with different values of m are encouraged. As $m \to \infty$, the empirical distribution of the m points rapidly converges to the uniform distribution. This was conjectured by S. Kakutani and subsequently proved by several authors. This partition of the unit interval has an extensive and advanced literature.

Problem 6.4.28: The elements P_{11} and P_{22} of a 2×2 stochastic matrix P are chosen independently and uniformly in (0, 1). Show that the steady-state probability π_1 has the density

$$\begin{aligned} g(u) &= \frac{1}{2(1-u)^2}, && \text{for } 0 \le u \le \frac{1}{2}, \\ &= \frac{1}{2u^2}, && \text{for } \frac{1}{2} \le u \le 1, \end{aligned}$$

and plot the density $g(\cdot)$. Next, fill a 3×3 array with independent uniform random variates. Normalize each row to obtain a 3×3 stochastic matrix P. Write an efficient subroutine to compute the steady-state probabilities of P. Repeat that 5,000 times and for each replication, store the computed values of π_1 and π_2. Construct histograms of the values of π_1 and π_2 after grouping each set of 5,000 values in classes of width 0.01. Also generate a three-dimensional plot of the frequencies of the pairs (π_1, π_2). Discuss the qualitative features of the outcome of that experiment. Are any of these to be expected on theoretical grounds?

Problem 6.4.29: The Growth of a Random Walk: The pairs $\{(X_n, Y_n)\}$, $n \ge 0$, are the coordinates of the points visited by a symmetric random walk on the lattice points in the plane. The walk starts at (0, 0). Consider the times τ_k when the distance from (0, 0) to (X_n, Y_n) *increases*.

Perform 100 replications of a simulation of the first 10,000 steps of the random walk. For each replication, record the times $\{\tau_k\}$ and the corresponding distances $\{\rho(\tau_k)\}$ to the origin. The numbers of times and distances recorded will vary from one simulation run to another. Study the empirical frequencies of the numbers of growth points in successive blocks of 1,000 steps of the walk. What do these suggest about the frequency of the occurrence of growth points over time? Analyze the resulting data to make inferences about the growth of the sequences $\{\tau_k\}$ and $\{\rho(\tau_k)\}$. Specifically, compute the regression lines of the sets of pairs $\{(k, \log \tau_k)\}$ and $\{(k, \log \rho(\tau_k))\}$. What do these suggest about the growth of these sequences?

Problem 6.4.30: The Growth of a Random Walk: Do the same study as in Problem 6.4.29 for the random walk in which, starting at (0, 0), the walk randomly moves to one of the points (1, 0), (–1, 0), (0, 1), or (0, –1). After that first step, the walk randomly moves one step to the right or to the left *perpendicular* to its previous step. Does your experiment suggest a different growth in this random walk from that in Problem 6.4.29?

Problem 6.4.31: Generate a random stochastic matrix P by filling an $m \times m$ array with independent uniform variates and normalizing each row to sum to 1. Denote the column sums by $Y_1, \cdots, Y_m$.

a. Empirically examine the distribution of the sample standard deviation

$$\sigma = \left[\frac{1}{m-1} \sum_{j=1}^{m} [Y_j - 1]^2 \right]^{1/2}$$

of the m column sums of such matrices. For $m = 2, 4, 10, 20, 50, 100$, plot informative histograms based on 10,000 samples. Any interesting observations? Note that the standard deviation of the column sums is zero if and only if the matrix is *doubly stochastic*. The distribution of σ therefore indicates of how close to doubly stochastic such randomly generated matrices are.

b. The quantities $Z_j = m^{-1} Y_j$, $1 \le j \le m$ form a probability vector. Its *entropy* is defined by $H(Z_1, \cdots, Z_m) = -\sum_{j=1}^{m} Z_j \log Z_j$. H is maximum when $Z_j = m^{-1}$ for $1 \le j \le m$. For each of the matrices generated in part *a*, also evaluate the ratio $V = H(Z_1, \cdots, Z_m)/\log m$. Empirically study the joint distribution of σ and V and the marginal distribution of V for the same orders m as in part *a*.

Problem 6.4.32: Do the same experiment as in Problem 6.4.31, but generate each row of the matrix P by taking m independent *exponential* variates and by normalizing the resulting matrix to be stochastic. The elements in each row are now the classical uniform spacings of $m - 1$ points placed independently in the interval (0, 1). Discuss the qualitative results suggested by your experiment.

Problem 6.4.33: In a further variant of Problems 6.4.31 and 6.4.32, randomly select the locations of n, $1 \le n \le m$, positive elements in each row of the matrix P. Fill these locations with n independent uniform or exponential variates. All other elements are zero. Normalize the resulting matrix to be stochastic. The case $n = m$ corresponds to the earlier problems. Carry out the same tasks and report your empirical findings for $m = 10$ and $5 \le n \le 10$.

The following five problems deal with discrete-time Markov processes on a continuous state space. The general theory of such processes is beyond the scope of this book. However, the models are sufficiently simple that they can be easily be investigated by simulation. A few special cases also are analytically tractable.

Problem 6.4.34: A Random Walk on the Unit Interval: Assume $\psi_1(\cdot)$ and $\psi_2(\cdot)$ are probability densities on [0, 1]. A random walk starts at a point $0 < X_0 < 1$ in that interval. At successive times n, it jumps to the left of its present position X_n with probability p, $0 < p < 1$, or to the right with probability $q = 1 - p$. If it jumps to the left, we generate a variate V_n from the density $\psi_1(\cdot)$ and place the point at $X_{n+1} = V_n X_n$. If it jumps to the right, V_n is generated from $\psi_2(\cdot)$ and we place the point at $X_{n+1} = X_n + (1 - X_n)V_n$.

a. Let $P(y \mid x)$ be the conditional density of the next position of the point after one jump from its current position x. Show that

$$P(y \mid x) = \frac{p}{x}\psi_1\left[\frac{y}{x}\right], \qquad \text{for } 0 < y \le x,$$

$$= \frac{q}{1-x}\psi_2\left[\frac{y-x}{1-x}\right], \qquad \text{for } 0 < y < x < 1.$$

The stationary probability density $\pi(\cdot)$ of the Markov chain satisfies the integral equation

$$\pi(x) = \int_0^1 P(x\,|u)\pi(u)\,du$$

$$= q\int_0^x \psi_2\left[\frac{x-u}{1-u}\right]\frac{1}{1-u}\pi(u)\,du + p\int_x^1 \psi_1\left[\frac{x}{u}\right]\frac{1}{u}\pi(u)\,du,$$

for $0 \le x \le 1$. For some special choices of the functions $\psi_1(\cdot)$ and $\psi_2(\cdot)$, that integral equation can be reduced to a differential equation which can be solved explicitly. The following is an example of the steps involved: If $\psi_1(\cdot)$ and $\psi_2(\cdot)$ are uniform densities on $[0, 1]$, then by differentiating in

$$\pi(x) = q\int_0^x \frac{\pi(u)}{1-u}\,du + p\int_x^1 \frac{\pi(u)}{u}\,du,$$

we obtain that

$$\pi'(x) = \left[\frac{q}{1-x} - \frac{p}{x}\right]\pi(x), \quad \text{for } 0 < x < 1.$$

That equation has the general solution $\pi(x) = Cx^{-p}(1-x)^{-q}$. The constant C needs to be chosen so that $\pi(\cdot)$ is a probability density and therefore $C = [B(q, p)]^{-1}$, where $B(q, p)$ is the beta function at q and p. In particular, for $p = 1/2$, $\pi(\cdot)$ is the *arc sine density*. Describe how a typical path of the random walk behaves and explain why a U-shaped density $\pi(\cdot)$ is to be expected.

b. Set $\psi_1(u) = \psi_2(1-u) = \alpha u^{\alpha-1}$, for $0 < u < 1$. The parameter α satisfies $\alpha \ge 1$. Derive the equation

$$\pi(x) = q\,\alpha(1-x)^{\alpha-1}\int_0^x (1-u)^{-\alpha}\pi(u)\,du + p\,\alpha x^{\alpha-1}\int_x^1 u^{-\alpha}\pi(u)\,du,$$

for $\pi(\cdot)$ on $(0, 1)$. Set $\alpha = 2$ and by twice differentiating in the resulting equation, show that

$$\pi''(x) = [2q(1-x)^{-1} - 2px^{-1}]\pi'(x), \quad \text{for } 0 < x < 1.$$

Conclude that

$$\pi'(x) = Cu^{-2p}(1-u)^{-2q},$$

where C is a constant. For $p = 1/2$, the expression for $\pi'(x)$ is integrable on $[0, 1]$ if and only if $C = 0$. Therefore, $\pi(x) \equiv 1$. For $0 < p < 1/2$,

$$\pi(x) = C\int_0^x u^{-2p}(1-u)^{-2q}\, du,$$

where C is chosen so that the integral from 0 to 1 of $\pi(x)$ is equal to 1. In that case, $\pi(x)$ is strictly increasing in $[0, 1]$. For $1/2 < p < 1$, we get the corresponding strictly decreasing function

$$\pi(x) = C\int_x^1 u^{-2p}(1-u)^{-2q}\, du.$$

c. Write a computer program to simulate the random walk for given densities $\psi_1(\cdot)$ and $\psi_2(\cdot)$ and values of p. Divide $[0, 1]$ into 200 intervals of equal length and keep track of the numbers of times N_j that each interval is visited in 200,000 transitions of the Markov process. Explain why the piecewise constant function with value $N_j/1{,}000$ on the jth interval is a plausible estimate of the stationary density $\pi(\cdot)$. Implement your code for $\psi_1(u) = 2u$, and $\psi_2(u) \equiv 1$ for $0 < u < 1$ and for four selected values of p. Plot the resulting estimated densities and discuss any qualitative conclusions.

d. For the case in part *b*, if α is an integer greater than 2, one can show that the density $\pi(\cdot)$ satisfies the differential equation

$$\pi^{(\alpha)}(x) = \left[\frac{q\alpha}{1-x} - \frac{p\alpha}{x}\right]\pi^{(\alpha-1)}(x),$$

which in some cases can be integrated explicitly. Use your simulation code as in part *c* to obtain estimated densities for $p = 1/4$ and for three selected values of α. What do these suggest about the behavior of the walk?

Problem 6.4.35: A Random Walk on the Unit Interval: For the random walk in Problem 6.4.34, set $\psi_1(u) = \psi_2(1-u)$ and define $\psi_1(u)$ by

$$\begin{aligned} \psi_1(u) &= 1, \quad \text{for } 0 \le u \le 0.1, \\ &= 9, \quad \text{for } 0.9 \le u \le 1, \end{aligned}$$

and zero elsewhere. Before you do simulations, describe verbally how the random walk will behave for $p = 1/2$ and make a thoughtful guess about how its stationary density will look. Next perform a simulation experiment to obtain an estimate of its stationary density as in Problem 6.4.34. How well do your experimental results agree with your intuition about the behavior of that process? The stationary density for this case is not analytically tractable.

Problem 6.4.36: A Random Walk on the Unit Interval: Consider the random walk in Problem 6.4.34, with $\psi_1(u) = 2u$ and $\psi_2(u) = 1$. Subdivide [0, 1] into 200 intervals of equal length. By simulating the random walk over 200,000 transitions, obtain estimates of the stationary densities $\pi(\cdot)$ for $p = 0.5$ and $p = 0.7$. You should find that their graphs should suggest that for some intermediate value p the density $\pi(\cdot)$ is close to symmetric on (0, 1). For this model, no explicit formula for $\pi(\cdot)$ is known. How would you construct an efficient search by simulation for a value p^* of p for which $\pi(\cdot)$ is nearly symmetric? To do this, you will need an operational definition of "close to symmetric." There are various ways if stating such a definition. See the discussion of Example 6.1.1.

Remember that in the search the results of each simulation run are themselves random. You are searching for the minimum or maximum of a function of which only noisy values are available. How would you approach this problem? Implement various methods and discuss the values for p^* that they produce. If that question had arisen in an important application (it has not!), decide which value of p^* you would declare. For that p^*, do an independent simulation over 1,000,000 transitions. Then, print and plot the 200 empirical cell frequencies for that case.

Problem 6.4.37: A Random Walk on the Unit Interval: Consider the case treated in part b of Problem 6.4.34. Recall that when $p = 1/2$ and $\psi_1(u) = \psi_2(1-u) = 2u$, the stationary density $\pi(\cdot)$ is uniform on (0, 1). Now set $p = 1/4$. The stationary density is now strictly increasing.

a. With the same density $\psi_1(u)$ as before, we would like to use a density

$\psi_2(u)$ of the form $\alpha(1-u)^{\alpha-1}$, with $\alpha > 2$, to counteract the drift of the steady-state probabilities to the right. Specifically, we would like to choose α (not necessarily integer) such that $\pi(\cdot)$ is close to symmetric. Use successive simulation experiments to search for that α. Count the fractions of time spent in each of 200 intervals of equal length as in Problem 6.4.36 and use a χ^2-criterion to measure how close to symmetric these are. Discuss the behavior of your search as well as any qualitative conclusions of the proposed experiment.

b. With $p = 1/2$ and with $\psi_1(u) = \psi_2(1-u) = 0.1(2u) + 0.9(8u^7)$, determine the estimated stationary density on 200 intervals of equal length and perform a χ^2-test for symmetric. Visualize a set of path functions of the random walk. Is the stationary density close to symmetric?

Problem 6.4.38: For $x \geq 0$, let $f(x)$ be a positive differentiable function whose integral

$$F(x) = \int_0^x f(u)\, du,$$

is finite for every finite x. In a discrete time random walk on the positive reals, a point jumps from a current position u to a new position v in $[0, u+1]$ according to the conditional density $P(v\,|u) = f(v)/F(u+1)$. If a stationary probability density $\pi(\cdot)$ exists for that Markov chain, it satisfies

$$\pi(x) = \int_0^\infty P(x\,|y)\pi(y)\, dy = f(x) \int_{L(x)}^\infty \frac{\pi(y)}{F(y+1)}\, dy,$$

for all $x \geq 0$. The lower integration limit is $L(x) = \max(0, x-1)$.

a. Show that a solution $\pi(\cdot)$ is differentiable and satisfies

$$\pi'(x) = \frac{f'(x)}{f(x)}\pi(x) - \frac{f(x)}{F(x)}\pi(x-1),$$

for $x \geq 1$ and that $\pi(x) = Cf(x)$ for $0 \leq x \leq 1$. By examining special cases such as $f(x) \equiv 1$ or $f(x) = \exp(-ax)$, it will be clear that the explicit form of $\pi(\cdot)$ is intractable. For $\pi(\cdot)$ to be a valid steady-state probability density, we should further verify the solution to the differential equation is positive and integrable, so that the normalizing constant C can be found. General conditions for the existence of such a solution are not known. The remainder of this problem deals with an experimental exploration of this model.

b. Define an array of counters $N(k)$, $1 \le k \le K$, and increase the counter $N(k)$ by 1 whenever the random walk visits the interval $[(k-1)h, kh)$, where h is a step size of your choice. Also keep track of the number of times that the position of the walk exceeds Kh. Write a subroutine to simulate 100,000 transitions of the random walk. From the contents of the counters, obtain and plot an informative empirical density of the average location of the walk. Use some initial runs with smaller sample size to determine useful values of h and K. Start all random walks at $x_0 = 5$.

c. For $f(x) \equiv 1$, $f(x) = \exp(-ax)$, and $f(x) = x^a$, with several choices of the exponent a, replicate your experiment a sufficient number of times to be able to make qualitative statements about the perceived recurrence or transience of the random walk.

Remark: This experiment can only suggest, not prove, that a particular version of this Markov process is recurrent. If several replications with different seeds yield substantially the same empirical density and if that density concentrates mostly on the lower values of the state space, that is definitely suggestive of recurrent behavior.

The following seven problems deal with basics of the calculation of *a posteriori* densities in Bayesian methodology. Some problems are adaptations of exercises in the important reference

Berger, J. O. *"Statistical Decision Theory and Bayesian Analysis,"* Second Edition, New York: Springer-Verlag, 1985.

In Bayesian analysis, a statistical model has an unknown parameter θ belonging to a set Θ. Prior information about θ is expressed by the (probability) density $\pi(\theta)$. A set of data denoted by the generic symbol x is obtained. For the parameter value θ, the probability density of the data x is $f(x \mid \theta)$. The joint density of θ and x is then $\pi(\theta) f(x \mid \theta) = h(x, \theta)$. The marginal density $m(x)$ of the data is found by integrating $h(x, \theta)$ over Θ. The posterior density $\pi(\theta \mid x)$ is then given by

$$\pi(\theta \mid x) = \frac{\pi(\theta) f(x \mid \theta)}{m(x)}.$$

The density $\pi(\theta|x)$ describes how, by taking the data x into account, the prior information about θ is updated.

Problem 6.4.39: The successive independent observations $X_1, \cdots, X_n$ have a normal density with mean θ and known variance σ^2. The prior density is normal $N(\mu, \tau^2)$.

a. Show that for a (single) observation x, $\pi(\theta|x)$ is normal with mean

$$\mu(x) = \frac{\tau^2}{\sigma^2+\tau^2}x + \frac{\sigma^2}{\sigma^2+\tau^2}\mu.$$

b. What is the posterior density for n observed values $x_1, \cdots, x_n$? Recall that $\bar{X}(n)$ is a sufficient statistic for θ.

c. Set $\sigma^2 = 1$, $\mu = 2$, and $\tau^2 = 4$. Generate a value of θ by using the prior density. Next, generate 25 observations $x_1, \cdots, x_{25}$ from the appropriate normal density and compute the means and standard deviations of the posterior densities $\pi(\theta|x_1, \cdots, x_j)$ for $1 \le j \le 25$. Plot these densities and explain what happens as more and more data are obtained.

Remark: This is one of the simplest cases of a posteriori analysis. Because the sample mean $\bar{X}(n)$ is sufficient statistic, the updating is particularly easy. In addition, the densities $\pi(\theta|x_1, \cdots, x_j)$ all remain normal. When the density $\pi(\theta|x)$ belongs to the same family as $\pi(\theta)$ then we say that $\pi(\theta)$ is a *conjugate* prior for the model $f(x|\theta)$. Further examples of conjugate priors are given in the next problems.

Problem 6.4.40: For the model in Problem 6.4.39, let $\sigma^2 = 1$. With n observations and letting x be the observed $\bar{X}(n)$, $f(x|\theta)$ is normal with mean θ and variance $1/n$. Some statisticians argue for specifying the variance τ^2 of the prior density as $1/n$. That amounts to making the sample mean $\bar{X}(n)$ as convincing an estimate of θ as the prior mean μ. For $n = 25$, $\tau^2 = 1/25$ and repeat the same numerical study as in part c of Problem 6.4.39. Any conclusions?

Problem 6.4.41: Let θ have the Pareto density

$$\pi(\theta) = \left[\frac{\alpha}{\theta_0}\right]\left[\frac{\theta_0}{\theta}\right]^{\alpha+1}, \qquad \text{for } \theta > \theta_0 > 0, \quad \text{with } \alpha > 0.$$

a. For given parameters θ_0 and α, how can we efficiently generate a variate with the density $\pi(\theta)$? Write a subroutine to do so.

b. Suppose that $x_1, \cdots, x_n$ are independently and uniformly chosen on $(0, \theta)$. What is the function $f(x_1, \cdots, x_n | \theta)$? Show that with the prior density $\pi(\theta)$ for θ, its posterior density $\pi(\theta | x_1, \cdots, x_n)$ is again Pareto with parameters $\theta^\circ = \max[\theta_0, x_1, \cdots, x_n]$, and $\alpha^\circ = \alpha + n$. Note that this shows that the Pareto density is the *conjugate* prior for the uniform density on $(0, \theta)$. Before doing any numerical examples, describe what happens to the graph of the posterior density as more and more observations are taken.

c. Set $\theta_0 = 0.1$, $\alpha = 0.8$, and generate a value of θ. Next, successively choose 20 independent random numbers on $(0, \theta)$. Compute and plot the posterior densities $\pi(\theta | x_1, \cdots, x_j)$ for successive $j \leq 20$.

d. The classical minimum variance unbiased estimator of θ based on n observations is

$$\hat{\theta}(n) = \frac{n+1}{n} \max[x_1, \cdots, x_n].$$

That estimator does not take any prior knowledge about θ into account. In the Bayesian setting, we can use quantities such as the mean, the median, or the mode of the posterior density as estimators for θ. Perform the experiment in part *c* 100 times but do not plot the posterior densities. For each replication and for $n = 20$ compute $\hat{\theta}(n)$, and the mean, median, and mode of the posterior density. Compare those estimates to the value of θ which is known for each replication. Based on your experimental results, which of the Bayesian estimates do you prefer? Does this experiment convince you that it is worthwhile to take the prior information on θ into account?

Remark: In Bayesian analysis, the appropriate estimator is chosen on the basis of a *loss function* chosen for the application at hand. The estimators to be examined here correspond to specific loss functions. A discussion of these would lead us too far afield.

Problem 6.4.42: For the model in Problem 6.4.41, suppose that the prior density is

$$\pi(\theta) = \alpha e^{-\alpha(\theta - \theta_0)}, \quad \text{for } \theta > \theta_0, \quad \text{with } \alpha > 0.$$

a. Show that the posterior density $\pi(\theta \mid x_1, \cdots, x_n)$ is given by

$$\frac{e^{-\alpha\theta}\theta^{-n}}{\int_{\theta^\circ}^{\infty} e^{-\alpha u} u^{-n}\, du}, \quad \text{for } \theta > \theta^\circ = \max\,[\theta_0, x_1, \cdots, x_n].$$

Using a careful numerical integration, write a subroutine to evaluate the posterior density and the Bayesian estimators proposed in Problem 6.4.41.

b. Carry out the same numerical experiment as in Problem 6.4.41. What is the qualitative difference between the two proposed prior densities? Does this affect your conclusions on the merits of Bayesian estimators?

c. Suppose that the prior information leading to the form of $\pi(\theta)$ with $\theta_0 = 0.1$ and $\alpha = 0.8$ is quite wrong and that in reality $\theta = 4$. Generate successive sets of 10 variates uniformly distributed in (0, 4). For $n = 10k$ and for successive values of k, compute the posterior densities and their means, modes, and medians. Compare these to the standard estimate $\hat{\theta}(n)$. Is the effect of wrong prior information evident from your numerical experiment? Discuss your numerical results in detail.

Remark: We see that the posterior densities are no longer shifted exponentials as is the prior. The advantage of using the conjugate prior is therefore lost. Even for this simple case, a numerical integration is required to compute the posterior densities.

Problem 6.4.43: The random variables X_1 and X_2 are independent and have the Cauchy density $f(u \mid \theta) = [\pi(1 + (\theta - u)^2]^{-1}$. A first statistician always estimates θ by $\hat{\theta} = (X_1 + X_2)/2$.

a. Show that the probability $P\{|\theta - \hat{\theta}| \leq 1\} = 1/2$.

b. A second statistician knows that θ has a uniform prior density on (−2, +2). For observations x_1 and x_2, she calculates the posterior density

$$\pi(\theta \mid x_1, x_2) = \frac{[1 + (\theta - x_1)^2]^{-1}[1 + (\theta - x_2)^2]^{-1}}{\int_{-2}^{2} [1 + (u - x_1)^2]^{-1}[1 + (u - x_2)^2]^{-1}\, du}.$$

Then, she numerically determines the mode $\hat{\theta}_1$ and the median $\hat{\theta}_2$ of $\pi(\theta \mid x_1, x_2)$ and uses these as estimates of θ. She is considered successful if the distance between her estimate and the true θ is at most 1. To assist that statistician in choosing between $\hat{\theta}_1$ and $\hat{\theta}_2$, generate 1,000 values of θ from the given prior. For each generate two observations X_1 and X_2 with the corresponding Cauchy density. Calculate the two estimates and determine the empirical frequencies of success for each of the two estimators. Based on your experimental results, which choice would you advise? Do you think she would be better off just using the estimator $\hat{\theta}$?

Hint: For part *a*, remember that $(X_1 + X_2)/2$ has the same Cauchy density as X_1 or X_2. To find the mode and median of the posterior density, just tabulate that density, find the mode of the computed values, and perform a simple numerical integration for the median. This problem does not call for high numerical accuracy.

Problem 6.4.44: Let x denote the number of successes in n independent Bernoulli trials with probability θ. The prior density $\pi(\theta)$ is the beta density with parameters α and β,

$$\pi(\theta) = \frac{1}{B(\alpha, \beta)}\theta^{\alpha-1}(1-\theta)^{\beta-1}, \quad \text{for } 0 < \theta < 1.$$

a. Show that the posterior density $\pi(\theta|x)$ is the beta density with parameters $\alpha^\circ = x + \alpha$, and $\beta^\circ = n - x + \beta$. (The beta density is therefore the conjugate prior for the estimation of the parameter θ of a binomial density).

b. Let $a(x)$ be an estimator of θ. If the loss because of an error in estimation is measured by $L[\theta, a(x)] = [\theta - a(x)]^2$, one can easily show that the estimator that minimizes the *a posteriori* expected loss

$$\int_0^1 L[\theta, a(x)]\pi(\theta|x)\, d\theta,$$

is the mean $(x + \alpha)/(n + \alpha + \beta)$ of the posterior density $\pi(\theta|x)$. That is the standard Bayes estimator

$$a(x) = \frac{x + \alpha}{n + \alpha + \beta} = \frac{x/n + \alpha/n}{1 + \alpha/n + \beta/n},$$

if θ has the prior density $\pi(\theta)$.

c. Set $\alpha = 1$ and $\beta = 9$. That corresponds to the prior 'belief' that θ is somewhere around 1/10. Generate a variate θ from that density. Do 1,000 sets of 25 Bernoulli trials with probability θ. For each, calculate the Bayes estimate $a(x)$. Plot histograms of these estimates and of the classical estimates x/n to see the effect of taking prior information into account.

Problem 6.4.45: As in Problem 6.4.44, consider the estimation of the parameter θ of the binomial density. There again are n observations and x, $0 \le x \le n$, successes are observed. However, the estimate must be one of the values $0, 0.1, \cdots, 0.9$, or 1. That is, based on the observations, we must declare that θ is 0, or 1, or one of the other nine values. Denote these 11 values in order by θ_j, $0 \le j \le 10$.

Subject only to their sum being 1, the prior probabilities $\pi(\theta_j)$ of these alternatives can be chosen arbitrarily. There is no need for restrictions on the loss function either. If the correct alternative is j and based on x successes we declare the alternative $a(x)$, our loss is $L[\theta_j, a(x)]$. It is convenient to think of these quantities as the elements of an 11×11 matrix. The posterior probabilities of the θ_j are given by

$$\pi(\theta_j \mid x) = \frac{\pi(\theta_j)\binom{n}{x}\theta_j^x(1-\theta_j)^{n-x}}{\sum_{k=0}^{10} \pi(\theta_k)\binom{n}{x}\theta_k^x(1-\theta_k)^{n-x}}, \quad \text{for } 0 \le j \le 10.$$

The *a posteriori* expected loss for the decision function $a(x)$ is

$$\rho[\boldsymbol{\pi}, a(x)] = \sum_{j=0}^{10} \pi(\theta_j \mid x) L[\theta_j, a(x)], \quad \text{for } 0 \le x \le n,$$

and for each x, the Bayes estimate $a(x)$ is the θ_r for which $\rho[\boldsymbol{\pi}, a(x)]$ is smallest.

a. Write subroutines to read the sample size n, the probabilities $\pi(\theta_j)$, and the loss matrix L. Write a further subroutine to compute the decision function $a(x)$ that minimizes the average *a posteriori* loss.

b. Suppose that the loss is 0 if the correct θ_j is identified, 1 if a neighboring θ_j is chosen, and 10 if any other incorrect choice is made. Set up the loss matrix and incorporate it into your program.

Generate a variate j from the discrete density $\{\pi(\theta_j)\}$. If $1 \le j \le 9$, generate the value of the probability θ *uniformly* in $[0.1j - 0.05, 0.1j + 0.05)$. If $j = 0$ or $j = 10$, draw θ uniformly from $[0, 0.05)$ or $[0.95, 1]$, respectively. Next, generate 1,000 sets of 25 Bernoulli trials all with that probability θ. For each set, find the Bayes estimate, compute your loss, and plot a histogram of the losses. Discuss the implications of your results.

Remarks: *a.* This problem is not artificial. In some situations, the estimates are limited to a specific set of values. For example, the estimate may have to be encoded on a chip and only three bits are available for its storage. The estimate must then be one of eight possible values. The particular valυes used in the formulating the problem are not important. The same Bayesian analysis applies when other values are specified. This problem should be compared to the classical Problem 6.4.44. Here there are no significant restrictions on the loss function or on the prior probabilities. The choice of the conjugate beta prior in Problem 6.4.44 is only for analytic tractability. The simple explicit formula for the Bayes estimator holds only for a special (squared error) loss function. Bayesian methods in statistics present a wealth of algorithmic problems. The statistics used in Bayesian analysis can be found explicitly only under special, often unrealistic assumptions.

b. In the problem, θ is chosen from a arbitrary, piecewise constant prior density. Some people may prefer another continuous density $\pi^*(\theta)$. That adds only a little extra work. The prior probabilities $\pi(\theta_j)$ are computed by integrating $\pi^*(\theta)$ over the intervals $[0, 0.05)$, $[0.05, 0.15)$, up to $[0.95, 1]$. For the numerical study in part b, θ is generated using $\pi^*(\theta)$.

Problem 6.4.46: Fictitious Play: This problem requires some acquaintance with the notion of a *zero-sum game* between two players, I and II. The payoff matrix of the game is the $m \times m$ matrix A whose rows correspond to the pure strategies for Player I. Its column indices are the pure strategies for Player II. A probability vector $\mathbf{p} = (p_1, \cdots, p_m)$ is called a *mixed strategy* for Player I. A probability vector $\mathbf{q} = (q_1, \cdots, q_n)$ is a mixed strategy for Player II. The return to Player I when the players respectively use the mixed strategies $\mathbf{p}$ and $\mathbf{q}$ is given by the expected value

$$A(\mathbf{p}, \mathbf{q}) = \sum_{i=1}^{m} \sum_{j=1}^{n} A_{ij} p_i q_j = \mathbf{p} A \mathbf{q}.$$

For the last equality, we think of $\mathbf{p}$ as a row vector and of $\mathbf{q}$ as a column vector. The *minimax* theorem of game theory states that there exists probability vectors $\mathbf{p}^*$ and $\mathbf{q}^*$ for which

$$\max_{\mathbf{p}} \min_{\mathbf{q}} A(\mathbf{p}, \mathbf{q}) = \min_{\mathbf{q}} \max_{\mathbf{p}} A(\mathbf{p}, \mathbf{q}) = A(\mathbf{p}^*, \mathbf{q}^*) = V. \tag{1}$$

V is called the *value* of the game, $\mathbf{p}^*$ and $\mathbf{q}^*$ are optimal strategies. If Player I uses a mixed strategy $\mathbf{p}$, a Bayes response to that strategy is for Player II to choose a pure strategy j for which $(\mathbf{p}A)_j$ is minimum. Similarly, a Bayes response of I to the mixed strategy $\mathbf{q}$ of II is to choose a pure strategy i for which $(A\mathbf{q})_i$ is maximum.

The method of *fictitious play* to solve a finite zero-sum two-person game proceeds as follows: Player I chooses a pure strategy i_1, to which II selects the Bayes response j_1. Next, player I selects the Bayes response i_2 to j_1. Thereupon II selects the Bayes response to the mixed strategy that puts weights 1/2 each on i_1 and i_2. I next uses the Bayes response i_3 against the mixed strategy that places weights 1/2 at j_1 and j_2.

The process continues in this manner. If respectively the pure strategies $i_1, \cdots, i_k$ and $j_1, \cdots, j_k$ have been used and the frequencies of the various strategies in these strings are $p_1(k), \cdots p_m(k)$ and $q_1(k), \cdots, q_n(k)$, then Player I will select i_{k+1} so as to maximize $(A\mathbf{q}(k))_i$, whereas II will choose j_{k+1} so as to minimize $(\mathbf{p}(k)A)_j$. If at any stage, several pure strategies are tied for the Bayes response, any of these may be chosen. A tie-breaking rule should be built into the algorithm to deal with this eventuality.

If $\underline{V}(k)$ and $\bar{V}(k)$ are defined by

$$\underline{V}(k) = \min_j (\mathbf{p}(k)A)_j, \quad \text{and} \quad \bar{V}(k) = \max_i (A\mathbf{q}(k))_i,$$

then for all $k \geq 1$, $\bar{V}(k) \geq V \geq \underline{V}(k)$, and $\bar{V}(k) \rightarrow V$, $\underline{V}(k) \rightarrow V$, as $k \rightarrow \infty$. Moreover, any limit point $\mathbf{p}^*$ of $\{\mathbf{p}(k)\}$ is an optimal strategy for Player I. Correspondingly, every limit point $\mathbf{q}^*$ of $\{\mathbf{q}(k)\}$ is optimal for II.

a. Write a subroutine to implement the method of fictitious play. After some trial runs, set the maximum allowable number of steps to 50,000. Stop when $\bar{V}(k) - \underline{V}(k) < 10^{-4}$. Visualize the behavior of your algorithm by plotting the sequences $\{\bar{V}(k)\}$ and $\{\underline{V}(k)\}$. Print the strategies $\mathbf{p}(k^\circ)$ and $\mathbf{q}(k^\circ)$ for the stopping index k°. Specifically, implement your algorithm for the game with payoff matrix

$$A = \begin{vmatrix} 4 & 0 & 2 & 1 \\ 0 & 4 & 1 & 2 \\ 1 & -1 & 3 & 0 \\ -1 & 2 & 0 & 3 \\ -2 & -2 & 2 & 2 \end{vmatrix} .$$

For that game, $V = 14/9$ and $\mathbf{p}^* = (1/9)(4, 4, 0, 0, 1)$ is the unique optimal strategy for Player I. A particular optimal strategy for II is $\mathbf{q}^* = (1/18)(1, 1, 8, 8)$, but Player II has other optimal strategies. Compare these results to those obtained by the method of fictitious play.

b. For each k, keep track of the indices k_1 and k_2, respectively, for which $\bar{V}(\nu)$, $1 \le \nu \le k$, is smallest and for which $\underline{V}(\nu)$, $1 \le \nu \le k$, is largest. Also save the strategies $\mathbf{p}(k_1)$ and $\mathbf{q}(k_2)$ and update these efficiently. Stop at the first index k' for which $\bar{V}(k_1) - \underline{V}(k_2) < 10^{-4}$. At the expense of a little extra computation, this version of the algorithm may stop sooner than that in part *a*. Implement that version of the algorithm for the examples as used for the original algorithm. Is this refinement worth the extra effort? Does it typically halt after significantly fewer steps?

Remark: At each step, the algorithm requires a sorting routine to determine the Bayes response for each player. To be useful for large matrices A, the code should therefore be written efficiently. Also try to store only essential information.

Reference: For a thorough discussion of zero-sum two-person games, see Karlin, Samuel, "*Mathematical Methods and Theory in Games, Programming and Economics,*" Reading, Mass: Addison-Wesley Publishing Company, 1959.

Problem 6.4.47: Symmetric Games: A zero-sum two-person game whose (square) payoff matrix satisfies $A_{ij} = -A_{ij}$ for all i and j is called a *symmetric* game. For such a game, $V = 0$ and Players I and II have the same optimal strategies. Efficiently modify the ficticious play algorithm in Problem 6.4.46 to compute an optimal strategy.

HARDER PROBLEMS

Problem 6.4.48: This experiment is aimed at obtaining estimates for the probability that an $m \times m$ matrix B of zeros and ones is the incidence matrix of an irreducible stochastic matrix. To that end, fill an $m \times m$ array with entries 0 or 1, according to m^2 independent Bernoulli trials with probability p for 1 and $1 - p$ for 0. Next, use the subroutine developed in Problem 4.4.6 to determine whether B is irreducible.

Choose k values of p in (0, 1), and repeat this experiment 1,000 times for each of the values of p. Record the corresponding empirical frequency of the matrices found to be irreducible. We suggest starting with the values 0.25, 0.50, 0.75 for p and adding $k - 3$ additional values based on the experimental results. Do this experiment for $m = 10$ and $m = 20$ and discuss any qualitative conclusions suggested by your results. Perform additional experiments to confirm your findings and report both results that confirm your earlier conclusions or cast doubt on them.

Problem 6.4.49: Problems 6.4.5, 6.4.6, and 6.4.27 suggest the construction of a statistical game. For a given integer m between 10 and 20, generate a number n, say, 100 partitions of the interval [0, 1] according to one of four procedures. The first procedure just places m points independently and randomly in [0, 1]. In the second, the next point is put randomly in the currently longest interval. For the third method, the next point is selected uniformly in one of the preceding intervals chosen at random.

The fourth procedure is a hybrid of the second and third. A first point is chosen at random in the unit interval. After $k \geq 2$ points have been chosen, an interval is chosen at random from among those whose length is at least $1/(k + 1)$. The $(k + 1)$st point is chosen at random in that interval.

For a given random number seed, each procedure is implemented by calling an appropriate subroutine. A friend or the computer chooses one of the four procedures. Given the n generated partitions, you win if you correctly identify the procedure that was used. You can use the criteria studied in the cited problems or you may propose others. The open-ended challenge of this problem is to find criteria under which the four procedures have very different signatures. When you are satisfied with your identification strategy for $n = 100$, make the game more challenging by reducing the value of n. Write a code for this game and enjoy playing it!

Remark: During 1993, this problem really fired the imagination of my students. Even for $n = 1$ and with m as small as 10, they came up with identification procedures with high success probabilities. An account of the various approaches we used is given in Neuts, M. F., Rauschenberg, D. E., and Li, J-M., "How did the cookie crumble? Identifying fragmentation procedures," Manuscript, 1994.

Problem 6.4.50: Consider the Markov chain with transition probability matrix $P(N)$ which we display for $N = 4$.

$$P(4) = \begin{vmatrix} a_0 & a_1 & a_2 & a_3 & a_4^\circ \\ a_0 & a_1 & a_2 & a_3 & a_4^\circ \\ 0 & a_0 & a_1 & a_2 & a_3^\circ \\ 0 & 0 & a_0 & a_1 & a_2^\circ \\ 0 & 0 & 0 & a_0 & a_1^\circ \end{vmatrix}$$

where the quantities a_k are the terms of a discrete probability density. The entries in the column with index N are defined by $a_k^\circ = \sum_{r=k}^{\infty} a_r = 1 - \sum_{r=0}^{k-1} a_r$, for $k \geq 1$. $P(N)$ is the transition probability matrix of the Markov chain embedded at departures in the $M/G/1$ queue, but in this problem, we allow the density $\{a_k\}$ to be general.

a. How would you simulate the behavior of that Markov chain by using the alias method with *only one alias table?*

b. For $N = 20$, simulate the behavior of the Markov chain with the initial state 0. Use four different probability densities $\{a_k\}$, respectively with means 0.5, 0.75, 1.0, and 1.25. For each density simulate 15 paths, each of 1,000 transitions. Display the paths of the Markov chain in an informative manner on a computer screen.

c. Compute the invariant probability vector π of $P(N)$. Repeat the task in part b, but now for each path choose the initial state as a variate from the steady-state density $\{\pi_i\}$. Is the difference in the behavior of the paths apparent from the graphs?

d. Does the behavior you see in the simulated paths correspond to your understanding of the behavior of a finite-capacity queue? If so, keep the means of each of the densities the same, but increase their variance, say by splitting the mass of some positive term a_i appropriately between a_0 and a_N. What new insights do you gain from that experiment?

Remark: The point of this problem is to write an efficient simulation code for an interesting Markov chain. The exploration of the effects of various densities $\{a_k\}$ is a worthwhile use of that code. It can serve to illustrate the behavior of a queue or storage model with various types of input.

Problem 6.4.51: The Borel - Tanner density

$$g_k = e^{-\lambda k}\frac{(\lambda k)^{k-1}}{k!}, \quad \text{for } k \geq 1,$$

is the probability density of the number of customers served during a busy period in an $M/D/1$ queue with arrival rate λ and service times of length one. When $\lambda < 1$, it is a proper density with mean $(1-\lambda)^{-1}$. Let $k_1, \cdots, k_n$ be n independent observations with common density $\{g_k\}$. Denote their sample mean by $\bar{k}$.

a. Show that the maximum likelihood estimator $\hat{\lambda}$ of λ based on the observations $k_1, \cdots, k_n$ is

$$\hat{\lambda} = \frac{\bar{k}-1}{\bar{k}}.$$

$\hat{\lambda}$ is also the first moment estimator found by setting $\bar{k} = (1-\lambda)^{-1}$.

b. Use the fact that the generating function $g(z)$ of $\{g_k\}$ satisfies

$$g(z) = z\exp\{-\lambda[1-g(z)]\},$$

to show that the variance σ^2 of the Borel - Tanner density is given by $\sigma^2 = \lambda(1-\lambda)^{-3}$. The second moment estimator λ° is defined as the unique solution in (0, 1) of

$$\frac{\lambda}{(1-\lambda)^3} = \frac{1}{n-1}\sum_{i=1}^{n}[k_i - \bar{k}]^2.$$

c. For $\lambda = 0.1\nu$, $1 \leq \nu \leq 9$, generate 1,000 samples of size 20 with the corresponding Borel - Tanner density. For each sample, compute the estimators $\hat{\lambda}$ and λ°. You now have two sets each of 1,000 estimates $\hat{\lambda}_j$ and λ_j°. Subtract the known value λ from each sample average of these 1,000 estimates. These give an idea of the size of the bias of each estimator. Similarly, calculate the sample standard deviation for each estimator. Plot the bias and the sample standard deviation as a function of λ. Which of the two estimators do you prefer? Give reasons for your choice.

Hint: Each sample is obtained by simulating 20 busy periods of the $M/D/1$ queue as in Problem 6.4.52.

Problem 6.4.52: Large Deviations: The random variables $X_1, \cdots, X_n$ are independent with a common $N(0, 1)$ distribution. Their sum is S_n. By the strong law of large numbers $n^{-1}S_n$ tends to 0 with probability 1. We say that a large deviation (to the right) occurs if for some positive constant γ, $\{S_n > \gamma n\}$ for a large value of n. In what follows, set $n = 10$, and $\gamma = 1.2$. The question of interest is whether the event $\{S_{10} > 12\}$, when it occurs, comes about because a few of the X_is are exceptionally large, or, because most of the summands X_i are larger than average.

To explore that question, do the following experiment: Generate successive samples of 10 independent normal variates. Retain only those samples for which $S_{10} > 12$ and stop when you have 20 such samples. What is the expected number of 10-tuples $X_1, \cdots, X_{10}$ needed to get 20 samples for which $S_{10} > 12$?

Denote the samples you have retained by $\{x_1(j), \cdots, x_{10}(j)\}$, $1 \le j \le 20$. Plot informative histograms of the 20 values $x_i(j)$ for $1 \le i \le 10$. Also compute the sample means and sample standard deviations for $1 \le i \le 10$. What does your experiment suggest is the answer to the question of interest?

Problem 6.4.53: Long Busy Periods: Consider the $M/D/1$ queue, the $M/G/1$ queue with constant service time. Every customer requires one unit of service time. The arrival rate λ is then also the traffic intensity of the queue. The probability density $\{g_k\}$ of the number of customers during a busy period starting with one customer is given by

$$g_k = e^{-\lambda k} \frac{(\lambda k)^{k-1}}{k!}, \quad \text{for } k \ge 1,$$

where $\{g_k\}$ is the Borel - Tanner density. For $\lambda \le 1$, it is a proper density. Its mean is finite if and only if $\lambda < 1$ and is then $(1 - \lambda)^{-1}$.

a. For $\lambda(r) = 0.9 + 0.01r$, $0 \le r \le 10$, determine the smallest value k_r of k

for which $\sum_{k=1}^{k_r} g_k \geq 0.99$.

b. Write an efficient simulation subroutine to generate the number of customers during a busy period for an $M/D/1$ queue with $\lambda \leq 1$.

c. For a given $\lambda(r)$ as defined in part *a*, we say that a busy period is long if it involves at least k_r customers. We are interested in what happens in the sequence of arrivals to cause a long busy period. In the $M/D/1$ queue, the number of arrivals during a service time has a Poisson density of mean λ. Is it more likely that long busy periods involve a small number of services during which unusually large numbers of arrivals occur or many services with a few more arrivals than expected?

To explore that question, perform the following experiment: For a given $\lambda(r)$, generate the numbers of arrivals ν_j during the successive services of a busy period. If the busy period involves fewer than k_r services, start over. If the busy period is long, plot the empirical frequencies of the values in the data string $\{\nu_j\}$, their sample mean, and sample standard deviation. Fit a Poisson density to the empirical frequencies. Repeat the experiment at least five times for each $\lambda(r)$ and draw qualitative inferences from your experimental results.

Hint: You need only a subroutine to generate variates from a Poisson density. Generate Poisson variates ν_j with parameter $\lambda(r)$ and denote their successive sums by S_j for $j \geq 1$. The number of services during a busy period is the first j for which $S_j - j + 1 = 0$.

Problem 6.4.54: Long Busy Periods: Consider a discrete distribution $H(\cdot)$ with masses p at a and $q = 1 - p$ at $b = q^{-1}(1 - pa)$, where $0 < a < 1$ and $0 < p < 1$. Its mean $\mu'_1 = 1$.

a. Verify that the n-fold convolution $H^{(n)}(\cdot)$ of $H(\cdot)$ is the discrete distribution with the binomial probability masses

$$p_k = \binom{n}{k} p^k q^{n-k},$$

at the (not necessarily distinct) points $c_k = ka + (n - k)b$, $0 \leq k \leq n$.

b. A formula of Takács states that the probability density $\{g_n\}$ of the number of customers served during a busy period starting with one customer in the $M/G/1$ queue with arrival rate λ and service time distribution $H(\cdot)$ is given by

$$g_n = \int_0^\infty e^{-\lambda u} \frac{(\lambda u)^{n-1}}{n!} \, dH^{(n)}(u), \quad \text{for } n \geq 1.$$

For the service time distribution in part *a*, verify that

$$g_n = \sum_{k=0}^{n} \binom{n}{k} p^k q^{n-k} e^{-\lambda[ka + (n-k)b]} \frac{[\lambda(ka + (n-k)b)]^{n-1}}{n!}, \quad \text{for } n \geq 1.$$

Develop an efficient subroutine to compute the probabilities g_n. Recall that for $\lambda \leq 1$, $\{g_n\}$ is a proper probability density and that with $\lambda < 1$, its mean equals $(1-\lambda)^{-1}$.

c. Do the same experimental study for this $M/G/1$ queue as described in Problem 6.4.53 for the $M/D/1$ queue. You now also have the option of varying the parameters a and p and there are several more interesting conditional empirical frequencies to be examined. Over a long period of time, the server will dispense services of lengths a and b approximately with frequencies p and q. Do long busy periods typically show a higher frequency of long services? What about the empirical frequencies of arrivals in long and short services during a long busy period?

Do your experiments for three well-chosen pairs a and p and for the values of λ specified in Problem 6.4.53. Compare the empirical frequencies of the number of arrivals during the services in a long busy period to the density

$$\phi_n = pe^{-a} \frac{a^n}{n!} + qe^{-b} \frac{b^n}{n!}, \quad \text{for } n \geq 0.$$

Should you find a close agreement, what does this suggest about the behavior of the queue during a long busy period?

Problem 6.4.55: The random variables X_i, $i \geq 1$, are independent with the common continuous probability distribution $F(x)$. N is the first index for which there is one and only one observation among $X_1, \cdots, X_{N-1}$ which is larger than X_N. The observation X_N is called a *near-record*.

a. Show that X_N has the distribution $F(x)$. Explain why it is sufficient to prove this for the uniform distribution on [0, 1].

b. Now verify this somewhat astonishing result empirically. Perform 1,000 independent experiments to generate values of N and X_N. Use the variates so obtained to test the X_N for uniformity, using the χ^2 and the Kolmogorov - Smirnov goodness of fit tests. If the observed fit is not very good, can you explain why this is so?

c. Can you derive the discrete density of N? Examine also the empirical conditional distributions of the X_N, given the various observed values for N.

Reference: A proof of the statement in *a* is found in *The American Mathematical Monthly,* Solution to Problem 6522, 95, 360 - 362, 1988.

Problem 6.4.56: For the closed queueing system with two servers in Problem 5.2.44, write a simulation code to visualize the behavior of the queue in front of Server I. First simulate the case where both service time distributions are exponential with the same mean. Next, run simulations with five different pairs of hyperexponential distributions with the same mean but different coefficients of variation. This experiment will be most informative if numerical results for Problem 5.2.44 are already available.

Problem 6.4.57: This problem deals with a special priority discipline for the $M/G/1$ queue operating under a gating procedure. The queue is initially empty. The first arriving customer starts service immediately. Any arrivals during that service are called the *offspring* of the first customer. At the end of the first service, the new arrivals are allowed inside a gate. If that customer has $n_1 \geq 1$ offspring we generate their service times. While these services are dispensed, new arrivals line up outside the gate. These make up the second-generation offspring of the initial customer. They are admitted inside the gate for service only when all first-generation offspring have been served. The procedure continues until there is a generation without arrivals during its time in service. At that time, the first busy period ends. Following an exponential idle period, the second busy period is generated in the same manner as the first and so on.

Now consider three priority rules. Under Rule 1 (first-come, first-served), customers are served in the order of arrival. Under Rule 2 (shortest service time first), *within each generation* the customers are served in *increasing* order of their service times. Under Rule 3 (longest service time first), the customers in each generation are processed in *decreasing* order of their service times.

The purpose of this problem is to compare the waiting times of the successive customers. A careful simulation study with proper estimators and confidence intervals is beyond what is intended here. We only ask the reader to generate the variates of interest correctly and to compute some summary statistics. In the experiment, simulate the queue for 10,000 busy periods. Number the customers in order of arrival, keep track of their arrival times, and generate their service times. For each customer n, calculate his waiting times $w_1(n)$, $w_2(n)$, and $w_3(n)$ if the queue were operating under each of the three rules. For each rule, report the sample mean of the waiting times. Also calculate the empirical frequency of those customers whose waiting time is shortest under Rules 1, 2, or 3, respectively.

Successively choose the service time distributions to be exponential, Erlang of order 5, and hyperexponential, in each case with mean 1. For each service time distribution, set the arrival rate successively equal to 0.4, 0.7, and 0.9. What conclusions do you draw from these nine experiments? As always, replications of the experiments are encouraged.

References: The priority rules in this problem were examined by rather involved transform methods in Nair, S. S. and Neuts, M. F., "A priority rule based on the ranking of the service times for the $M/G/1$ queue," *Operations Research,* 17, 466 - 477, 1969, and "An exact comparison of the waiting times under three priority rules," *Operations Research,* 19, 414 - 423, 1971. No numerical examination by exact computations of these rules has yet been undertaken. To do so remains of some interest.

Problem 6.4.58: The Galton - Watson Process: This problem deals with an experimental investigation of aspects of the Galton - Watson process discussed in Section 4.2.4. With probability a_k each individual is replaced in the next generation by k offspring. We assume that $a_k = 0$ for $k > K$ for some $K \geq 3$ and that a_K is positive. The mean of the density $\{a_k\}$ is α. The population is started by one individual in generation zero.

The offspring of each individual are numbered and $N_j(n)$ is the number of individuals in the nth generation that bear the number j. In other words, an individual in generation $n - 1$ with i progeny contributes one unit to each of $N_1(n), \cdots, N_i(n)$. The sum $N_1(n) + \cdots + N_K(n) = N(n)$ is the size of the nth generation. Now set $K = 5$,

$$a_k = \frac{(1-p)p^k}{1-p^6}, \quad \text{for } 0 \leq k \leq 5,$$

and choose p such that $\alpha = 0.9$, 1.0, or 1.1. For each of these p-values, simulate the process $\{N_1(n), \cdots, N_K(n)\}$ for successive n. Stop when the number of individuals in a generation exceeds 10,000 or when the population becomes extinct. For successive generations, plot the quantities $N_j(n)$ in a manner that brings out how many individuals in the total population have $j-1$ siblings. As long as $N(n)$ is positive, display also the ratios $N_j(n)/N(n)$ in an informative manner. Do 20 replications of this experiment. Which of the graphical displays do you find most useful in understanding the effects of the individual probabilities a_k on the behavior of the Galton - Watson process?

Problem 6.4.59: The Galton - Watson Process: For the process in Problem 6.4.58, successively number the individuals when they are *born*. Record the numbers of offspring of each individual in a linear array. For example, the array 1, 3, 2, 0, 3, 0, 2, 4, 1, 2, $\cdots$ indicates that the progenitor has three descendants. These respectively have 2, 0, and 3 offspring. These 5 individuals in turn produce 0, 2, 4, 1, and 2 offspring, respectively. That array permits us to reconstruct the family tree of the branching process. In particular, the counts $N_j(n)$ in Problem 6.4.58 can be recovered from it. However, to store that information for a process with realistic parameters, a very long array may be needed, or else, the array must be stored by segments on an external storage device.

Assuming that you can store up to 200,000 integers, generate the array for the density in Problem 6.4.58. If the population becomes extinct before 200,000 individuals have ever lived, there is no storage problem. Otherwise stop at the last *complete* generation for which all the information can be stored. First, discuss how the counts $N_j(n)$ in Problem 6.4.58 can be efficiently extracted from the array. Write a program to do that and generate the same plots as called for in Problem 6.4.58.

Next, suppose that all individuals with $K = 5$ progeny were allowed to have only four. How can we modify the array to take that into account? For each fifth descendant of an individual, not only that descendant but also his entire line of descent must be deleted. Write a subroutine to make that modification *without* using a new storage array. Obtain the counts $N_j(n)$ and the corresponding plots for that array. Based on several replications, discuss the qualitative effects of such a population control policy for each of the three values of p.

Remark: This problem offers an example of an experiment with a random graph, in this case, the family tree of a Galton - Watson process. It illustrates the storage problems common in such investigations. The reader may wish to explore storage and processing algorithms to overcome the limitation to 200,000 individuals that was imposed here.

Problem 6.4.60: The Poisson Process in the Plane: The homogeneous Poisson process in the plane is the translation invariant point process with independent increments for which the number of events N_A in a (nice) set A of area V_A has a Poisson distribution with mean λV_A. The parameter λ is a positive real number.

a. Verify that the area between two circles of radii $u_1^{1/2}$ and $(u_1 + du_1)^{1/2}$, centered at the origin, is $\pi\, du_1 + o(du_1)$. For a Poisson process in the plane, number the points in increasing distance from the origin. For $i \geq 1$, Z_i is the distance of the ith point to (0, 0). Prove that $\{Z_i^2\}$ is a Poisson process of rate $\lambda\pi$ on $(0, \infty)$. Show that, given Z_i, the ith point is uniformly distributed on the circle of radius Z_i centered at the origin.

b. How would you efficiently generate the cartesian coordinates (x_1, x_2) of a random point on the circle $x_1^2 + x^2 = u^2$? For given positive values of λ and r, how would you generate the coordinates of all points inside a circle of radius r for the Poisson process of parameter λ in the plane?

Problem 6.4.61: The Poisson Process in R^m: Derive (or look up) the formula for the volume of a sphere of radius r in R^m, $m \geq 3$. Extend the results in Problem 6.4.60 to an m-dimensional Poisson process of parameter λ. Develop an efficient subroutine to generate the coordinates of a random point on the surface of an m-dimensional sphere centered at the origin.

Problem 6.4.62: Random Points in a Sphere: If U_i, $1 \leq i \leq m$, are independent uniform variates then the point $(X_1, \cdots, X_m)$ with $X_i = 2U_i - 1$ is uniformly distributed in a cube of volume 2^m. If, in addition, $\sum_{i=1}^m X_i^2 \leq 1$, the point is uniformly distributed over the sphere of radius 1 centered at the origin.

a. Say that a point in the cube is *accepted* if it also lies in the sphere. Show that a random point in the cube is accepted with probability

$$p(m) = \frac{\pi^{m'}}{2^m m'!}, \qquad \text{for } m = 2m',$$

$$= \frac{m'!\pi^{m'}}{m!}, \qquad \text{for } m = 2m' + 1, \quad m' \text{ integer.}$$

b. This suggests a simple acceptance - rejection method to generate random points in the unit sphere. For $m \le 5$, tabulate the probabilities $p(m)$ and the expected number $E(m)$ of uniform random variates to generate one accepted point.

c. For $3 \le m \le 5$, we can save the generation of some random numbers by rejecting as soon as a partial sum in $\sum_{i=1}^{m} X_i^2$ exceeds one. For each of the three values of m, generate 10,000 random points in the unit sphere. Form estimates of the expected numbers $E'(m)$ of uniform variates used if you encode that extra test. Are the savings appreciable? Why is the method in this problem not practical for larger values of m?

Remark: Notice how rapidly the ratio of the volumes of the euclidean unit sphere and the cube in R^m tends to zero.

Problem 6.4.63: Random Points in a Sphere: In two dimensions, if you choose R according to the density $2r$, $0 \le r \le 1$, and independently Θ uniformly in $[0, 2\pi]$, the point with coordinates

$$X_1 = R \cos \Theta, \quad \text{and} \quad X_2 = R \sin \Theta$$

has a uniform density over the circle of radius 1. A friend asserts that the same procedure extended to *spherical coordinates* will generate a random point in the unit sphere in R^m. The transformation from cartesian coordinates $(x_1, \cdots, x_m)$ to spherical coordinates $(r, \theta_1, \cdots, \theta_{m-1})$ is given by

$$\begin{aligned}
x_1 &= r \cos \theta_1 \sin \theta_2 \sin \theta_3 \cdots \sin \theta_{m-1}, \\
x_2 &= r \sin \theta_1 \sin \theta_2 \sin \theta_3 \cdots \sin \theta_{m-1}, \\
x_3 &= r \cos \theta_2 \sin \theta_3 \cdots \sin \theta_{m-1}, \\
x_4 &= r \sin \theta_2 \sin \theta_3 \cdots \sin \theta_{m-1}, \\
&\cdots \\
x_{m-1} &= r \cos \theta_{m-2} \sin \theta_{m-1}, \\
x_m &= r \cos \theta_{m-1},
\end{aligned}$$

where $r \ge 0$, $0 \le \theta_1 \le 2\pi$, and $0 \le \theta_i \le \pi$, for $2 \le i \le m-1$. The Jacobian J of that transformation is given by

$$J = (-1)^m r^{m-1} \sin\theta_2 (\sin\theta_3)^2 (\sin\theta_4)^3 \cdots (\sin\theta_{m-1})^{m-2}.$$

For $m = 3$, your friend generates R, Θ_1, and Θ_2, with the joint density

$$\psi(r, \theta_1, \theta_2) = \frac{3r^2}{2\pi^2},$$

of independent random variables, respectively with marginal densities $3r^2$, $1/(2\pi)$ and $1/\pi$, on $[0, 1]$, $[0, 2\pi]$, and $[0, \pi]$. Thereupon he computes

$$X_1 = R\cos\Theta_1 \sin\Theta_2, \quad X_2 = R\sin\Theta_1 \sin\Theta_2, \quad X_3 = R\cos\Theta_2.$$

a. Generate independent 25,000 points with coordinates (X_1, X_2, X_3) by that procedure. By criteria of your choice, examine your friend's claim empirically.

b. Show analytically that the actual joint density of (X_1, X_2, X_3), is

$$\phi(u_1, u_2, u_3) = \frac{3}{2\pi^2}\left[\frac{u_1^2 + u_2^2 + u_3^2}{u_1^2 + u_2^2}\right]^{1/2},$$

over the set (u_1, u_2, u_3): $u_1^2 + u_2^2 + u_3^2 \leq 1$.

Problem 6.4.64: Random Points in a Sphere: Generating random points in a sphere by the method of Problem 6.4.62 is clearly not efficient for $m > 5$. After proving the necessary preliminary results, explore the following method and discuss its performance.

a. If $(X_1, \cdots, X_m)$ is uniformly distributed over the sphere $\sum_{i=1}^m u_i^2 \leq 1$ in euclidean m-space, show that the marginal density of X_1 is

$$\phi(u) = \frac{1}{B[1/2, (m+1)/2]}(1 - u^2)^{(m-1)/2}, \qquad \text{for } -1 \leq u \leq 1.$$

Deduce that X_1^2 has the beta density with parameters 1/2 and $(m + 1)/2$. Explain why, given that $X_1 = u$, $-1 \leq u \leq 1$, the $(m - 1)$-tuple $X_2, \cdots, X_m$ is uniformly distributed over the sphere of radius $(1 - u^2)^{1/2}$ in $(m - 1)$-space.

b. Look up and code an efficient procedure to generate variates with a beta distribution. Generate m beta variates, respectively, with parameters 1/2, $(r + 1)/2$, for $1 \leq r \leq m$ and use these in a routine to generate random vectors uniformly distributed over the unit sphere.

Problem 6.4.65: Random Points in a Sphere: The following are two methods for generating random points in the m-dimensional unit sphere starting with m independent $N(0, 1)$ variates $X_1, \cdots, X_m$.

a. Verify that X_1^2 has a gamma density with parameters 1/2 and 1/2 and conclude that $L^2 = \sum_{i=1}^m X_i^2$ has the gamma distribution

$$G_m(x) = \frac{1}{\Gamma(m/2)} \int_0^{x/2} e^{-v} v^{m/2-1}\, dv, \qquad \text{for } x \geq 0.$$

b. Show that the point with cartesian coordinates

$$Y_i = \frac{X_i}{L} \left[G_m(L^2) \right]^{1/m}, \qquad \text{for } 1 \leq i \leq m,$$

is uniformly distributed in the unit sphere.

c. Explain why the quantities $X_i L^{-1}$, $1 \leq i \leq m$, are the directional cosines of a random direction in R^m. Let T be a random variable with probability distribution $F(x) = x^m$ on [0, 1], independent of $X_1, \cdots, X_m$. Show that the point with coordinates $Y'_i = X_i L^{-1} T$, $1 \leq i \leq m$, is uniformly distributed in the unit sphere.

The results in parts *b* and *c* can be used to generate random points in the unit sphere. Which method would you prefer? Write efficient codes to implement both methods. Compare the execution times of the generation of 10,000 random points in the 10-dimensional unit sphere.

Hint: The joint distribution of the $X_1, \cdots, X_m$ and the uniform distribution over the unit sphere are both spherically symmetric. Given $L = r$, the point $(X_1, \cdots, X_m)$ is uniformly distributed on the boundary of the sphere of radius r centered at the origin.

Problem 6.4.66: The Unit Sphere for Various Norms: $S_p(m)$ is the set of m-tuples of real numbers $x_1, \cdots, x_m$ for which $\sum_{i=1}^m |x_i|^p \leq 1$, for a given $p > 0$. $S_\infty(m)$ is the cube of m-tuples satisfying $|x_i| \leq 1$, for $1 \leq i \leq m$.

Notice that the sequence of sets $\{S_p(m)\}$ is increasing to the limit $S_\infty(m)$.

a. Show that the probability $P_p(m)$ that a random point in $S_\infty(m)$ belongs to $S_p(m)$ is given by

$$P_p(m) = \int \cdots \int I\{E_p(m)\}\, du_1 \cdots du_m,$$

where $I\{E_p(m)\}$ is the indicator function of the set of m-tuples $(x_1, \cdots, x_m)$ with $x_i \geq 0$, for $1 \leq i \leq m$ and $\sum_{i=1}^m |x_i|^p \leq 1$.

b. For $m = 2$, calculate $P_p(2)$ explicitly in terms of the beta function. Plot a graph of $P_p(2)$.

c. For a point $\mathbf{x}$ in $E_\infty(m)$, the largest p for which $\mathbf{x} \in E_p(m)$ is the solution to the equation $\sum_{i=1}^m |x_i|^p = 1$. Write an *efficient* subroutine to solve that equation. For $m = 5$, obtain a numerical estimate of $P_p(5)$ by generating a large number of random points in $E_\infty(5)$. Using the solution to Problem 6.4.62, we know that $P_2(5) = \pi^2/60$. Can you find $P_1(5)$ explicitly?

Remark: In developing your code for part *c*, use the sample size 10,000, but to get reasonably accurate estimates of the function $P_p(5)$ you will need much larger samples sizes. It is therefore essential that the solution to the equation in part *c* be very efficiently written. This task also illustrates the computational effort needed to evaluate multivariate integrals by simulation methods. Compare the graphs you obtain experimentally to that for $m = 2$ and discuss the qualitative implications of your results.

Problem 6.4.67: Spherically Symmetric Densities: An m-variate probability density $\phi(u_1, \cdots, u_m)$ of m random variables $X_1, \cdots, X_m$ is *spherically symmetric* if it is a function of $\sum_{i=1}^m u_i^2$ only.

a. Write $R = [\sum_{i=1}^m X_i^2]^{1/2}$. Explain why, conditional on $R = r > 0$, the vector $(R^{-1}X_1, \cdots, R^{-1}X_m)$ is uniformly distributed over the sphere of radius r centered at the origin. If $\phi(u_1, \cdots, u_m) = \psi([\sum_{i=1}^m u_i^2]^{1/2})$, show that $\psi(u)$ is the probability density of R.

b. Show that a vector $\mathbf{X} = (X_1, \cdots, X_m)$ with the density $\phi(u_1, \cdots, u_m)$ is obtained by generating a vector $\mathbf{Y} = (Y_1, \cdots, Y_m)$ uniformly distributed on the unit sphere in R^m and then forming $\mathbf{X} = mV_m(1)R^{m-1}\mathbf{Y}$, where R is an independent variate with the density $\psi(\cdot)$. $V_m(1)$ is the volume of the unit sphere in R^m.

c. Set $m = 2$ and let $\psi_k(u)$ be the Erlang density of order k with mean one. To visualize how a bivariate spherically symmetric density looks, obtain three-dimensional plots of the density $\phi(u_1, u_2)$ for $k = 1, 3, 5$. Next, let $\psi(u)$ be a bimodal mixture of two Erlang densities (you get to choose its parameters). Obtain a plot of the corresponding density $\phi(u_1, u_2)$.

Problem 6.4.68: How would you generate the equation of an n-dimensional hyperplane whose normal has a random direction and such that the distance from the origin to the plane has a given probability distribution $F(\cdot)$ on $[0, \infty)$? Write a subroutine for that purpose. Test it for the Erlang distribution of order 3 with mean 5 and for $n = 2$, and $n = 3$.

Problem 6.4.69: The coordinates of N points $[x(i), y(i)]$ are independent exponential random variables with parameter $\lambda = 10$. All points are independently chosen. Their *convex hull* C is the smallest convex set that contains all N points. C is a polygon whose vertices are a subset of the N given points. These vertices are called the *extreme points* of the convex hull. The purpose of this problem is to explore a search procedure (definitely not the best possible) for the extreme points of C. It uses the fact that the minimum and maximum of a linear function defined on C are attained at extreme points.

a. Write a subroutine to pick θ at random in $(0, 2\pi)$ and to find at which of the N given points $x \sin \theta - y \sin \theta$ is largest or smallest. With probability one, these points are uniquely determined. They are extreme points of C. In successive calls of the subroutine, more and more extreme points will be found but eventually some that were found earlier will be repeated. When the subroutine finds two extreme points that are already known say that a *repetition* occurs. Let α be a positive constant. Call the subroutine until $M = [\alpha N]$ successive repetitions occur.

b. Repeated calls to the subroutine may not find all extreme points. Next, write a second subroutine to identify which of the N given points belong to the convex hull of the extreme points you have found. The following is a method to do that. Find the two points among the given N closest to the origin. They are extreme points (why?) If one or both of these were not located earlier, move them from the list of given points to a list of failed points. Continue until you find two points nearest to the origin that earlier have been identified as extreme points.

Consider the line through these two points. Rotate it counterclockwise about the point with the largest abscissa until another point is found to lie on it. If that point was found earlier, rotate the line about that point. If not, move it to the list of failed points. Continue in this manner until the line again passes through the first two extreme points. All points remaining in the list are now either the extreme points found by your random search procedure or they belong to the convex hull of these points. Keep track of the proportion δ of failed points.

c. Execute your program for different sets of $N = 500$ points and for $\alpha = 0.50, 0.75, 1.00, 1.25$. Compare the proportions δ for the various values of α. Are they consistent for the various sets of N points you generate?

Remark: A proper procedure to find the extreme points of C is to apply the rotating line method directly to the N points without doing the random search. See Problem 6.4.70. Finding the extreme points of the convex hull of N given points in n-space is a classical problem in computational geometry. See Preparata, F. P. and Shamos, M. I., *"Computational Geometry, An Introduction,"* New York: Springer-Verlag, 1985.

Problem 6.4.70: Generate N points according to the procedure described in Problem 6.4.69, and apply the rotating line method directly to those N points. Let $X(N)$ be the number of extreme points among the N given points. Obtain an empirical distribution of $X(N)$ by doing this for 1,000 samples of $N = 300$ points.

Remark: The distribution of the number of extreme points among N points generated by random procedures is studied in geometric probability. Few theoretical results are known and their proofs are difficult.

Problem 6.4.71: Consider the $H_2/M/1$ queue with interarrival time distribution

$$F(x) = 1 - pe^{-\lambda_1 x} - qe^{-\lambda_2 x}, \quad \text{for } x \geq 0.$$

For given λ_1 and λ_2 choose p such that the mean of $F(\cdot)$ is one. The queue is then stable whenever the service rate $\mu > 1$. We shall construct an experiment to visualize the effect of various service rates on the behavior of the queue length of the $H_2/M/1$ queue.

a. Write an efficient subroutine to generate 40 sets, each of 5,000 independent interarrival times. Also generate 5,000 independent exponential variates with mean one.

b. Successively set the service rate μ equal to 2, 1.75, 1.5, 1.25, and 1. Write a program to visualize the set of 40 paths described by the queue length with the simulated input streams. The service time of the ith customer in each stream is obtained by multiplying the ith exponential variate by μ^{-1}. Suppose that the queue is initially empty and, for each value of μ, trace all 40 paths on a common graph.

c. For μ equal to 2, 1.75, 1.5, and 1.25, successively compute the parameters η of the geometric steady-state queue length density prior to an arrival. For each of the 40 input streams, choose the initial queue length as a variate from the corresponding geometric density. For each μ, again trace all 40 paths on a common graph.

Remark: In this experiment, we visualize the effect of the rate μ and of two choices of the initial conditions on the queue for 5,000 jobs with given exponential service times but with different arrival patterns. That is quite different from 40 independent simulations of the $H_2/M/1$ queue. The effect of the arrival patterns is more apparent if the same service times are used.

Problem 6.4.72: Consider n independent Markov-modulated Poisson processes with the same $m \times m$ parameter matrices D and Λ. $\boldsymbol{\theta}$ is the stationary probability vector of D. The diagonal elements of Λ make up the column vector $\boldsymbol{\lambda}$ of the rates λ_i.

a. Explain why by choosing the initial state of the Markov chain D according to $\boldsymbol{\theta}$ we obtain the stationary version of each *MMPP*. Why does division of each element of D by the quantity $\boldsymbol{\theta\lambda}$ cause the stationary process to have arrival rate 1?

b. How would you *efficiently* simulate the *superposition* of the n stationary *MMPP* s, each normalized to have unit rate?

c. To visualize the behavior of such a superposition, generate the numbers $N_1, N_2, \cdots$ of arrivals in successive intervals of length 1. Generate these counts for 10,000 intervals; display them in an informative manner on a computer screen. Implement your code for $n = 25$ and for the *MMPP* s with (unnormalized) parameter matrices

$$D(1) = \begin{vmatrix} -1. & 0.2 & 0.8 \\ 2.5 & -10. & 7.5 \\ 99. & 1. & -100. \end{vmatrix}, \qquad \lambda(1) = \begin{vmatrix} 1. \\ 10. \\ 100. \end{vmatrix},$$

and

$$D(2) = \begin{vmatrix} -1. & 0.2 & 0.8 \\ 2.5 & -10. & 7.5 \\ 99. & 1. & -100. \end{vmatrix}, \qquad \lambda(2) = \begin{vmatrix} 10. \\ 10. \\ 1. \end{vmatrix}.$$

Reference: Neuts, M. F. and M. E. Pagano, "Generating random variates from a distribution of phase type," *Proceedings 1981 Winter Simulation Conference,* T. I. Oren, C. M. Delfosse, C. M. Shub (eds.), 381 - 387.

Problem 6.4.73: The following is an example of a point process in which the marginal density of any interarrival time is *exponential* with parameter λ, but the successive interarrival times are *dependent.* For $1 \le i \le N$, let $G_i(\cdot)$ be the generalized Erlang distribution with Laplace - Stieltjes transform

$$g_i(s) = \prod_{j=i}^{N} \left[\frac{j\lambda}{s + j\lambda} \right].$$

P is an $N \times N$ *doubly stochastic* matrix. The initial state of a Markov chain with matrix P is chosen according to the invariant probability vector $\boldsymbol{\pi} = N^{-1}(1, \cdots ,1)$. The point process, a particular *MAP*, is constructed as follows: If the chain successively visits the states i_0, i_1, $\cdots$, the times $X_1, X_2, \cdots$ between the successive arrivals are conditionally independent and have the distributions $G_{i_0}(\cdot)$, $G_{i_1}(\cdot)$, $\cdots$.

a. Explain why the probability distribution $F(\cdot)$ of an arbitrary interarrival interval has the Laplace - Stieltjes transform

$$f(s) = \frac{1}{N} \sum_{j=1}^{N} g_j(s).$$

Verify that $f(s) = \lambda(s + \lambda)^{-1}$, so that $F(\cdot)$ is exponential.

b. Show that, for $k \ge 1$, the joint Laplace - Stieltjes transform $f_k(s_1, s_2)$ of

X_1 and X_{k+1} is given by

$$f_k(s_1, s_2) = \frac{1}{N} \sum_{j=1}^{N} \sum_{r=1}^{N} g_j(s_1)[P^k]_{jr} g_r(s_2).$$

Derive a formula for the correlation coefficient $\rho(k)$ of X_1 and X_{k+1}. Conclude that, in general, the interarrival times are dependent.

c. To construct specific examples, we choose matrices P that are convex combinations of particular permutation matrices K. For a given permutation $(\phi_1, \cdots, \phi_N)$ of $1, \cdots, N$, we write $K(\phi_1, \cdots, \phi_N)$ for the matrix with elements $K_{i,\phi_i} = 1$ and $K_{ij} = 0$ otherwise. Consider the examples obtained by setting

$$P_1 = 0.9K(1, 2, 3, 4, 5, 6, 7, 8, 9, 10) + 0.1K(6, 5, 1, 2, 8, 4, 9, 7, 10, 3),$$

and

$$P_2 = 0.9K(5, 7, 1, 9, 8, 10, 2, 3, 4, 6) + 0.1K(6, 5, 1, 2, 8, 4, 9, 7, 10, 3).$$

Compute the correlation coefficients $\rho(k)$, $1 \leq k \leq 10$, for the interarrival times corresponding to these two matrices P.

d. Set $\lambda = 1$. Plot 1,000 interarrival times of a Poisson process of rate 1 and of the point processes corresponding to P_1 and P_2. Do these plots reflect the behavior suggested by the values of the correlation coefficients?

e. The three sets of interarrival times in part *d* are used as input streams to a single-server queue with exponential service times. Generate 1,000 exponential variates with mean one. As in Problem 6.4.71, successively set the service rate μ equal to 2, 1.75, 1.5, 1.25, and 1. Write a program to visualize the paths described by the queue length. The service time of the ith customer in each stream is obtained by multiplying the ith exponential variate by μ^{-1}. Let the queue be empty initially. For each μ, trace all three paths on a common graph. Is the effect of the dependence in the second and third arrival streams evident in the behavior of the queues?

Reference: This problem is based on results in Latouche, G., "An exponential semi-Markov process, with applications to queueing theory," *Stochastic Models,* 1, 137 - 169, 1985, where also various analytic comparisons of queues are discussed.

Problem 6.4.74: A Queue with Maintenance: In an $M/G/1$ queue a maintenance period is inserted after the completion of each mth service. That makes the model a queue with semi-Markov services, an $M/SM/1$ queue. Its analytic treatment is beyond the scope of this book. For a special case, this problem deals with the selection of m by experimentation. The service time distribution $H(\cdot)$ has mean one. The arrival rate of customers is λ. If maintenance is done after every mth service, each maintenance period lasts $a + cm^{\alpha}$ units of time. The parameters satisfy $a > 0$, $c > 0$, and $\alpha \geq 1$. From the theory of the $M/SM/1$ queue it is known that the queue is stable if and only if

$$\frac{1}{\lambda} > 1 + \frac{a}{m} + cm^{\alpha-1}.$$

a. For $\lambda = 0.75$, $a = 2$, $c = 0.1$, and $\alpha = 1.25$, compute the smallest and largest values m° and m^{*} of m for which the queue is stable. For $m^{\circ} \leq m \leq m^{*}$, plot the traffic intensity $\rho(m) = \lambda[1 + m^{-1}a + cm^{\alpha-1}]$.

b. Generate 20,000 arrival times in a Poisson process of rate 0.75 and equal number of variates from the distribution $H(\cdot)$ of mean one. Let the queue be empty at time zero. Customers are served in order of arrival. Compute the waiting times of 20,000 customers with these arrival and service times. Determine the 19,800th largest waiting time $W_{0.99}$. Obviously, 1% of the customers wait longer than $W_{0.99}$.

c. For the same simulated arrival and service times, find the value of m^{+}, $m^{\circ} \leq m^{+} \leq m^{*}$, for which, in the queue with maintenance, the number of customers waiting longer than $W_{0.99}$ is as small as possible. Denote by $p_{0.99}(m^{+})$ the proportion of customers whose waiting time exceeds $W_{0.99}$.

d. Using the value m^{+} in part c, do five independent replications of a simulation of the waiting times of 20,000 customers in the queue with maintenance. For each replication, print the proportion of customers who wait longer than $W_{0.99}$.

e. Implement this experimental determination of m^{+} for three different service time distributions $H(\cdot)$. Discuss the qualitative merits of this method. In a real situation, would you use it to schedule maintenance periods to minimize their effect on the delays of regular jobs?

Remark: For an algorithmic solution of this problem the steady-state waiting time distribution of a customer must be computed for various values of m. For the $M/SM/1$ queue, efficient algorithms for that purpose have only recently been developed.

Problem 6.4.75: We consider N identical, independent sources of data packets. The output of each source is a sequence of strings of $b \geq 1$ successive packets alternating with geometrically distributed silent periods. The lengths of the silent periods are independent. Their common density is $s_\nu = c^{\nu-1}(1-c)$, for $\nu \geq 1$. J_n, the number of packets submitted at time n, is an integer between 0 and N. We are interested in simulating the sequence of counts $\{J_n\}$ in *steady-state*. Let I_r, $1 \leq r \leq b$, be the number of sources that submitted the rth packet of a string at time n. Clearly, the I_rs are integers whose sum is the number I^* of sources active at time n.

a. Explain why, to generate J_{n+1}, it suffices to remember the numbers I_r. Discuss how you would generate J_{n+1} and update the counts $\{I_r\}$ for the time slot $n+1$ as efficiently as possible.

b. Explain why the following initialization produces a steady-state realization of the sequence $\{J_n\}$. Set all $I_r = 0$. For each of the N sources, determine by a Bernoulli trial of probability $\pi = (1-c)(1-c+b)^{-1}$ whether it is active at time zero or not. If a source is active, choose an integer r at random from $\{1, \cdots, b\}$ and increase I_r by 1.

c. Show that the expected number of packets submitted at an arbitrary time is $\lambda^* = N\pi$. Next, do the following experiments first with $N = 50$ and next with $N = 100$. Fix λ^* at the value 20. For each N, that determines the value of the probability π. Now successively set $b = 1, 4, 10, 25$ and adjust the parameter c accordingly. For each case, simulate the behavior of the J_n over 10,000 time slots. Informatively display the resulting traffic on a computer screen. How does the observed traffic differ for the various values of b? How does it appear to depend on N?

Remark: For $b = 1$, the J_ns are independent. What is their probability density? For the other cases, you should strikingly see the effect of dependence in the generated sequences.

Problem 6.4.76: The data traffic simulated in Problem 6.4.75 is the input to a queue. Its server processes a packets per time slot in the following manner. If at time n, ξ_n packets are waiting, first any arrivals at time n join the queue and then $\min(\xi_n + J_n, a)$ packets are processed. The number of packets present at time $n+1$ is therefore given by

$$\xi_{n+1} = \max [0, \xi_n + J_n - a].$$

a. Use the simulated traffic data of Problem 6.4.75 to visualize the behavior of the queue. As is to be expected, the queue will exhibit instability when $a < \lambda^*$. Choose some values of a that are moderately larger than λ^*, say, $a = 21, 25, 30$ and visualize how the different types of traffic respond to faster service.

b. Now suppose that the queue has a buffer of size 150. There are $N = 50$ sources and $\lambda^* = 20$. In order to be processed, a packet must be admitted to the buffer. Any packets that arrive when the buffer is full are lost. We are interested in finding the smallest value of a for which fewer than 1% of the packets will be lost. That value of a will depend on the parameter b. Why should we expect that a will increase with b?

Organize an experimental search as follows: Generate a string $\{J_n\}$ of length 100,000. With $b = 1$ and $a = 21$, determine the numbers $\{L_n\}$ of each of the successive J_ns that would be lost. If the sum of the L_n exceeds 999, successively increase a and, *as efficiently as possible*, determine the next $\{L_n\}$. Stop when the total number of packets is less than 1,000 and record the corresponding value of a.

Replicate this part of the experiment five times. If, in the first round, the starting value $a = 21$ resulted in far too many lost packets, have your code choose a better starting value of a. Which of the five values of a found so far would you propose for use? Give reasons for your choice. Starting with the a-value chosen for $b = 1$, now successively set $b = 4, 10, 25$. For each case, repeat the procedure used for $b = 1$ to select a based on five replications with 100,000 J_n each.

c. The experiment in part b can be used to illustrate another feature of traffic overflow. You can add an appropriate code to report statistics on the *persistence* of overflows once they start. To that end, use the counts $\{L_n\}$ for the a-values selected by each search. If packet losses are few and far between, we would expect that *runs* of positive L_n will be short. Alternatively, if overload conditions are persistent, we would anticipate longer runs of positive L_n. For each type of traffic, i.e., for each value of b, report the number of observed runs of positive L_n as well as the lengths of such runs and the total numbers of packets lost in each. Can you draw qualitative conclusions on the persistence of overload (and packet loss) for the various values of b? If so, do your conclusions agree with what you expected?

Remark: This problem is a simplified version of a matter of concern to communications engineering. Data traffic that is not easily described by familiar point process models is called *bursty*. That includes arrival streams with long persistent behavior as is seen here for increasing values of b. The design of buffers, or in this case, the choice of processing speed, poses difficult problems that often can be settled only by extensive simulation experiments.

Problem 6.4.77: Suppose that, in a certain competition, the skill of an individual i is measured by a number $\pi(i)$ in [0, 1]. There are N individuals competing in a "tournament by pairwise elimination," that is, individuals are paired off and compete in a first round. The winners of the first round are again paired off and so on until only one person remains. That player is the champion. The last two competitions are the semifinals and the finals, respectively. If some intermediate round involves an odd number of competitors, one individual is selected to sit out that round. He moves on to the next round. There is a function $\phi[\pi(i), \pi(j)]$, that measures the probability individual i wins a match against j. That function should be chosen to give the individual with the greater skill a higher probability of winning and $\phi(x, y) = 1/2$, if and only if $x = y$.

In scheduling the successive rounds, consider two possible rules. Under Rule 1, the competitors involved in that round are ranked in order of decreasing skills $\pi(x)$. If the number of persons is odd, the one with the median skill sits out that round. The highest ranked player competes against the highest ranked in the lower half; the second ranked player competes against the second ranked player in the lower half, and so on. Under Rule 2, the pairs are drawn at random. That is done by forming a random permutation of the players in that round and by pairing off the first two, the next two, and so on. If the number of players is odd, the last one remaining sits out the round.

The purpose of the experiment in this problem is to compare several choices of $\phi(x, y)$ and to assess the impact of Rules 1 and 2. We are interested in finding functions ϕ for which there is a high probability that the competitors with the highest skills appear in the semifinals and finals, whereas those with lower ranks will not be completely excluded.

Select the skill parameters $\pi(x)$ for the N individuals as independent observations with common density $\theta(u) = (\alpha + 1)u^{\alpha}$, for $0 \le u \le 1$, and with a parameter $\alpha \ge 0$ of your choice. Which qualitative feature of the population of competitors do you model by taking a high value for α? Define five plausible functions $\phi(x, y)$ and explain the reasons for your choices. For example, one choice could be

$$\phi[\pi(i), \pi(j)] = \frac{\pi(i)}{\pi(i) + \pi(j)}.$$

With the skills assigned and the functions $\phi(x, y)$ defined, simulate 100 replications of the tournament, using both Rule 1 and Rule 2. Keep track of how many of the four (or three) best players compete in the last two rounds and how often the highest ranking player wins the tournament.

Eventually, you will have to make a recommendation on the choice between Rules 1 and 2 and also to justify your use of a particular function $\phi(x, y)$. Calculate statistics that summarize your experimental findings and offer clear statistical information in support of your recommendations. Write the necessary code to perform the experiments and implement it for a tournament with $N = 50$.

CHALLENGING PROBLEMS

Problem 6.4.78: This experiment is inspired by the assignment of "slots" in packetized data transmission. Suppose that there are 500 boxes placed at the vertices of a regular polygon with 500 sides. The boxes are numbered counterclockwise from 1 to 500. Each box can hold one item. There are 300 single items (Type 1) and 200 items bunched in 10 groups of 20 (Type 2.) The items of Type 1 are placed first according to one of the following two random procedures:

In the *random assignment method* (*RAM*), the 300 items are assigned so that all $\binom{500}{300}$ possible patterns are equally likely.

In the *first available method* (*FAM*), the 300 items are taken one at a time. A random integer i between 1 and 500 is chosen and the item is placed in box i if it is empty, or if it is not, the boxes are scanned counter clockwise and the item is placed in the first available empty box.

By looking at all possible outcomes, say, for five boxes and three items, convince yourself that *RAM* and *FAM* assign different probabilities to the various outcomes. After the 300 items of Type 1 have been placed, we try to fit in the items of Type 2. There are two possible ways to do so. Under *contiguous placement CP*, the bunches must be placed in successive empty boxes. Under *regular placement* (*RP*), the items in each bunch must be placed at the vertices of a regular 20-sided polygon of empty boxes.

It may not be possible to fit all 10 bunches of size 20 into the remaining free spaces. We are interested in which of the methods, *RAM* or *FAM*, is more favorable to the assignments *CP* and *RP*. We are also interested, for any combination of a placement and an assignment method, in the probability densities of the number of bunches fitted.

Write four subroutines, two to place the 300 single items under *RAM* and *FAM* and two to determine, for a given pattern obtained, the numbers of bunches that can be fit under *CP* and *RP*. Perform 10,000 replications of an experiment to study the empirical probability distribution of the number of bunches placed. Select appropriate statistics to summarize your experimental results and state qualitative conclusions. In particular, on intuitive grounds, one might guess that *RAM* will often allow more bunches under *RP*. *FAM*, on the other hand, will leave larger gaps and should therefore be more favorable to *CP*. Is that borne out by your experimental results? Replicate the entire experiment several times to see whether your conclusions are consistent.

Problem 6.4.79: In this variant of the experiment in Problem 6.4.78, the number of single items is not fixed but has a binomial distribution. Perform 500 independent binomial trials with probability $p = 0.60$. The number of Type I items is the number of successes of which there are 300 on average. Modify the code for Problem 6.4.78 to perform the corresponding experiment. Compare the experimental results to those found there.

Problem 6.4.80: Consider an $M/G/1$ queue with arrival rate $\lambda = 1$. The service times have the Pareto density

$$\phi(u) = \left[\frac{\alpha}{c}\right]\left[\frac{c}{u}\right]^{\alpha+1}, \qquad \text{for } u > c > 0, \qquad \text{with } \alpha > 0.$$

a. Verify that the queue is stable if and only if $\alpha > (1-c)^{-1}$ and that for $1 < \alpha \leq 2$, the variance of the service time is infinite. With $c < 1/2$ and $(1-c)^{-1} < \alpha \leq 2$, we have a stable $M/G/1$ queue for which the variance of the service times is infinite.

b. Set $c = 0.4$ and $\alpha = 11/6$. Generate 25,000 service times $\{s_j\}$ with the corresponding distribution. Next, generate the numbers of arrivals $\{n_j\}$ in a Poisson process of rate 1 for each of these service times. Denote the sum of the 25,000 counts n_j by S. Print a frequency chart of the $\{n_j\}$. It is very likely that during several service times there will be many arrivals, say, more than 5, and that also there will be a few very large counts.

c. Starting with an empty queue, generate the queue lengths $\{q_j\}$ at the ends of successive services. Print a frequency chart of the $\{q_j\}$. Sort the array $\{q_j\}$ applying the same permutation to an array containing the integers 1 to 25,000. Consider the 250 largest elements of $\{q_j\}$. If q° is such an element with corresponding index j°, print the 30 elements of the array $\{n_j\}$ just before and including n_{j°. Call these strings of 30 elements the *antecedents* of the long queue lengths. Print or visualize these antecedents on a screen. Can you conclude that the long queues are caused by the (long) services with many arrivals?

d. Let the cutoff value for the 250 largest queue lengths be q^*. Now examine a control rule under which during any service at most K arrivals are admitted to the queue. By adjusting your array $\{n_j\}$, search for the largest value of K for which *all* elements of the adjusted array $\{q_j\}$ are less than q^*. What fraction of the total number of arrivals S have been turned away? Calculate the sample mean of the adjusted array $\{n_j\}$. That is an estimate of the traffic intensity of the "controlled" queue. How does it compare to the traffic intensity $\rho = 0.88$ of the original queue?

e. Replicate the experiment in parts *b–d* nine more times. Print the values of K, q^*, the fraction of lost customers, and the estimate of the traffic intensity for each replication. What do you think of this experimental approach to the design of a service system?

Hint: If U is a uniform variate, then $Y = cU^{-1/\alpha}$ has the given Pareto distribution. If s is a service time, the corresponding number n of arrivals during a Poisson process of rate 1 is generated by finding the first index ν for which

$$\prod_{i=1}^{\nu} U_i > e^{-s},$$

and setting $n = \nu - 1$.

Reference: For a queueing theoretic discussion of this control rule, see Neuts, M. F., "The $M/G/1$ queue with a limited number of admissions or a limited admission period during each service time," *Stochastic Models,* 1, 361 - 391, 1985.

Problem 6.4.81: For the model in Problem 6.4.80, consider a different control rule. For a positive constant T we admit customers only during the initial portion $\min[s, T]$ of each service time s. How would you determine the largest T for which the 250 longest queue lengths are reduced (as in Problem 6.4.80)? Note that it is now necessary also to store the arrival times of customers during each service. Implementation of the algorithm requires much storage. Write an efficient code to implement this experimental study. Report the same descriptors of the queue as in Problem 6.4.80.

Problem 6.4.82: This problem deals with a simple *stochastic difference equation,* known as a first-order autoregressive scheme. You are asked to examine three related versions of that equation by a combination of analysis and experimentation. $\{U_n, n \geq 1\}$ is a sequence of independent random variables with common distribution $F(\cdot)$ of mean μ'_1 and variance σ^2. The random variable X_0 with mean $\mu(0)$ and finite variance $\sigma^2(0)$ is independent of the U_n. We consider the difference equation

$$X_n = \theta_n X_{n-1} + U_n, \quad \text{for } n \geq 1, \tag{1}$$

with three different specifications for the random coefficients θ_n.

In Case 1, $\{\theta_n, n \geq 1\}$ is a sequence of independent, identically distributed random variables with values in the interval (0, 1). Their common probability distribution $\Psi(\cdot)$ has mean θ^* and variance η^2. X_0 and the sequences $\{U_n\}$ and $\{\theta_n\}$ are mutually independent.

In Case 2, $\theta_n = \theta$, for all $n \geq 1$. θ is independent of X_0 and of the sequence $\{U_n\}$. Its probability distribution $\Psi(\cdot)$ is the same as in Case 1.

In Case 3, $\theta_n = \theta^*$, for all $n \geq 1$. That is the simplest case. It has an extensive literature and some practical applications. For example, a stock of radioactive material is depleted by spontaneous decay from one time unit to the next. It is also replenished by random increments U_n. X_n is then the amount of the isotope in stock at time n.

We note that Case 3 is a degenerate case of 1 and 2, but that 2 is not a particular form of Case 1. In Case 2 the random discount factor θ is chosen once and for all; in Case 3 different discount factors are chosen for each successive time point.

a. Show that for all three cases

$$EX_n = \mu(n) = \theta^{*n}\mu(0) + (1 - \theta^*)^{-1}(1 - \theta^{*n})\mu'_1, \quad \text{for } n \geq 0. \tag{2}$$

b. Show that for all cases the variance $\sigma^2(n)$ of X_n satisfies the recurrence relation

$$\sigma^2(n + 1) = (\eta^2 + \theta^{*2})\,\sigma^2(n) + \eta^2(EX_n)^2 + \sigma^2, \tag{3}$$

where $\eta = 0$ for Case 3.

c. Multiply both sides of equation (3) by $(\eta^2 + \theta^{*2})^{-n-1}$ and write the resulting equality for $n = 0, \cdots, m - 1$. Sum from 0 to $m - 1$ to obtain that

$$\sigma^2(m) = (\eta^2 + \theta^{*2})^m \sigma^2(0)$$

$$+ \eta^2 \sum_{n=0}^{m-1} (\eta^2 + \theta^{*2})^{m-n-1}[EX_n]^2 + \sigma^2 \sum_{n=0}^{m-1} (\eta^2 + \theta^{*2})^{m-n-1}.$$

From (2) and the preceding equation deduce that, for $m \geq 0$,

$$\sigma^2(m) = (\eta^2 + \theta^{*2})^m \sigma^2(0) + [\mu(0) - (1 - \theta^*)^{-1}\mu'_1]^2[(\eta^2 + \theta^{*2})^m - \theta^{*2m}]$$

$$+ 2(1 - \theta^*)^{-1}\mu'_1[\mu(0) - (1 - \theta^*)^{-1}\mu'_1]\eta^2 \frac{(\eta^2 + \theta^{*2})^m - \theta^{*m}}{\eta^2 + \theta^{*2} - \theta^*}$$

$$+ [\sigma^2 + \eta^2(1 - \theta^*)^{-2}\mu'^2_1]\frac{1 - (\eta^2 + \theta^{*2})^m}{1 - \eta^2 - \theta^{*2}}. \tag{4}$$

Explain why $\eta^2 + \theta^{*2} < \theta^* < 1$ and derive expressions for the limits $\mu(\infty)$ and $\sigma^2(\infty)$ of $\mu(n)$ and $\sigma^2(n)$ as $n \to \infty$.

d. Write general subroutines to compute $\mu(n)$ and $\sigma(n)$ for successive n until respectively $|\mu(n) - \mu(\infty)| < 0.01\,|\mu(\infty)|$, or $|\sigma(n) - \sigma(\infty)| < 0.01\sigma(\infty)$, for ten successive values of n. Let N_1 and N_2 be the smallest values of n for which the first, respectively the second, of these criteria holds.

e. For the experimental part of the problem, choose X_0 to be exponential with mean 50. Let the distribution $F(\cdot)$ of the U_n be exponential with $\mu'_1 = 1$. Successively set θ^* equal to 0.5, 0.75, 0.9, and 0.98. Remembering that the variance of $\Psi(\cdot)$ is less than $\theta^*(1 - \theta^*)$, select for each θ^*, a "small" and a "large" value of the standard deviation η. The first two moments of $\Psi(\cdot)$ have now been chosen in eight different ways.

Select a distribution $\Psi(\cdot)$ with these first two moments. Generate and plot realizations of the sequences $\{X_n\}$ for the three cases and for n up to $\max(N_1, N_2)$. Although the $\{\mu(n)\}$ agree for all three cases and the $\{\sigma^2(n)\}$ agree for Cases 1 and 2, you will find that, for the various cases, typical realizations are quite different.

Try expressing these differences in precise quantitative terms. By replicating your experiment, see whether the differences are real. As an example of a quantitative measure of the difference, consider the fraction of indices n satisfying $0 \le n \le \max(N_1, N_2)$ for which $|X_n - \mu(n)| > 1.5\sigma(n)$. For all three cases and for the various choices of θ^* and η, obtain numerical estimates of that fraction. Another measure is obtained by computing the straight lines fitted by least squares to the data sets $\{n, \log X_n\}$ and by comparing and interpreting scatter plots of their intercepts and slopes.

Remark *a*: A nice way of visualizing the different behavior of the three cases is to concatenate, for each of the cases, say, 10 or 20 realizations of length $\max(N_1, N_2)$. These can then be displayed as a single graph on a screen or printed on a strip. By comparing these displays for the three cases, some features will stand out that may not be apparent in single realizations.

Remark *b*: The scope of this problem can be broadened in many ways. We suggest writing a general code in which the choices of $F(\cdot)$, of the probability distribution of X_0, and of $\Psi(\cdot)$ are all left up to the user. By adding subroutines to compute analytically tractable features or to simulate those that are not, the code can be made into an exploratory tool for the study of a model of considerable interest to applications.

Problem 6.4.83: The object of this problem is to develop an efficient code to generate a large number of events in a Markovian arrival process (*MAP*) with single arrivals. Your code should start by offering options to simulate a realization from a discrete-time or from a continuous-time *MAP*. After the coefficient matrices D_0 and D_1 (of order m) have been read, perform checks to verify that they are valid parameter matrices.

Initialization: Offer the option to specify an initial state for the Markov chain with transition probability matrix (or generator) D, or to select the initial state from the steady-state probability vector. If that option is chosen, compute the vector $\boldsymbol{\theta}$ and write a subroutine to generate a variate with values in $1, \cdots, m$ with the probabilities $\theta_1, \cdots, \theta_m$. (I would not use the alias method here. Do you see why?)

Efficiency: For the discrete case, suppose that the current state of the Markov chain is i and that we wish to generate the time until the next event in the *MAP*. Except for the final transition, that time clearly only involves transitions dictated by the matrix D_0. Let $d(k)$ be the sum of the elements in the kth row of D_0. Set up the alias tables for each of the $2m$ discrete densities

$$\delta(k) = [d(k)]^{-1}[(D_0)_{k1}, \cdots, (D_0)_{km}]$$

and

$$\delta'(k) = [1 - d(k)]^{-1}[(D_1)_{k1}, \cdots, (D_1)_{km}],$$

unless $d(k)$ is either 0 or 1. For the state index i, do a Bernoulli trial with probability $d(i)$ to determine whether the next time epoch is not an event. If it is not, generate the next state i_1 using the density $\delta(i)$. Continue this for the successive states visited until an event occurs. Keep track of the elapsed number n of transitions; that is the time until the next event. Generate the state at the event using the density $\delta'(i_n)$ and repeat the procedure until either the required number of events or the total specified time of the process is reached.

A little extra efficiency is attained by setting the kth element of $\delta(k)$ equal to 0 and, if that element δ_k was positive to begin with, by dividing by it. If you do that, keep track of the number of times each state was visited *before* the event occurs. Suppose that the numbers of visits to the various states are $v_1, \cdots, v_m$. To figure out the total time until the event, we now accumulate n_1 geometric holding times with parameter $1 - \delta_1$, n_2 with parameter

$1 - \delta_2$, and so on.

Examine a similarly efficient procedure to generate the time until the next event for a continuous-time *MAP*. Generate the counts of the numbers of states visited in the embedded discrete-parameter Markov chain and calculate the total elapsed time by generating and summing appropriate exponential holding times. Test your code by simulating runs of 1,000 events each for *MAP* s of your choice.

The following five problems deal with the visualization of the behavior of important Markov chains, known as *quasi-birth-and-death* or *QBD* processes. Their state space consists of the set of pairs $\{(i, j)\}$ with $i \geq 0$ and $1 \leq j \leq m$ and the transition probability matrix P (or the infinitesimal generator Q for Markov chains in continuous time) is of the form

$$P = \begin{vmatrix} B_0 & A_0 & 0 & 0 & 0 & 0 & \cdots \\ B_1 & A_1 & A_0 & 0 & 0 & 0 & \cdots \\ 0 & A_2 & A_1 & A_0 & 0 & 0 & \cdots \\ 0 & 0 & A_2 & A_1 & A_0 & 0 & \cdots \\ 0 & 0 & 0 & A_2 & A_1 & A_0 & \cdots \\ & \cdots & & \cdots & & \cdots & \end{vmatrix}.$$

In the discrete-time case, B_0, B_1, A_0, A_1, and A_2 are substochastic matrices. The sums $B_0 + A_0$ and $A = A_0 + A_1 + A_2$ are stochastic matrices. Here we assume that the Markov chain P is irreducible and that A is an *irreducible* stochastic matrix. π is the invariant probability vector of A. The Markov chain with transition probability matrix P is then *positive recurrent* if and only if $\pi A_0 e < \pi A_2 e$.

For the continuous time case, A is an irreducible generator; the only difference is that the diagonal elements of B_0 and A_1 are now negative and all row sums of Q are zero. The Markov chain with generator Q is also positive recurrent if and only if $\pi A_0 e < \pi A_2 e$.

With fewer restrictions than stated here, such Markov chains and many of their applications are discussed in Neuts, M. F. "*Matrix-Geometric Solutions in Stochastic Models: An Algorithmic Approach,*" Baltimore: The Johns Hopkins University Press, 1981.

Problem 6.4.84: Simulation of Quasi-Birth-and-Death Processes: Write a general purpose code to simulate the behavior of a quasi-birth-and-death process allowing coefficient matrices B_0, B_1, A_0, A_1, and A_2 up to order 10. Give the user the option of a discrete- or a continuous-time model. Check that the input matrices meet the required conditions. Allow specification of arbitrary initial conditions. Strive for maximum efficiency in generating the successive states visited. For example, use the *alias method* to generate the next state, but compute and store as few alias tables as possible.

Plot a grid of $200 \times m$ points on the screen. Start the process at some point $(0, j)$, $1 \le j \le m$, and, by highlighting the current state, trace its path over 1,000 transitions. That should be useful in visualizing the behavior of the chain and in comparing the different behaviors of chains with different parameter matrices. For continuous chains, try to achieve a dynamic display by updating the screen after every transition (so that the time differences between state changes are apparent). If the Markov chain visits states outside of the grid, indicate this in your visualization by printing those states in a separate window. Resume depicting the state in the grid when the chain returns to that set. Use your code to visualize the behavior of the Markov chains with the following coefficient matrices and explain that behavior.

1. The matrices A_0, A_1, and A_2 are given by

$$A_0 = \begin{vmatrix} 0.20 & 0.22 \\ 0.01 & 0.42 \end{vmatrix}, \quad A_1 = \begin{vmatrix} 0.02 & 0.40 \\ 0.02 & 0.30 \end{vmatrix}, \quad A_2 = \begin{vmatrix} 0.06 & 0.10 \\ 0.05 & 0.20 \end{vmatrix},$$

and $B_0 = A_1 + A_2$, and $B_1 = A_2$.

2. With the same matrices A_0, A_1, A_2, and B_1, as in 1, replace B_0 by

$$B_0 = \begin{vmatrix} 0 & 0.58 \\ 0.57 & 0 \end{vmatrix}.$$

What does that change amount to? Is its effect apparent in a visualization of 1,000 transitions?

3. The matrices A_0, A_1, and A_2 are given by

$$A_0 = \begin{vmatrix} 4 & 0 \\ 0 & 5 \end{vmatrix}, \quad A_1 = \begin{vmatrix} -7 & 0 \\ 16 & -35 \end{vmatrix}, \quad A_2 = \begin{vmatrix} 2 & 1 \\ 3 & 11 \end{vmatrix},$$

and $B_0 = A_1 + A_2$, and $B_1 = A_2$.

Problem 6.4.85: Simulation of Quasi-Birth-and-Death Processes: Let $N_{ij}(n)$ be the number of times the Markov chain visits the state (i, j) in the first n transitions. Add a feature to the code of Problem 6.4.84 to visualize how the empirical frequencies $n^{-1}N_{ij}(n)$ for the states in the grid change with n. That can be done by plotting three-dimensional graphs of the empirical densities. To get a lively and informative display, generate consecutive batches of 100 transitions and display the plots of the current empirical frequencies after every batch. Be sure to take the frequencies of visits to states outside the grid into account.

The display should start with a unit mass at the specified initial state. You will see the empirical frequencies spread out as the chain evolves. Eventually, the plots will settle on a stable pattern corresponding to the steady-state probabilities of the states in the grid. You may want to place a high upper limit on the number of transitions generated so that you can stop when the displays cease to change noticeably (or when you get tired of watching it). What do you think will happen to the successive displays if the Markov chain is transient or null-recurrent? Try out some examples of such chains. What is the precise interpretation of the plots for continuous-time Markov chains?

Problem 6.4.86: Simulation of Quasi-Birth-and-Death Processes: For the continuous-time option in Problem 6.4.83, keep track of the total length of time $L_{ij}(t)$ spent in state (i, j) in the grid. As in Problem 6.4.85, create displays of the quantities $t^{-1}L_{ij}(t)$ at time points kT, where T is a step size to be specified. Organize the bookkeeping carefully as only part of the last sojourn time is to be included at each display epoch. For a positive recurrent chain, these displays will also settle down eventually. What is the significance of the limit configuration? Generate some examples to illustrate that the limit configurations can be quite different from those in Problem 6.4.85 for the same model.

Problem 6.4.87: Simulation of a Queue in Random Environment: A is the generator of an m-state, irreducible Markov chain. Let $\boldsymbol{\lambda}$ and $\boldsymbol{\mu}$ be nonzero, nonnegative vectors. Λ and M are diagonal matrices with the components of $\boldsymbol{\lambda}$ and $\boldsymbol{\mu}$ as diagonal elements. The QBD-process with coefficient matrices $A_0 = \Lambda$, $A_1 = A - \Lambda - M$, $A_2 = M$, $B_0 = A_1 + A_2$, and $B_1 = A_1$ describes an $M/M/1$ queue in *random environment*. The arrival and service rates vary randomly according to the states visited by an m-state Markov chain with generator A. The condition for positive recurrence of the QBD-process, $\boldsymbol{\pi\lambda} < \boldsymbol{\pi\mu}$, states that the arrival rates averaged by the steady-state probabilities π_j, should be less than the corresponding average of the service rates. The quantity $\rho = (\boldsymbol{\pi\lambda})(\boldsymbol{\pi\mu})^{-1}$ is the traffic intensity of the queue.

Particularly if $\lambda_j > \mu_j$ for some indices j, the behavior of the $M/M/1$ queue in random environment is strikingly different from that of the $M/M/1$ queue with the same traffic intensity.

a. Let $(\boldsymbol{\alpha}, T)$ and $(\boldsymbol{\beta}, S)$ be irreducible representations of PH-distributions. Define the matrix A by

$$A = \begin{array}{c|cc|} 1 & T & \mathbf{T}^{\circ}\boldsymbol{\beta} \\ 2 & \mathbf{S}^{\circ}\boldsymbol{\alpha} & S \end{array} .$$

The vector $\boldsymbol{\lambda}$ has identical components $\lambda(1)$ for the indices in set **1** and identical components $\lambda(2)$ for the indices in **2**. The vector $\boldsymbol{\mu}$ is of a similar form with identical components $\mu(1)$ on **1** and $\mu(2)$ on **2**. What queueing model does the corresponding QBD-process describe? For that case, what is the explicit form of ρ? Write an efficient special purpose routine to simulate the behavior of that QBD-process and to visualize its path functions.

b. Choose a value of $\lambda(1)$ such that with $\lambda(2) = \lambda(1)$, the queue would be unstable. Then choose $\lambda(2)$ so that $\rho = 0.80$. Start the queue in a state $(0,j)$, where $j \in \mathbf{1}$ and generate 10,000 arrivals and services. Visualize the path behavior and the evolution of the state probabilities on the grid defined in Problems 6.4.85 and 6.4.86. For purpose of comparison, also simulate and visualize 10,000 arrivals and services in an $M/M/1$ queue with $\rho = 0.80$.

c. Repeat the experiment in part *b* for several pairs of PH-distributions with different representations $(\boldsymbol{\alpha}, T)$ and $(\boldsymbol{\beta}, S)$ but with the same means μ'_1 and μ'_2. Discuss how the variability of these distributions affects the behavior of the queues, all of which have the same traffic intensity.

Problem 6.4.88: Simulation of a Voice and Data Traffic: A communication system has $m \geq 2$ channels. Voice calls arrive to it according to a Poisson process of rate λ. If such a call finds a free channel, it is accepted and has an exponentially distributed processing time with parameter μ. If no channel is free, the call is lost.

a. Derive explicit expressions for the steady-state probabilities π_j, $0 \leq j \leq m$, that j channels are occupied. For $m = 10$ and $\lambda = 1$, find the value of μ for which $\sum_{r=0}^{7} \pi_r = 0.75$. Use these parameter values for the numerical study in part *c*.

b. The m communication channels are also used for data transmission. Data messages arrive according to a Poisson process of rate λ° and have exponential service times with parameter μ°. Such messages can wait. They are processed by the free processors. If r channels are free and there are at least r data messages, these are processed at a rate $r\mu^\circ$. Voice calls have priority. If all channels are busy, an arriving voice call is assigned to a channel at random. Any data job in process there is repeated later. Show that the communication system can be described by an $M/M/m$ queue in a random environment. Its state (i, j) signifies that there are i data jobs and j voice calls in the system. Write its generator Q for $m = 3$ and note that it is block tridiagonal, also that for $i \geq m$, blocks A_0, A_1, and A_2 are repeated as in the generator of a QBD-process. Verify that the queue is stable if and only if $\lambda^\circ < \mu^\circ \sum_{j=0}^{m} \pi_j (m - j)$. The ratio ρ of these quanitities is the traffic intensity of the queue.

c. In general, the arrival and service rates of data jobs are much higher than for voice calls. Even with $\rho < 1$, the queue of data jobs fluctuates rapidly. Using the parameters for the voice calls computed in part *a*, set $\mu^\circ = 200$ and calculate λ° for which $\rho = 0.75$. As efficiently as possible, simulate the behavior of the system immediately after the arrivals of the first 20,000 voice calls. Start the simulation with $i = 0$ and $j = m$. Visualize the behavior of the queue of data jobs at these 20,000 arrival epochs.

d. Suppose that you would like to reconstruct a stochastically equivalent continuous-time path of the system from information saved at the arrival epochs in part *c*. What (minimal) additional information would you need to store and how would you reconstruct the path in continuous time from it?

REFERENCES

The following are references to works that are highly relevant to the subject of this book. They deal with probability, statistics, stochastic processes, simulation methodology, matrix theory, or numerical analysis. In them, the reader will find theoretical developments and additional algorithmic problems.

Berger, J. O., *"Statistical Decision Theory and Bayesian Analysis."* Second Edition, New York: Springer-Verlag, 1985.

David, F. N. and Barton, D. E., *"Combinatorial Chance."* London: Ch. Griffin & Company Ltd., 1962.

Devroye, L., *"Non-Uniform Random Variate Generation."* New York: Springer-Verlag, 1986.

Feller, W., *"An Introduction to Probability Theory and Its Applications, Vol. I."* Third Edition; New York: J. Wiley & Sons, 1967.

Fishman, G. S., *"Principles of Discrete Event Simulation."* New York: J. Wiley and Sons, 1978.

Gantmacher, F. R., *"The Theory of Matrices."* New York, Chelsea, 1959.

Grenander, U., *"Mathematical Experiments on the Computer."* New York: Academic Press, 1982.

Johnson, N. L., Kotz, S., and Kemp, A., *"Univariate Statistical Distributions."* (Second Edition), New York: John Wiley & Sons, 1992.

Karlin, S. and Taylor, H. M., *"A First Course in Stochastic Processes."* 2d. ed. New York: Academic Press, 1975.

Muirhead, Robb J., *"Aspects of Multivariate Statistical Theory."* New York: John Wiley & Sons, 1982.

Neuts, M. F., *"Matrix-Geometric Solutions in Stochastic Models: An Algorithmic Approach."* Baltimore: The Johns Hopkins University Press, 1989.

Neuts, M. F., *"Structured Stochastic Matrices of M/G/1 Type and Their Applications."* New York: Marcel Dekker Inc., 1989.

Press, William H.; Flannery, Brian P.; Teukolsky, Saul A. and Vetterling, William T., *"Numerical Recipes: The Art of Scientific Programming."* Cambridge, England: Cambridge University Press, 1986. (Versions of this book with computer programs in Fortran, Pascal or C are available. Many computer centers have electronic libraries of the programs.)

Resnick, S. I., *"Adventures in Stochastic Processes."* Boston: Birkhäuser, 1992.

Riordan, J., *"Stochastic Service Systems."* New York: John Wiley & Sons.

Saad, Y., *"Numerical Methods for Large Eigenvalue Problems."* Manchester, U.K: Manchester University Press, 1992.

Scott, David R., *"Multivariate Density Estimation: Theory, Practice and Visualization."* New York: John Wiley & Sons, 1992.

Seneta, E., *"Nonnegative Matrices and Markov Chains."* Second Edition, New York: Springer-Verlag, 1981.

Tijms, H. C. *"Stochastic Models: An Algorithmic Approach."* New York: John Wiley & Sons, 1994.

Varga, R. S., *"Matrix Iterative Analysis."* Englewood Cliffs, New Jersey: Prentice-Hall, 1962.

APPENDIX 1

SOME TOPICS FROM MATRIX ANALYSIS

In this appendix, we summarize some properties of matrices that are frequently used in probability modelling. The summaries provide the factual information necessary to do the problems on Markov chains. However, the reader is encouraged to study the proofs and further developments in the sources that are listed.

A.1.1: FUNCTIONS OF A MATRIX

Suppose $f(z)$ is a function that can be represented by a convergent power series $f(z) = \sum_{n=0}^{\infty} a_n z^n$, in an open disk $|z| < r$. If A is an $m \times m$ matrix whose eigenvalues η_j, $1 \le j \le m$, satisfy $|\eta_j| < r$, we can define the function $f(A)$ of the matrix A. First suppose that A is *diagonalizable,* i.e., A can be written as $A = H^{-1}\Xi H$, where Ξ is an $m \times m$ diagonal matrix with the eigenvalues η_j of A as its diagonal elements. For such matrices, $f(A)$ is defined by

$$f(A) = H^{-1} f(\Xi) H,$$

where $f(\Xi)$ is the diagonal matrix with the values $f(\eta_j)$ as its corresponding diagonal elements. For matrices A that are not diagonalizable, $f(A)$ is defined by continuity. A is then approximated by a sequence of diagonalizable matrices A_n and $f(A)$ is defined as the limit of the sequence of matrices $f(A_n)$. There are minor technical difficulties in showing that such a definition by continuity is consistent, that is, that the required limit exists and does not depend on the sequence $\{A_n\}$.

The two most important functions of a matrix argument used in this book are inverses of the form $(I - A)^{-n}$, defined by

$$(I - A)^{-n} = \sum_{k=0}^{\infty} \binom{n + k - 1}{k} A^k, \quad \text{for } n \ge 1.$$

The series converges for all matrices A whose eigenvalues satisfy $|\eta_j| < 1$; and the *matrix exponential function,* defined by the series

$$\exp(A) = \sum_{k=0}^{\infty} \frac{1}{k!} A^k,$$

which converges for all matrices A.

As a general rule, all familar formulas for scalar analytic functions extend to the corresponding functions of a matrix argument unless their derivations depend on nonsingularity or commutativity. This is illustrated by the following examples which are used in some of the chapters.

Example A.1.1.1: Some Properties of the Matrix-Exponential: If the $m \times m$ matrices A and B *commute,* the equality $\exp(A + B) = \exp(A)\exp(B)$ holds. For noncommutative matrices, that equality is in general not valid. In calculations for Markov chains, we frequently use that equality with $B = aI$ to write that

$$e^a \exp(A) = \exp(aI + A).$$

Matrix integrals of the form

$$\int_0^x f(u)\exp(Au)\, du$$

are common in probability. When the matrix A is *nonsingular,* they can frequently be calculated by using the same formal steps as used to evaluate the scalar integrals

$$\int_0^x f(u)\exp(au)\, du.$$

For example, if the function $f(\cdot)$ is differentiable, we have the usual formula for integration by parts

$$\int_0^x f(u)\exp(Au)\, du = f(x)\exp(Ax)A^{-1} - f(0)A^{-1} - A^{-1}\int_0^x f'(u)\exp(Au)\, du.$$

Similarly, for s with $\mathrm{Re}(s) \geq 0$, we evaluate the matrix-Laplace transform

$$\int_0^\infty e^{-su}\exp(Au)\, du = \int_0^\infty \exp[-(sI - A)u]\, du = (sI - A)^{-1}.$$

That formula is valid for any square matrix A with all eigenvalues in the complex left half-plane. It is frequently used in calculations for continuous-time Markov chains. To illustrate the care needed when A is singular, we consider the integral

$$M(x) = \int_0^x \exp(Qu)\, du,$$

where Q is an irreducible generator. The element $M_{ij}(x)$ is the conditional expected value of the time spent by the Markov chain in the state j during the interval $(0, x)$, given the chain starts in state i.

If $\boldsymbol{\theta}$ is the stationary probability vector of Q, the matrix $\mathbf{e}\boldsymbol{\theta} - Q$ is nonsingular (see Problem 5.3.17.) We write:

$$M(x)(\mathbf{e}\boldsymbol{\theta} - Q) = \int_0^x \exp(Qu)(\mathbf{e}\boldsymbol{\theta} - Q)\, du = \mathbf{e}\boldsymbol{\theta}x - \int_0^x \exp(Qu)Q\, du$$

$$= \mathbf{e}\boldsymbol{\theta}x - [\exp(Qx) - I],$$

since the integrand in the last integral is the derivative of $\exp(Qu)$. It follows that

$$M(x) = \mathbf{e}\boldsymbol{\theta}x + [I - \exp(Qx)](\mathbf{e}\boldsymbol{\theta} - Q)^{-1}, \qquad \text{for } x \geq 0.$$

Derivatives: Because of noncommutativity, the calculation of *derivatives* of functions of a matrix argument requires particular attention. For matrices $A(z)$ whose elements depend on z, we consider the calculation of derivatives of matrix functions of the form:

$$f[A(z)] = \sum_{n=0}^{\infty} a_n [A(z)]^n.$$

If $A(z)$ and $B(z)$ are matrices whose elements are differentiable in z, the usual product rule yields that

$$[A(z)B(z)]' = A'(z)B(z) + A(z)B'(z),$$

so that, for example,

$$\{[A(z)]^n\}' = \sum_{k=0}^{n-1} [A(z)]^k A'(z)[A(z)]^{n-k-1},$$

because, in general, $A(z)$ and $A'(z)$ do not commute. If the matrix $V(z) = [A(z)]^{-1}$ exists in a neighborhood of z, its derivative is given by

$$V'(z) = \{[A(z)]^{-1}\}' = -[A(z)]^{-1}A'(z)[A(z)]^{-1}.$$

To show this, we differentiate in the equation $V(z)A(z) = I$ and solve for $V'(z)$. Combining the preceding two formulas, we obtain that

$$\{[A(z)]^{-n}\}' = -\sum_{k=0}^{n-1} [A(z)]^{-k-1}A'(z)[A(z)]^{k-n}, \quad \text{for } n \geq 1.$$

In the discussion of finite Markov chains we often encounter matrices of the form $K_n(z) = (I - zT)^{-n}$. In this case, the matrix $I - zT$ and its derivative $-T$ commute, so that the derivative $K'_n(z)$ is given by the usual formula

$$K'_n(z) = n(I - zT)^{-n-1}T.$$

Higher derivatives can also be calculated in the usual manner.

In general, the derivative of $f[A(z)]$ is given by

$$\frac{d}{dz} f[A(z)] = \sum_{n=1}^{\infty} a_n \sum_{k=0}^{n-1} [A(z)]^k A'(z)[A(z)]^{n-k-1}.$$

For example, the derivative of the matrix exponential $\exp[A(z)]$ is

$$\{\exp[A(z)]\}' = \sum_{n=1}^{\infty} \frac{1}{n!} \sum_{k=0}^{n-1} [A(z)]^k A'(z)[A(z)]^{n-k-1}.$$

The second derivative, obtained by a further term-by-term differentiation, leads to a double summation which, in general, cannot be simplified.

A.1.2: THE KRONECKER PRODUCT

For two matrices A and B of dimensions $k_1 \times k_2$ and $k_1' \times k_2'$, their *Kronecker product* $A \otimes B$ is the matrix of dimensions $k_1 k_1' \times k_2 k_2'$, written in partitioned form as

$$\begin{vmatrix} A_{11}B & A_{12}B & \cdots & A_{1k_2}B \\ \cdots & & & \cdots \\ A_{k_1 1}B & A_{k_1 2}B & \cdots & A_{k_1 k_2}B \end{vmatrix}.$$

In working with Markov chains or *MAP* s, the Kronecker product is used to express probabilities for systems consisting of two or more independent processes. A useful property, of frequent use in calculations with Kronecker products, is the following: If A, B, C, and D are rectangular matrices whose ordinary matrix products AC and BD are defined, then

$$(A \otimes B)(C \otimes D) = AC \otimes BD.$$

For $m \times m$ and $n \times n$ matrices A and B, the matrix

$$A \oplus B = A \otimes I_n + I_m \otimes B,$$

where I_r denotes an identity matrix of order r, is called the *Kronecker sum,* of A and B. That operation is useful in studying two (or more) simultaneous, independent continuous-time Markov chains. An important property of the matrix exponential sum is the equality

$$\exp[A \oplus B] = \exp(A) \otimes \exp(B),$$

which holds for any square matrices A and B. A simple proof of that equality goes as follows: Consider the matrix $V(t) = \exp[(A \oplus B)t]$, which is the unique solution of the differential equation

$$V'(t) = V(t)[A \oplus B], \quad \text{with} \quad V(0) = I_m \otimes I_n = I_{mn}.$$

The derivative of the matrix $\exp(At) \otimes \exp(Bt)$ is successively given by

$$\exp(At)A \otimes \exp(Bt) + \exp(At) \otimes \exp(Bt)B =$$

$$= [\exp(At) \otimes \exp(Bt)](A \otimes I) + [\exp(At) \otimes \exp(Bt)](I \otimes B)$$

$$= [\exp(At) \otimes \exp(Bt)](A \oplus B),$$

so that the matrix $\exp(At) \otimes \exp(Bt)$ satisfies the differential equation for $V(t)$. Since the solution to the differential equation is unique, that implies the stated equality.

Operations with Kronecker products frequently lead to large matrices. In many computations, the following operation is useful in reducing the order of matrices to be stored or in avoiding unnecessary arithmetic operations.

To multiply the Kronecker product $A \otimes B$ of matrices A and B of dimensions $k_1 \times k_2$ and $k_1' \times k_2'$, by the column vector **u** with $k_2 k'_2$ components, we partition **u** into k_2 vectors $\mathbf{u}(1), \cdots, \mathbf{u}(k_2)$ of dimension k'_2. We next form the $k_2 \times k'_2$ matrix U with columns $\mathbf{u}(1), \cdots, \mathbf{u}(k_2)$. The vector **u** is said to be the *direct sum* of the columns of U. The product $(A \otimes B)\mathbf{u} = \mathbf{v}$ is now a vector partitioned into k_1 vectors $\mathbf{v}(i)$ of dimension k'_1, given by $\mathbf{v}(i) = \sum_{r=1}^{k_2} A_{ir} (BU)_r$, where $(BU)_r$ is the rth column of BU. We see that $\mathbf{v}(i)$ is the ith column of the matrix BUA^T, where A^T is the transpose of A. Forming the matrix V of which the direct sum of the columns is **v**, we see that $V = BUA^T$. The matrix V is obtained as a product of smaller matrices and there is no need to store $A \otimes B$. The vector **v** is obtained by unwrapping V into the direct product of its columns. By a similar construction, we obtain the row vector $\mathbf{u}(A \otimes B) = \mathbf{v}$ from the matrix product $V = A^T UB$. In this case, **u** and **v** are the direct products of the rows rather than the columns of U and V.

A.1.3: THE PERRON - FROBENIUS EIGENVALUE

Irreducible nonnegative matrices play an important role in many problems of applied probability. The following is a summary of their most basic properties. For future reference, we shall label them by [P-F X].

[P-F 1] If A is an irreducible nonnegative $m \times m$ matrix, it has a positive eigenvalue η which is simple. All its other eigenvalues η_j, $1 \le j < m$, satisfy $| \eta_j | \le \eta$. η is called the *Perron - Frobenius eigenvalue* of A.

[P-F 2] In addition, the left and right eigenvectors **u** and **v**, corresponding to η, have all components of the same sign. With the normalization

$$\mathbf{ue} = 1, \quad \text{and} \quad \mathbf{uv} = 1,$$

these vectors are uniquely determined and positive.

[P-F 3] If an irreducible nonnegative matrix A has a nonnegative (left or right) eigenvector, that eigenvector corresponds to η.

[P-F 4] If A is irreducible and nonnegative and B is a (complex) matrix satisfying $|B_{ij}| \leq A_{ij}$ for all (i,j), then for all eigenvalues λ_j of B, $\lambda_j \leq \eta$.

[P-F 5] **Periodicity:** If there are $r > 1$ eigenvalues η_k, $1 \leq k \leq r$, with the same modulus as η, they lie at the vertices of a regular r–gon, inscribed in the circle of radius η with center at (0, 0). These eigenvalues are then given by

$$\eta_k = \eta \exp[\frac{2\pi i \ k}{r}], \quad \text{for } 1 \leq k \leq r.$$

In that case, a permutation of both row and column indices allows the matrix to be written in the form:

$$A = \begin{vmatrix} 0 & A_1 & 0 & 0 & \cdots & 0 & 0 \\ 0 & 0 & A_2 & 0 & \cdots & 0 & 0 \\ 0 & 0 & 0 & A_3 & \cdots & 0 & 0 \\ \cdots & & & \cdots & & & \cdots \\ 0 & 0 & 0 & 0 & \cdots & 0 & A_{r-1} \\ A_r & 0 & 0 & 0 & \cdots & 0 & 0 \end{vmatrix},$$

where $A_1, \cdots, A_r$ are square matrices of order m/r. For finite stochastic matrices, that corresponds to *periodicity* of the corresponding Markov chain.

Remark: Irreducible nonnegative matrices are sometimes called *indecomposable*. We do not use that term in this text. An irreducible matrix for which the moduli of all other eigenvalues are strictly smaller than the Perron eigenvalue η is sometimes called a *primitive* matrix. In discussions of Markov chains, we call such a matrix *aperiodic*.

[P-F 6] **The Limiting Behavior of A^n:** If all other eigenvalues have a *smaller* modulus than η (the aperiodic case), the following very important limit result holds

$$\lim_{n \to \infty} \eta^{-n} A^n = \mathbf{vu}.$$

The right-hand side is an $m \times m$ matrix with elements $v_i u_j$. A more refined statement of that limit is that as $n \to \infty$,

$$A^n = \eta^n \mathbf{vu} + O[n^{r-1} \eta_1^n].$$

where η_1 is the next largest modulus of an eigenvalue of A and r is the multiplicity of that eigenvalue. This says that, the smaller the ratio of the modulus η_1 to η, the faster the difference between $\eta^{-n} A^n$ and $\mathbf{vu}$ tends to zero.

Principal Submatrices: A principal submatrix of A is a matrix obtained by deleting a set of rows and columns with the same indices. The principal submatrix $A(i_1, \cdots, i_r)$ is obtained by deleting the rows and columns with indices not in the set $\{i_1, \cdots, i_r\}$.

[P-F 7] The modulus of all eigenvalues of $A(i_1, \cdots, i_r)$ is less than η, the Perron - Frobenius eigenvalue of A.

In probability, the preceding results are most commonly applied to the *(sub)stochastic* matrices encountered for discrete-time Markov chains. The principal submatrices of the generator Q for continuous-parameter Markov chains are a subclass of the *semi-stable* matrices. Their properties are analogous to those of nonnegative matrices. Statements about the Perron - Frobenius eigenvalue and the corresponding eigenvectors are easily translated to properties of the eigenvalues of semi-stable matrices.

Semi-Stable Matrices: A (real) $m \times m$ matrix A is *semi-stable* if its diagonal elements A_{ii} are negative, all off-diagonal elements are nonnegative and for all $1 \le i \le m$,

$$\sum_{j=1}^{m} A_{ij} \le 0.$$

Such a matrix is called *stable,* if at least one inequality is strict. It is *irreducible* if and only if the nonnegative matrix obtained by setting all diagonal elements equal to zero, is irreducible.

Let θ satisfy $\theta \geq \max\{-A_{ii}\}$, then $A + \theta I$ is an irreducible nonnegative matrix. We denote its Perron - Frobenius eigenvalue by $\eta(\theta)$ and its (normalized) left and right eigenvectors by **u** and **v**. If η' is an eigenvalue of $A + \theta I$, then clearly, $\eta' - \theta$ is an eigenvalue of A; the eigenvalues of A are obtained by subtracting θ from those of $A + \theta I$. Therefore $\eta(\theta) - \theta$ is the *eigenvalue with largest real part* of A (and does not depend on θ). The matrix $I + \theta^{-1}A$ is substochastic, so that its Perron - Frobenius eigenvalue $\theta^{-1}\eta(\theta)$ is at most one. Therefore, $-\eta^* = \eta(\theta) - \theta$ is nonpositive. It follows that

[P-F 8] The eigenvalue $-\eta^*$ of an irreducible semi-stable matrix A is real, simple, and nonpositive. ($-\eta^*$ is also called the Perron - Frobenius eigenvalue.) The eigenvectors **u** and **v** satisfy [P-F 2].

[P-F 9] If an irreducible semi-stable matrix A has a nonnegative (left or right) eigenvector, that eigenvector corresponds to $-\eta^*$.

[P-F 10] The exponential matrix $\exp(At)$ has the following limit:

$$\lim_{t \to \infty} e^{\eta^* t} \exp(At) = \mathbf{vu}.$$

A refined statement of that limit is that, as $t \to \infty$,

$$\exp(At) = e^{-\eta^* t}\mathbf{vu} + O[t^{r-1}e^{-\eta_1 t}].$$

The quantity $-\eta_1$ is the real part of the eigenvalue with the next smallest real part, and r is the multiplicity of that eigenvalue.

Reference: Gantmacher, F. R., *"The Theory of Matrices,"* New York, Chelsea, 1959, and Seneta, E., *"Nonnegative Matrices and Markov Chains,"* Second Edition, New York: Springer-Verlag, 1981.

APPENDIX 2

PHASE-TYPE DISTRIBUTIONS

This appendix deals with properties of *PH*-distributions that are useful in solving problems in this book. Discrete and continuous *PH*-distributions have analogous properties proved by similar techniques. We shall sketch proofs for either case and leave the other to the initiative of the reader. In some instances, we give proofs by insight rather than by formal calculation. Such insight is important when *PH*-distributions are used as building blocks in more involved Markovian models.

A.2.1: *PH*-DISTRIBUTIONS

The definitions are given, for the discrete case, in Example 4.1.2 and, for the continuous case, in Example 5.1.2. A *PH*-distribution is specified by the row vector $\boldsymbol{\alpha}$ and the matrix T. The pair $(\boldsymbol{\alpha}, T)$ is called a *representation* of the distribution. Representations are *not unique*. For example, the pairs $\alpha = 1$, $T = -\lambda$ and $\boldsymbol{\alpha} = (\theta, 1-\theta)$ and

$$T = \begin{vmatrix} -\lambda-\mu & \mu \\ \mu & -\lambda-\mu \end{vmatrix}$$

both represent the exponential distribution. In fact, every *PH*-distribution can be represented in infinitely many ways. That raises interesting and quite difficult mathematical questions about minimal representations, that is, the representations requiring the smallest number of phases. For computations, such issues are not of immediate concern. The most commonly used *PH*–distributions have simple, natural representations. For example, the *hyperexponential* distributions

$$F(x) = \textstyle\sum_{\nu=1}^{m} \alpha_\nu (1 - e^{-\lambda_\nu x}),$$

are represented by $\boldsymbol{\alpha} = (\alpha_1, \ldots, \alpha_m)$, $\alpha_0 = 0$, and $T = -\,diag(\lambda_1, \ldots, \lambda_m)$, and the (mixed) *Erlang* distributions $F(x) = \sum_{\nu=1}^{m} p_\nu E_\nu(\lambda; x)$, by $\boldsymbol{\alpha} = (p_m,$

$p_{m-1}, \cdots, p_1)$, $\alpha_0 = 0$, and

$$T = \begin{vmatrix} -\lambda & \lambda & 0 & \cdots & 0 & 0 & 0 \\ 0 & -\lambda & \lambda & \cdots & 0 & 0 & 0 \\ & \cdots & & & \cdots & & \\ 0 & 0 & 0 & \cdots & 0 & -\lambda & \lambda \\ 0 & 0 & 0 & \cdots & 0 & 0 & -\lambda \end{vmatrix} .$$

Irreducible Representations: In computations with *PH*-distributions, we should avoid representations with superfluous phases. As a simple example, the hyperexponential distribution

$$F(x) = 1 - \alpha e^{-\lambda_1 x} - (1 - \alpha) e^{-\lambda_2 x},$$

has a superfluous phase when α is 0 or 1. In general, a phase i is superfluous if the ith component of the vector $\boldsymbol{\alpha}\exp(Tx)$, or of $\boldsymbol{\alpha}T^n$ in the discrete case, is identically zero in x or in n.

It may be shown there are no superfluous phases if and only if in the continuous (discrete) case, the matrix $T + (1 - \alpha_0)^{-1}\mathbf{T}^{\circ}\boldsymbol{\alpha}$ is an irreducible generator (stochastic matrix). The representation $(\boldsymbol{\alpha}, T)$ is then called *irreducible*. In this book, we always assume that the given representations of *PH*-distributions are irreducible. This causes no loss of generality, but the issue is important. Using a reducible representation in a computer code can induce errors.

Testing Code with Hidden Exponential Distributions: For models involving *PH*-distributions, the special case where all distributions are exponential is always much simpler and is often analytically tractable. We test a general code (and implicitly the matrix derivations) as follows:

The exponential case is solved explicitly and a separate (simpler) code is written for its numerical solution. The general code is tested by entering representations $(\boldsymbol{\alpha}, T)$ with $m > 1$ phases for the same exponentials. Of course, any numerical results that do not depend on the chosen representation must agree.

To do such tests it is useful to maintain a data file with a few representations of hidden exponential distributions. These are constructed by choosing T to be an irreducible stable matrix for which $T\mathbf{e} = -\lambda\mathbf{e}$. Alternatively, we may calculate the Perron eigenvalue $-\eta^*$ and the corresponding left eigenvector $\mathbf{u}$ of any irreducible stable matrix T. We normalize $\mathbf{u}$ by $\mathbf{ue} = 1$ and set $\boldsymbol{\alpha} = \mathbf{u}$. The PH-distribution is then exponential with parameter $-\eta^*$. Codes for models with discrete PH-distributions can similarly be tested by hidden *geometric* distributions.

Closure Properties: When applied to PH-distributions, many standard constructions of probability theory again yield PH-distributions. We say that the family of PH-distributions is *closed* under these constructions. While closure properties of great generality are known, the practically useful ones are those for which a convenient representation for the resulting distributions can be written down. Closure properties often yield representations with much larger matrices. Exploiting their structure in numerical computations then becomes essential.

In what follows, $F_1(\cdot)$ and $F_2(\cdot)$ will both be continuous or discrete PH-distributions with representations $[\boldsymbol{\alpha}(1), T(1)]$ and $[\boldsymbol{\alpha}(2), T(2)]$ and m_1 and m_2 phases, respectively. The probabilities of instantaneous absorption are $\alpha_0(1)$ and $\alpha_0(2)$. The vectors $\mathbf{T}^\circ(1)$ and $\mathbf{T}^\circ(2)$ are defined in the usual manner and depending on whether $F_1(\cdot)$ and $F_2(\cdot)$ are continuous or discrete PH-distributions. The following are the principal closure properties of PH-distributions:

[PH-1] The mixture $pF_1(\cdot) + (1-p)F_2(\cdot)$ is a PH-distribution with (α, T) given by $\boldsymbol{\alpha} = [p\boldsymbol{\alpha}(1), (1-p)\boldsymbol{\alpha}(2)]$ and

$$T = \left| \begin{matrix} T(1) & 0 \\ 0 & T(2) \end{matrix} \right| .$$

[PH-2] The convolution $F_1(\cdot) * F_2(\cdot)$ is a PH-distribution with representation (α, T) given by $\boldsymbol{\alpha} = [\boldsymbol{\alpha}(1), \alpha_0(1)\boldsymbol{\alpha}(2)]$ and

$$T = \left| \begin{matrix} T(1) & \mathbf{T}^\circ(1)\boldsymbol{\alpha}(2) \\ 0 & T(2) \end{matrix} \right| .$$

Properties [PH-1] and [PH-2] readily extend to more than two distributions. They can easily be proved by interpretation. To establish [PH-2], we start the absorbing Markov chain for F_1 in one of its transient states according to the vector $\boldsymbol{\alpha}(1)$ unless instantaneous absorption occurs. Upon absorption in that first Markov chain, we immediately start the absorbing chain for F_2 in one of its transient states unless that chain is instantaneously absorbed. The stated representation is obvious by considering the combined process. The term $\alpha_0 = \alpha_0(1)\alpha_0(2)$ is the probability that both chains are immediately absorbed. The vector $\mathbf{T}^\circ$ is given by

$$\mathbf{T}^\circ = \begin{vmatrix} \alpha_0(2)\mathbf{T}^\circ(1) \\ \\ \mathbf{T}^\circ(2) \end{vmatrix}.$$

[PH-3] If X_1 and X_2 are independent with PH-distributions $F_1(\cdot)$ and $F_2(\cdot)$, the random variables $Z = \min(X_1, X_2)$ and $U = \max(X_1, X_2)$ have PH-distributions. Representations are given in Example 5.1.4.

Let $F(\cdot)$ be a continuous PH-distribution with representation $(\boldsymbol{\alpha}, T)$. Since we assume the representation to be irreducible, the matrix $Q^* = T + (1 - \alpha_0)^{-1}\mathbf{T}^\circ\boldsymbol{\alpha}$ is an irreducible generator. Its stationary probability vector $\boldsymbol{\theta}^*$ satisfies the equation

$$\boldsymbol{\theta}^* = (1 - \alpha_0)^{-1}(\boldsymbol{\theta}^*\mathbf{T}^\circ)\boldsymbol{\alpha}(-T)^{-1}.$$

Since $\boldsymbol{\theta}^*$ is a probability vector, we obtain that $\boldsymbol{\theta}^*\mathbf{T}^\circ = (1 - \alpha_0)(\mu'_1)^{-1}$, where $\mu'_1 = \boldsymbol{\alpha}(-T)^{-1}\mathbf{e}$ is the mean of $F(\cdot)$. It follows that

$$\boldsymbol{\theta}^* = \frac{1}{\mu'_1}\boldsymbol{\alpha}(-T)^{-1}.$$

[PH-4] The probability distribution $F^*(\cdot)$ defined by

$$F^*(x) = \frac{1}{\mu'_1}\int_0^x [1 - F(u)]\, du,$$

is a PH-distribution with representation $(\boldsymbol{\theta}^*, T)$.

Proof: A proof by calculation of [PH-4] proceeds as follows:

$$F^*(x) = \frac{1}{\mu'_1}\int_0^x \boldsymbol{\alpha}\exp(Tu)\, du\ \mathbf{e}$$

$$= \frac{1}{\mu'_1}\boldsymbol{\alpha}(-T)^{-1}[I - \exp(Tx)]\mathbf{e} = 1 - \boldsymbol{\theta}^*\exp(Tx)\mathbf{e}.$$

In the discussion of the PH-renewal process we give an argument by interpretation. •

The discrete analogue of [PH-4] states that if $\{\phi_r\}$ is a PH-density with representation $(\boldsymbol{\alpha}, T)$, the sequence with terms

$$\phi_n^* = \frac{1}{\mu'_1}[1 - \sum_{r=0}^{n} \phi_r], \quad \text{for } n \geq 0,$$

is a PH-density represented by $\boldsymbol{\theta}^* = (\mu'_1)^{-1}\boldsymbol{\alpha}(I - T)^{-1}$ and T.

For the next closure property, we consider infinite mixtures of the form

$$G(x) = \sum_{r=0}^{\infty} \phi_r F^{(r)}(x),$$

where $\{\phi_r\}$ is a discrete PH-density with representation $(\boldsymbol{\beta}, S)$, $F(\cdot)$ is a continuous PH-distribution with representation (α, T), and $F^{(r)}(\cdot)$ is the r-fold convolution of $F(\cdot)$.

[PH-5] $G(\cdot)$ is a PH-distribution with representation $(\boldsymbol{\gamma}, L)$ given by

$$\gamma = \alpha \otimes \boldsymbol{\beta}(I - \alpha_0 S)^{-1}, \qquad L = T \otimes I + \mathbf{T}^{\circ}\alpha \otimes (I - \alpha_0 S)^{-1} S.$$

The height of the jump at 0 is given by $\gamma_0 = \beta_0 + \alpha_0 \boldsymbol{\beta}(I - \alpha_0 S)^{-1}\mathbf{S}^{\circ}$, and the vector $\mathbf{L}^{\circ}$ by $\mathbf{L}^{\circ} = \mathbf{T}^{\circ} \otimes (I - \alpha_0 S)^{-1}\mathbf{S}^{\circ}$.

Proof: A proof by calculation of property [PH-5] is quite lengthy and involves extensive manipulations with Kronecker products. The following is an argument by interpretation: Consider a discrete-time clock for the density $\{\phi_r\}$ and a continuous-time clock for the distribution $F(\cdot)$. We shall show that $G(\cdot)$ is the distribution of the absorption time τ in a Markov chain with the stated parameters.

With probability β_0, there is only the term for $r = 0$ in the sum for $G(\cdot)$. It is also possible that every time we advance the discrete-time clock, there is instant absorption in the Markov chain that emulates $F(\cdot)$. That may go on up to the absorption time in the discrete Markov chain for $\{\phi_r\}$. The height of the jump at 0 for $G(\cdot)$ is therefore $\gamma_0 = \beta_0 + \sum_{r=1}^{\infty} \boldsymbol{\beta} S^{r-1}\mathbf{S}^{\circ}\alpha_0^r$, which is equivalent to the stated formula.

If τ is positive, there may still be a number of instantaneous absorptions, but eventually the continuous-time clock is started. That yields the expression for $\boldsymbol{\gamma}$. The matrix $T \otimes I$ in L corresponds to the case where that clock is running and the discrete-time clock remains fixed. When absorption occurs in the continuous chain for $F(\cdot)$, we first check whether there is also absorption in the discrete chain for $\{\phi_r\}$. If not, we advance the discrete-time clock and instantaneously try restarting the continuous clock. There could be a number of instantaneous absorptions in the continuous chain, but eventually that chain is restarted in one of its transient states. That alternative contributes the second term in the matrix L.

The form of the vector $\mathbf{L}^\circ$ corresponds to the various ways in which the random duration τ may end. At the end of a nonzero lifetime of the continuous clock, the discrete clock may terminate. Alternatively, that clock may stop with an instantaneous absorption in the continuous Markov chain. •

In thinking through that argument, the reader may want to set $\alpha_0 = \beta_0 = 0$ first. For that case, the formulas and the argument are simpler. A useful special case arises when $\{\phi_r\}$ is *geometric,* and $\alpha_0 = 0$. To obtain the density $\phi_r = pq^r$, for $r \geq 0$, we set $\beta_0 = p$, $\boldsymbol{\beta} = q$, $S = q$, and $\mathbf{S}^\circ = p$. For that case, the parameters of the geometric mixture $G(\cdot)$ of the successive convolutions of $F(\cdot)$ are given by

$$\gamma_0 = p, \quad \boldsymbol{\gamma} = q\boldsymbol{\alpha}, \quad L = T + q\mathbf{T}^\circ\boldsymbol{\alpha}, \quad \mathbf{L}^\circ = p\mathbf{T}^\circ.$$

Property [PH-5] also holds when the distribution $F(\cdot)$ has a discrete PH-density. Formally, the representation is the same but T is now a substochastic matrix.

The following property states that the time spent in a subset of the transient states in an absorbing finite Markov chain has a PH-distribution. We state and prove that property for the discrete case. The corresponding result for the continuous case is left as an exercise (Problem 5.2.36). Consider a Markov chain with a single absorbing state that is reached from every initial state. Its transition probability matrix which is of the form

$$P = \begin{vmatrix} T & \mathbf{T}^\circ \\ 0 & 1 \end{vmatrix},$$

is further partitioned as

$$P = \begin{array}{c} 1 \\ 2 \\ * \end{array} \left| \begin{array}{ccc} T(1,1) & T(1,2) & \mathbf{T}^{\circ}(1) \\ T(2,1) & T(2,2) & \mathbf{T}^{\circ}(2) \\ 0 & 0 & 1 \end{array} \right| .$$

The square submatrices $T(1, 1)$ and $T(2, 2)$ correspond to transitions *within* subsets 1 and 2 of transient states. The submatrices $T(1, 2)$ and $T(2, 1)$ correspond to transitions from set 1 to 2 and vice versa. The initial probability vector is written in the corresponding partioned form $[\boldsymbol{\alpha}(1), \boldsymbol{\alpha}(2), \alpha_0]$.

[PH-6] The number N of time units spent by the Markov chain in the subset 1 prior to absorption in the state * has a PH-distribution equal to that of the absorption time in the Markov chain with

$$P^* = \left| \begin{array}{cc} S & \mathbf{S}^{\circ} \\ 0 & 1 \end{array} \right| ,$$

and initial probability vector $[\boldsymbol{\gamma}, \gamma_0]$, where

$$S = T(1, 1) + T(1, 2)[I - T(2, 2)]^{-1}T(2, 1),$$

$$\mathbf{S}^{\circ} = \mathbf{T}^{\circ}(1) + T(1, 2)[I - T(2, 2)]^{-1}\mathbf{T}^{\circ}(2),$$

$$\boldsymbol{\gamma} = \boldsymbol{\alpha}(1) + \boldsymbol{\alpha}(2)[I - T(2, 2)]^{-1},$$

$$\gamma_0 = \alpha_0 + \boldsymbol{\alpha}(2)[I - T(2, 2)]^{-1}\mathbf{T}^{\circ}(2).$$

Proof: By counting only visits to set 1, we see that the generating function of the random variable N is $P(z) = \alpha_0 + [z\boldsymbol{\alpha}(1), \boldsymbol{\alpha}(2)][I - T(z)]^{-1}\mathbf{T}^{\circ}$, where the matrix $I - T(z)$ has the form

$$I - T(z) = \left| \begin{array}{cc} I - zT(1, 1) & -T(1, 2) \\ -zT(2, 1) & I - T(2, 2) \end{array} \right| .$$

We write that matrix as the sum of two matrices. The first consists of all the blocks that do not contain z; the second of those that do. Factoring out the first matrix, which is clearly nonsingular, $I - T(z)$ may be written as

$$\begin{vmatrix} I & -T(1, 2) \\ 0 & I - T(2, 2) \end{vmatrix} \times \begin{vmatrix} I - zT(1, 1) - zLT(2, 1) & 0 \\ -zKT(2, 1) & I \end{vmatrix},$$

where the matrices K and L are defined by

$$K = [I - T(2, 2)]^{-1}, \quad \text{and} \quad L = T(2, 1)[I - T(2, 2)]^{-1}.$$

Clearly, $T(1, 1) + LT(2, 1) = S$. The inverse of the first matrix in the product is

$$\begin{vmatrix} I & L \\ 0 & K \end{vmatrix}.$$

Next, we take the inverse of the matrix product and multiply it on the right by the column vector

$$\begin{vmatrix} \mathbf{T}^{\circ}(1) \\ \mathbf{T}^{\circ}(2) \end{vmatrix}$$

and on the left by the row vector $[z\alpha(1), \alpha(2)]$. After routine matrix calculations we find that $P(z)$ is given by $P(z) = \gamma_0 + z\gamma(I - S)^{-1}\mathbf{S}^{\circ}$, which is the generating function of the PH-density with representation (γ, S). It is easy to verify that $S\mathbf{e} + \mathbf{S}^{\circ} = \mathbf{e}$, and $\gamma_0 + \gamma\,\mathbf{e} = 1$. •

The following two closure properties hold for discrete PH-densities only.

> Every probability density on a finite number of nonnegative integers is a PH-density.
> If the random variable N has a discrete PH-density, so does $N + n$ for every positive integer n.

Proofs of these properties are straightforward exercises.

Avoiding Numerical Integrations: In many cases, probability formulas that require numerical integrations can be evaluated by matrix-iterative procedures when the underlying distributions are of phase type. We illustrate this by a few useful examples.

The probability a_k of k arrivals in a Poisson process of rate λ during an (independent) interval of duration X with distribution $F(\cdot)$ is

$$a_k = \int_0^\infty e^{-\lambda u} \frac{(\lambda u)^k}{k!} \, dF(u), \quad \text{for } k \geq 0.$$

[PH-7] When $F(\cdot)$ is a PH-distribution with representation $(\boldsymbol{\alpha}, T)$, $\{a_k\}$ is a discrete PH-density with representation

$$\boldsymbol{\beta} = \lambda \boldsymbol{\alpha} (\lambda I - T)^{-1}, \quad S = \lambda (\lambda I - T)^{-1}.$$

Proof: We prove property [PH-7] by using generating functions. In terms of the Laplace - Stieltjes transform $f(s)$ of $F(\cdot)$, the generating function $A(z)$ of $\{a_k\}$ is

$$A(z) = f(\lambda - \lambda z) = \alpha_0 + \boldsymbol{\alpha}[(\lambda(1 - z)I - T]^{-1}\mathbf{T}^\circ$$

$$= \alpha_0 + \boldsymbol{\alpha}(\lambda I - T)^{-1}\mathbf{T}^\circ + z\,\boldsymbol{\alpha}\lambda(\lambda I - T)^{-1}[I - z\lambda(\lambda I - T)^{-1}]^{-1}(\lambda I - T)^{-1}\mathbf{T}^\circ$$

$$= \beta_0 + z\,\boldsymbol{\beta}(I - zS)^{-1}\mathbf{S}^\circ.$$

That is clearly the generating function of a discrete PH-density. •

In the next example, the evaluation of the convolution of a PH-distribution $F(\cdot)$ and a general distribution function $H(\cdot)$ on $[0, \infty)$ is reduced to solving a linear differential equation. The steps in that reduction can be applied in many other contexts. To avoid uninteresting details, we assume that $\alpha_0 = 0$. For $x \geq 0$, the convolution $F * H(\cdot)$ is given by

$$F * H(x) = \int_0^x F(u)\, \boldsymbol{\alpha} \exp[T(x - u)]\mathbf{T}^\circ \, du = \boldsymbol{\alpha} \int_0^x F(u) \exp[T(x - u)] \, du \; \mathbf{T}^\circ.$$

Set

$$\boldsymbol{\phi}(x) = \boldsymbol{\alpha}\int_0^x F(u) \exp[T(x - u)]\, du,$$

then

$$\boldsymbol{\phi}(x) \exp(-Tx) = \boldsymbol{\alpha}\int_0^x F(u) \exp(-Tu)\, du.$$

Differentiating both sides of that equality and multiplying both sides of the resulting equation by $\exp(Tx)$, we obtain the differential equation

$$\boldsymbol{\phi}'(x) = \boldsymbol{\phi}(x)T + F(x)\boldsymbol{\alpha},$$

with initial condition $\boldsymbol{\phi}(0) = 0$, and clearly $F * H(x) = \boldsymbol{\phi}(x)\mathbf{T}^{\circ}$.

A.2.2: THE *PH*-RENEWAL PROCESS

A renewal process has an underlying continuous *PH*-distribution $F(\cdot)$ with representation $(\boldsymbol{\alpha}, T)$. For convenience, set $\alpha_0 = 0$, so that there are no instantaneous renewals. That renewal process is simply related to the irreducible Markov chain with generator $Q^* = T + \mathbf{T}^{\circ}\boldsymbol{\alpha}$. That chain is obtained by instantaneously restarting the absorbing chain with the same initial probability vector $\boldsymbol{\alpha}$, upon each absorption. Its state is that in which the absorbing chain is restarted (equivalently its path functions are defined to be right-hand continuous). The epochs of restarts are clearly the *renewals* in the *PH*-renewal process.

The stationary probability vector of Q^* is the vector $\boldsymbol{\theta}^*$ introduced earlier. We now see that property [PH-4] is obvious. At an arbitrary time the state probabilities of the Markov chain Q^* are the components of $\boldsymbol{\theta}^*$. The time until the next renewal, or the forward recurrence time, is simply the time to the next restart. It therefore has a *PH*-distribution with representation $(\boldsymbol{\theta}^*, T)$ which, from renewal theory, is also $F^*(\cdot)$.

The *renewal function* $H(\cdot)$ is the expectation of the number $N(t)$ of renewals in $(0, t)$. It satisfies the *renewal equation*

$$H(t) = F(t) + \int_0^t H(t - u)\, dF(u), \qquad \text{for } t \geq 0.$$

Its derivative $H'(t)$ is the *renewal density*. $H'(t)dt$ can be interpreted as the elementary probability of a renewal in $(t, t + dt)$. The graph of $H'(t)$ is therefore quite informative. It depicts the *rate* at which renewals occur at time t if we start with a new item at 0. For the PH-renewal process, $H(\cdot)$ and $H'(\cdot)$ may be computed together by a simple algorithm. Before showing that, we develop some other material of interest.

$N(t)$ is the number of renewals in $(0, t)$. $J(t)$ is the state of the Markov chain Q^* at t. We define the conditional probabilities

$$P\{N(t) = n, J(t) = j \mid J(0) = i\} = P_{ij}(n; t),$$

and write the matrix with elements $P_{ij}(n; t)$ as $P(n; t)$. It is now easy to show that the matrices $P(n; t)$, $n \geq 0$, satisfy the differential equations

$$P'(0; t) = P(0; t)T, \quad \text{and} \quad P'(n; t) = P(n; t)T + P(n-1; t)\mathbf{T}^{\circ}\boldsymbol{\alpha}, \quad \text{for } n \geq 1,$$

with initial conditions $P(0; 0) = I$, $P(n; 0) = 0$, for $n \geq 1$.

For $T = -\lambda$, $\mathbf{T}^{\circ} = \lambda$, and $\boldsymbol{\alpha} = 1$, we obtain the familiar equations for the density of the number of events in a Poisson process. Their solution is of course

$$P(n; t) = e^{-\lambda t}\frac{(\lambda t)^n}{n!}, \quad \text{for } n \geq 0.$$

The matrix equations do not have a concise explicit solution but they are well-suited for numerical computation. Defining the matrix generation function

$$P^*(z; t) = \sum_{n=0}^{\infty} P(n; t)z^n,$$

the differential equations lead to

$$\frac{\partial}{\partial t}P^*(z; t) = P^*(z; t)[T + z\mathbf{T}^{\circ}\boldsymbol{\alpha}], \quad \text{with } P^*(z; 0) = I.$$

The solution to that equation is $P^*(z; t) = \exp[(T + z\mathbf{T}^{\circ}\boldsymbol{\alpha})t]$, for $t \geq 0$. That formula is the generalization of $P^*(z; t) = \exp[-\lambda(1 - z)t]$, for the Poisson process. It is useful in deriving formulas for the moments of $N(t)$.

Differentiation of $P^*(z; t)$ with respect to z and setting $z = 1$ leads to

$$M_1(t) = \sum_{n=1}^{\infty} nP(n; t) = \sum_{n=1}^{\infty} \frac{t^n}{n!} \sum_{r=0}^{n-1} (Q^*)^r \mathbf{T}^\circ \alpha (Q^*)^{n-r-1}.$$

Multiplying on the right by $\mathbf{e}$, we obtain that

$$M_1(t)\mathbf{e} = \sum_{n=1}^{\infty} \frac{t^n}{n!} (Q^*)^{n-1} \mathbf{T}^\circ = \int_0^t \exp(Q^* u)\, du\, \mathbf{T}^\circ.$$

By using the integration formula in Appendix 1, we see that

$$\int_0^t \exp[Q^* u]\, du = \mathbf{e}\boldsymbol{\theta}^* t + [I - \exp(Q^* t)](\mathbf{e}\boldsymbol{\theta}^* - Q^*)^{-1},$$

so that, since $\boldsymbol{\theta}^* \mathbf{T}^\circ = (\mu'_1)^{-1}$,

$$M_1(t)\mathbf{e} = \frac{t}{\mu'_1}\mathbf{e} + [I - \exp(Q^* t)](\mathbf{e}\boldsymbol{\theta}^* - Q^*)^{-1}\mathbf{T}^\circ.$$

Due to the special form of the matrix Q^*, the vector $\mathbf{d} = (\mathbf{e}\boldsymbol{\theta}^* - Q^*)^{-1}\mathbf{T}^\circ$ can be written in a more explicit form. The preceding equation and the definition of Q^* lead to

$$\mathbf{T}^\circ = (\boldsymbol{\theta}^* \mathbf{d})\mathbf{e} - T\mathbf{d} - (\alpha\mathbf{d})\mathbf{T}^\circ.$$

Multiplication on the left by $\boldsymbol{\theta}^*$ yields $\boldsymbol{\theta}^* \mathbf{d} = (\mu'_1)^{-1}$. By substitution and multiplication by T^{-1} we obtain that

$$\mathbf{d} = \frac{1}{\mu'_1}\boldsymbol{\theta}^* T^{-1}\mathbf{e} + (1 + \alpha\mathbf{d})\mathbf{e}.$$

The constant $\alpha\mathbf{d}$ can easily be determined, but we do not need it. Upon substitution the term in which it occurs cancels and we get that

$$M_1(t)\mathbf{e} = \frac{t}{\mu'_1}\mathbf{e} + [I - \exp(Q^* t)](\mu'_1)^{-1}T^{-1}\mathbf{e}.$$

The ith component of that vector is the expected number of renewals in $(0, t)$ if the Markov chain Q^* is started in state i. The renewal function $H(\cdot)$ is therefore given by

$$H(t) = \boldsymbol{\alpha} M_1(t)\mathbf{e} = \frac{t}{\mu'_1} + \boldsymbol{\alpha}[I - \exp(Q^* t)](\mu'_1)^{-1}T^{-1}\mathbf{e}.$$

By using the interpretation of $H'(t)\,dt$ or by direct calculation we find that $H'(t) = \boldsymbol{\alpha}\exp(Q^* t)\mathbf{T}^\circ$. To compute $H(t)$ and $H'(t)$, it suffices to evaluate the matrix $\exp(Q^* t)$.

After some calculation, $H(\cdot)$ may be rewritten as

$$H(t) = \frac{t}{\mu'_1} + \frac{\sigma^2 + \mu'^2_1}{2\mu'^2_1} + \boldsymbol{\alpha}[\mathbf{e}\boldsymbol{\theta}^* - \exp(Q^* t)](\mu'_1)^{-1}T^{-1}\mathbf{e}.$$

The last term tends to 0 as $t \to \infty$. We recognize the first two terms on the right as the linear asymptote of the renewal function.

If the initial state of the Markov chain Q^* is $\boldsymbol{\theta}^*$, we see that

$$\boldsymbol{\theta}^* M_1(t)\mathbf{e} = \frac{t}{\mu'_1}, \qquad \text{for } t \geq 0.$$

For the stationary renewal process, $EN(t) = \mu'^{-1}_1 t$, and μ'^{-1}_1 is the rate of renewals in that process.

Formulas for the second (and higher) moments of $N(t)$ can be derived by further differentiations of $P^*(z; t)$, but the calculations become quite belabored. The variance Var $N(t)$ of $N(t)$ for the *stationary* renewal process is useful in some applications. Its graph, or that of the *dispersion function* $D(t) = \text{Var } N(t)/EN(t)$, are used as descriptors of the variability of renewal (and other point) processes. The variance Var $N(t)$ is given by

$$\text{Var } N(t) = \frac{\sigma^2}{\mu_1'^2} t + 2\alpha[I - \exp(Q^* t)] \, [\mu_1'^{-2} T^{-2}\mathbf{e} + \frac{1}{2}\mu_1'^{-3}\mu_2' T^{-1}\mathbf{e}],$$

for $t \geq 0$. Here σ^2 and μ'_2 are respectively the variance and the second moment of $F(\cdot)$. Moreover,

$$\text{Var } N(t) = \frac{\sigma^2}{\mu_1'^2} t + \frac{1}{6} + \frac{\sigma^4}{2\mu_1'^4} - \frac{\mu_3^*}{3\mu_1'^3} + o(1), \quad \text{as } t \to \infty,$$

where μ_3^* is the third central moment of $F(\cdot)$. The graph of the variance has a linear asymptote which involves only the first three moments of $F(\cdot)$.

Remark: The PH-renewal process is a particular case of the Markovian arrival process discussed in Appendix 3.

Reference: Chapter 2 of Neuts, M. F., ''*Matrix-Geometric Solutions in Stochastic Models: An Algorithmic Approach*,'' Baltimore: The Johns Hopkins University Press, 1989, is devoted to the basic properties of PH-distributions.

APPENDIX 3

THE MARKOVIAN ARRIVAL PROCESS

We discuss properties of the *Markovian arrival process, (MAP),* a useful model for point processes. As for *PH*-distributions, there are discrete- and continuous-time versions of the *MAP*. For both the matrix formulas are similar. We consider only *MAP*s with *single* arrivals; the extension to group arrivals is entirely similar but the formulas are more complicated. The discrete case is treated in Section 4.3, so here we deal with the continuous case. Throughout, A is an irreducible $m \times m$ generator, the sum of matrices A_0 and A_1. A_1 is a nonnegative matrix. The diagonal elements of A_0 are negative, its off-diagonal elements are nonnegative, and the inverse of A_0 exists. The initial probability vector is $\boldsymbol{\alpha}$.

A.3.1: DESCRIPTION OF THE MAP

To describe informally how the *MAP* is constructed, we suppose that the Markov chain A has just entered state i. It then stays in i for an exponentially distributed time with parameter $-A_{0,ii}$. The transition at the end of the sojourn may be labeled or unlabeled. With conditional probability $A_{1,ij}[-A_{0,ii}]^{-1}$, a *labeled* transition occurs and the state changes to j. In that case, j may be equal to i. Alternatively, with conditional probability $A_{0,ij}[A_{0,ii}]^{-1}$, the sojourn ends with an *unlabeled* transition and the process moves to the state $j \neq i$. Following a transition of either type, the future of the process depends only on the (now current) state j.

The following is another way of looking at the transition mechanism. In $(t, t + dt)$, the process may move to state $j \neq i$ with elementary conditional probability $A_{0,ij}dt$ via an unlabeled transition, or to state j (which may be the same as i) with elementary conditional probability $A_{1,ij}dt$ via a labeled transition, or it stays in i with probability $1 - \sum_{h \neq i} A_{0,ih}\, dt - \sum_h A_{1,ih}\, dt$.

We are interested in the random point process of the labeled transitions. That process is often used to model the input to a queue. The labeled transitions are usually called *arrivals*. The point process is a Markovian arrival process. A_0 and A_1 are its *parameter matrices*. Formally the *MAP* is defined as the point process of the transition epochs of the m-state Markov renewal process with the transition probability matrix

$$R(x) = \int_0^x \exp(A_0 u) A_1 \, du, \qquad \text{for } x \geq 0.$$

To work with the *MAP*, it is useful but not essential to be acquainted with the theory of Markov renewal processes.

Example A.3.1.1: Let the matrices A_0 and A_1 be given by

$$A_0 = \begin{array}{c|cc|} 1 & -10 & 2 \\ 2 & 2 & -5 \end{array}, \qquad A_1 = \begin{array}{c|cc|} 1 & 4 & 4 \\ 2 & 0 & 3 \end{array},$$

with the initial probability vector (1/3, 2/3). If the initial state 1 is selected, the chain stays there for an exponential sojourn with mean 1/10. At the end of that sojourn, it produces an arrival and stays in state 1 with probability 4/10; with probabilities 2/10 and 4/10, it moves to state 2, respectively without or with an arrival. In state 2, its mean sojourn time is 1/5. It moves to state 1 without producing an arrival with probability 2/5 or it produces an arrival with probability 3/5 and stays in state 2.

Example A.3.1.2: The *PH*-Renewal Process: For the *PH*-renewal process in Appendix 2, the Markov chain with irreducible generator Q^* is obtained by restarting the absorbing chain for the *PH*-distribution according to the probability vector $\boldsymbol{\alpha}$. Conversely, we can start with a Markov chain with generator $A = Q^*$, written as the sum of $A_0 = T$ and $A_1 = \mathbf{T}^{\circ}\boldsymbol{\alpha}$, and construct the *MAP* with those parameter matrices. The special form of A_1 guarantees that the intervals between successive arrivals are *independent*. The *PH*-renewal process is a particular case of the *MAP*.

Example A.3.1.3: Departures from Markovian Queues: Consider a single-server queue with a finite waiting room of capacity K. Arrivals form a Poisson process of rate λ; services are exponential of rate μ. That queue is described by the Markov chain with generator

$$A = \begin{vmatrix} -\lambda & \lambda & 0 & 0 & \cdots & 0 & 0 & 0 \\ \mu & -\lambda-\mu & \lambda & 0 & \cdots & 0 & 0 & 0 \\ 0 & \mu & -\lambda-\mu & \lambda & \cdots & 0 & 0 & 0 \\ & \cdots & & & \cdots & & & \\ 0 & 0 & 0 & 0 & \cdots & \mu & -\lambda-\mu & \lambda \\ 0 & 0 & 0 & 0 & \cdots & 0 & \mu & -\mu \end{vmatrix}.$$

Departures correspond to transitions from a state i to $i-1$. If all such transitions are labeled, we obtain the *departure process* of the queue. The departure process is a *MAP*. The matrix A_1 inherits all the elements μ of A and $A_0 = A - A_1$. Similarly, the point process of the times of *admissions* to the queue is a *MAP* whose matrix A_1 inherits all the elements λ of A. Many finite capacity queues are described by Markov chains with a generator of the form

$$A = \begin{vmatrix} A_{00} & A_{01} & 0 & 0 & \cdots & 0 & 0 & 0 \\ A_{10} & A_{11} & A_{12} & 0 & \cdots & 0 & 0 & 0 \\ 0 & A_{20} & A_{21} & A_{22} & \cdots & 0 & 0 & 0 \\ & \cdots & & & \cdots & & & \\ 0 & 0 & 0 & 0 & \cdots & A_{K-1,0} & A_{K-1,1} & A_{K-1,2} \\ 0 & 0 & 0 & 0 & \cdots & 0 & A_{K,0} & A_{K,1} \end{vmatrix},$$

where the elements A_{rh} themselves are matrices. The departure process of such a queue is a *MAP*. Its coefficient matrices are constructed in the same manner as for the elementary $M/M/1$ queue. The *MAP* is obtained by labeling all transitions corresponding to departures from the queue. Other useful *MAP* s arise from labeling selected types of transitions in Markov chains.

Example A.3.1.4: The Markov Modulated Poisson Process: With the coefficient matrices $A_0 = A - \Lambda$, $A_1 = \Lambda$, where Λ is a diagonal matrix with nonnegative diagonal elements $\lambda_1, \cdots, \lambda_m$, we get a nice generalization of the Poisson process. It is called the *Markov modulated Poisson process (MMPP)*, and describes a point process that behaves like a Poisson process of rate λ_i whenever the Markov chain is in its state i. In applications, the *MMPP* is used to model arrivals with a randomly varying rate.

A.3.2: PROBABILITY DISTRIBUTIONS FOR THE MAP

Several choices of the initial probability vector $\boldsymbol{\alpha}$ produce interesting versions of the *MAP*. The *stationary version* of the process is obtained by setting $\boldsymbol{\alpha} = \boldsymbol{\theta}$, the stationary probability vector of the generator A. Given that there is an arrival at a time point in the stationary version, the probabilities of the various states immediately after that arrival are the components of the vector $\boldsymbol{\theta}_{arr} = [\boldsymbol{\theta} A_1 \mathbf{e}]^{-1} \boldsymbol{\theta} A_1$. By choosing $\boldsymbol{\alpha}$ to be $\boldsymbol{\theta}_{arr}$, we start the *MAP* at an arbitrary arrival epoch in the stationary version.

The probability distribution of the time τ_1 to the first arrival is

$$P\{\tau_1 \le x\} = \boldsymbol{\alpha}\int_0^x \exp(A_0 u)A_1\, du\, \mathbf{e} = \boldsymbol{\alpha}[I - \exp(A_0 x)](-A_0)^{-1}A_1\mathbf{e}, \quad \text{for } x \ge 0.$$

By setting $\boldsymbol{\alpha} = \boldsymbol{\theta}$, we obtain the corresponding distribution for the stationary version. For $\boldsymbol{\alpha} = \boldsymbol{\theta}_{arr}$, we get the distribution $F_1(\cdot)$ of the time between two successive arrivals in the stationary version. It is readily seen that all these distributions are of phase type. The mean of $F_1(\cdot)$ is

$$\mu'_1 = \frac{\boldsymbol{\theta}A_1(-A_0)^{-1}\mathbf{e}}{\boldsymbol{\theta}A_1\mathbf{e}} = \frac{1}{\boldsymbol{\theta}A_1\mathbf{e}}.$$

The quantity $\lambda^* = \boldsymbol{\theta}A_1\mathbf{e}$ is therefore the *stationary rate of arrivals* for the *MAP*. For numerical computations, it is often advisable to rescale the parameter matrices A_0 and A_1 to give λ^* a desired value. The variance σ^2 of $F_1(\cdot)$ is

$$\sigma^2 = \frac{2}{\lambda^*}\boldsymbol{\theta}(-A_0)^{-1}\mathbf{e} - \frac{1}{\lambda^{*2}}.$$

The Laplace - Stieltjes transform of the joint distribution of the first n intervals between arrivals is

$$\phi(s_1, \cdots, s_n) = \boldsymbol{\alpha}(s_1 I - A_0)^{-1}A_1 \cdots (s_n I - A_0)^{-1}A_1\mathbf{e}.$$

With an arbitrary choice of $\boldsymbol{\alpha}$, there is not necessarily an arrival at time zero. With $\boldsymbol{\alpha} = \boldsymbol{\theta}_{arr}$, we have the joint transform of n successive interarrival times $\tau_1, \cdots, \tau_n$, in the stationary version of the *MAP*. By differentiating the transform $\phi(s_1, \cdots, s_n)$, we can readily calculate the serial correlations $\rho(\tau_1, \tau_n)$ of τ_1 and τ_n. We obtain that

$$\rho(\tau_1, \tau_n) = \frac{\dfrac{1}{\lambda^*}\boldsymbol{\theta}[(-A_0)^{-1}A_1]^{n-1}(-A_0)^{-1}\mathbf{e} - \dfrac{1}{\lambda^{*2}}}{\dfrac{2}{\lambda^*}\boldsymbol{\theta}(-A_0)^{-1}\mathbf{e} - \dfrac{1}{\lambda^{*2}}}.$$

The Counting Process: The probability distributions of the numbers of arrivals during nonoverlapping time intervals are of particular interest. Let $N(t)$ be the number of arrivals during the interval $(0, t]$. To exploit the

Markovian properties of the *MAP*, we need to keep track of the states $J(0)$ and $J(t)$ of the Markov chain A at 0 and at t. We define the matrices $P(n;\, t)$ with elements

$$P_{ij}(n;\, t) = P\{N(t) = n;\, J(t) = j \mid J(0) = i\}.$$

By a standard argument, we show that the matrices $\{P(n;\, t)\}$ satisfy the differential equations

$$P'(0;\, t) = P(0;\, t)A_0, \quad \text{and} \quad P'(n;\, t) = P(n;\, t)A_0 + P(n-1;\, t)A_1, \quad \text{for } n \geq 1,$$

with the initial conditions $P(0;\, 0) = I$, $P(n;\, 0) = 0$, for $n \geq 1$. If $A_1 = -A_0 = \lambda$, and $m = 1$, these are the standard equations for the Poisson process and the scalar sequence $\{P(n;\, t)\}$ is then the Poisson density with parameter λt. The equations for the matrix case do not have explicit solutions, but they are usually well-suited for numerical computation.

As in Section A.2.2, the matrix generating function $P^*(z;\, t) = \sum_{n=0}^{\infty} P(n;\, t)z^n$ is given by

$$P^*(z;\, t) = \exp[(A_0 + zA_1)t], \quad \text{for } t \geq 0.$$

By differentiating $P^*(z;\, t)$ with respect to z, setting $z = 1$, and repeating the same calculations as in Section A.2.2, we obtain that

$$M_1(t)\mathbf{e} = A_1\mathbf{e}t + [I - \exp(At)](\mathbf{e}\boldsymbol{\theta} - A)^{-1}A_1\mathbf{e}.$$

With a general initial probability vector $\boldsymbol{\alpha}$, we obtain $E[N(t)]$ by forming the inner product $\boldsymbol{\alpha}M_1(t)\mathbf{e}$. For the stationary version of the *MAP*,

$$\boldsymbol{\theta}M_1(t)\mathbf{e} = (\boldsymbol{\theta}A_1\mathbf{e})t = \lambda^* t, \quad \text{for } t \geq 0.$$

The counts N_1 and N_2 of the arrivals in the nonoverlapping intervals $(0, t_1]$ and $(t_1 + u, t_1 + u + t_2]$, with $u \geq 0$ are *conditionally independent,* given the states $J(0)$, $J(t_1)$, $J(t_1 + u)$, $J(t_1 + u + t_2)$, of the Markov chain A at their endpoints. That also holds for any number of disjoint intervals. That property accounts for the similarity of the matrix formulas for the *MAP* to the simple scalar formulas for the Poisson process. For example, the joint probability generating function $\phi(z_1,z_2)$ of N_1 and N_2 is given by

$$\phi(z_1, z_2) = \alpha P^*(z_1; t_1)\exp(Au)P^*(z_2; t_2)\mathbf{e}.$$

Using the same care in differentiating the matrix functions as in the preceding calculations, we obtain a computationally useful formula for the correlation coefficient of N_1 and N_2.

References: An overview of the *MAP* with many examples is found in Neuts, M. F., "Models based on the Markovian arrival process," *IEICE Transactions on Communications,* E75-B, 1255 - 1265, 1992. For a clear exposition leading to applications in queueing theory, see Lucantoni, D. M., "New results on the single-server queue with a batch Markovian arrival process," *Stochastic Models,* 7, 1 - 46, 1991. Algorithms and asymptotic expansions for the moments of the *MAP* are discussed in Narayana S. and Neuts, M. F., "The first two moment matrices of the counts for the Markovian arrival process," *Stochastic Models,* 8, 459 - 477, 1992.

SOLUTIONS TO SELECTED PROBLEMS

Chapter 1

Solution to Problem 1.2.5: We discuss this basic method for the binomial density. The largest term of the binomial density occurs at the index $k^* = [np]$. We first compute the corresponding probability $p(k^*)$ by means of logarithms. To guard against loss of significance, we try to evaluate as few logarithms as possible. To that end, we rewrite $p(k^*)$ as

$$q^{n-k} \times \left[\frac{np}{k}\right] \times \left[\frac{(n-1)p}{k-1}\right] \cdots \left[\frac{(n-k+1)p}{1}\right],$$

when $k \leq n - k$, and as

$$p^k \times \left[\frac{nq}{n-k}\right] \times \left[\frac{(n-1)q}{n-k-1}\right] \cdots \left[\frac{(k+1)q}{1}\right],$$

when $k > n - k$. The sum of the logarithms of the factors is formed and $p(k^*)$ is computed as the antilogarithm.

At this stage, identifiers *kleft* and *kright* are set equal to k^* and the value of $p(k^*)$ is entered into identifiers *uleft* and *uright*. We now alternatingly reduce *kleft* by one and increase *kright* by one. The new terms of the binomial density are computed by means of the recurrence relations

$$\begin{bmatrix} n \\ k-1 \end{bmatrix} p^{k-1} q^{n-k+1} = \left[\frac{kq}{(n-k+1)p}\right] \times \begin{bmatrix} n \\ k \end{bmatrix} p^k q^{n-k},$$

and

$$\begin{bmatrix} n \\ k+1 \end{bmatrix} p^{k+1} q^{n-k-1} = \left[\frac{(n-k)p}{(k+1)q}\right] \times \begin{bmatrix} n \\ k \end{bmatrix} p^k q^{n-k}.$$

If *kleft* reaches 0 or *kright* reaches n, no further terms are computed on the left, or on the right. We accumulate the sum of terms and stop when it exceeds $1 - 10^{-10}$, say. We store the lower and upper indices of the non-negligible terms of binomial density and the values of these and all intermediate terms.

Solution to Problem 1.2.7: For $0 \le m \le n$, denote by $V_n(r, m)$, the number of nonnegative n-tuples of integers j_k, $1 \le k \le n]$, that satisfy the equations $\sum_{k=1}^{n} j_k = r$, and $\sum_{k=1}^{n} kj_k = m$, for given integers $m \ge r \ge 0$. If we fix the value of j_1, the remaining $n - 1$ integers must satisfy

$$\sum_{k=2}^{n} j_k = r - j_1, \quad \text{and} \quad \sum_{k=2}^{n} (k-1)j_k = m - r,$$

a system of equations of the same type, but with one unknown less. The number of its solutions is $V_{n-1}(r - j_1, m - r)$. It follows that

$$V_n(r, m) = \sum_{j_1=0}^{K(r)} V_{n-1}(r - j_1, m - r),$$

where $K(r) = \min(r, 2r - m)$, since, clearly, for the second system to have solutions, we must have that $r - j_1 \le m - r$. The quantities $V_n(r, m)$ can be computed using a single two-dimensional array. It is initialized by setting $V_1(1, 1) = 1$. The array is updated for successive n and only for indices satisfying $0 \le r \le m \le n$. At each stage, the elements $V_n(r, n)$, which provide the answers to the problem, are printed.

Solution to Problem 1.3.3: Each outcome is an assignment of birthdays to an ordered list of m persons. There are 365^m such lists. The event A is the set of lists in which k distinct pairs of birthdays are duplicated and all $m - 2k$ other people have distinct birthdays.

The second factor in the first expression for $P(k; m)$ is the number of ways we can choose the k duplicated dates; the second is the number of ways the $2k$ corresponding persons can be chosen. The k duplicated dates can be assigned to these $2k$ persons in $2^{-k}(2k)!$ ways. Finally, the remaining people can be assigned distinct birthdays from among the $365 - k$ remaining days in $(365 - k)(365 - k - 1) \cdots (365 - m + k + 1)$ ways.

Solution to Problem 1.3.4: Same reasoning as in Problem 1.3.3. The factor m is the number of choices for the student with the same birthday as the instructor.

Solution to Problem 1.3.14: The denominator is the number of equally likely repartitions of n items in m boxes. If there are to be k boxes between the first and the last of the nonempty boxes, the label i of the first nonempty box can be chosen in $m-k$ ways. Now place one item in each of the boxes i and $i+k$. The $n-2$ remaining items can be distributed among the $k+1$ allowable boxes in

$$\binom{n+k-2}{n-2},$$

different ways. The formula for $p(k)$ follows. Since

$$\frac{p(k)}{p(k-1)} = \frac{(m-k)(n+k-2)}{(m-k+1)k},$$

$p(k)$ is nondecreasing as long as $k \le k^*$. Therefore, k^* is the index of the largest $p(k)$. We notice that

$$P\{m-R=j\} = p(m-j)$$

$$= jn(n-1)\frac{(m-1)(m-2)\cdots(m-j+1)}{(n+m-1)(n+m-2)\cdots(n+m-j-1)}$$

$$= j\left(1-\frac{1}{n}\right)\frac{\left(\frac{m-1}{n}\right)\cdots\left(\frac{m-j-1}{n}\right)}{\left(\frac{n+m-1}{n}\right)\cdots\left(\frac{n+m-j-1}{n}\right)},$$

upon dividing numerator and denominator by n^{j+1}. For fixed j, this last fraction tends to $j\alpha^{j-1}(1+\alpha)^{-j-1}$, when m and n tend to infinity in such a manner that $m/n \to \alpha > 0$. However,

$$\beta_j = j\left(\frac{1}{1+\alpha}\right)^2\left(1-\frac{1}{1+\alpha}\right)^{j-1}, \quad \text{for } j \ge 1,$$

is the jth term of the stated negative binomial density.

Solution to Problem 1.3.15: The stated formula for $P\{X = n\}$ follows by considering the number of games resulting in a tie. To prove the convexity in part c, we evaluate the generating function

$$P^*(z) = \sum_{n=0}^{\infty} P\{X = n\} \frac{z^{2n}}{2n!},$$

which yields that

$$P^*(z) = \sum_{k=0}^{\infty} \frac{(xz)^{2k}}{(2k)!} \sum_{\nu=0}^{\infty} \frac{1}{(\nu!)^2} \left[\frac{(1-x)^2 z^2}{4}\right]^{\nu} = \frac{1}{2}(e^{xz} + e^{-xz})\, I_0[(1-x)z],$$

where $I_0(z)$ is the modified Bessel function of order zero. Using the integral representation

$$I_0(z) = \frac{1}{\pi} \int_0^{\pi} e^{-z\cos\theta}\, d\theta,$$

of $I_0(z)$ we obtain that

$$P^*(z) = \frac{1}{2\pi} \int_0^{\pi} [e^{-z[(1-x)\cos\theta - x]} + e^{-z[(1-x)\cos\theta + x]}]\, d\theta$$

$$= \frac{1}{2\pi} \sum_{n=0}^{\infty} \frac{(-1)^n z^n}{n!} \int_0^{\pi} \Big[[(1-x)\cos\theta - x]^n + [(1-x)\cos\theta + x)]^n \Big]\, d\theta.$$

But, the integrand may be written as

$$\sum_{r=0}^{n} \binom{n}{r} (\cos\theta)^{n-r} (1-x)^{n-r} x^r [1 + (-1)^r] = 2 \sum_{r=0}^{[n/2]} \binom{n}{2r} (\cos\theta)^{n-2r} (1-x)^{n-2r} x^{2r},$$

and for n odd,

$$\int_0^{\pi} (\cos\theta)^{n-2r}\, d\theta = 0,$$

so that the coefficients of z^{2n+1} vanish. We therefore obtain that

$$P^*(z) = \frac{1}{\pi}\sum_{n=0}^{\infty} \frac{z^{2n}}{(2n)!} \sum_{r=0}^{n} \binom{2n}{2r} x^{2r}(1-x)^{2(n-r)} \int_0^{\pi} (\cos\theta)^{2(n-r)}\, d\theta,$$

so that for $n \geq 0$,

$$P\{X = n\} = \frac{1}{\pi}\sum_{r=0}^{n} \binom{2n}{2r} x^{2r}(1-x)^{2(n-r)} \int_0^{\pi} (\cos\theta)^{2(n-r)}\, d\theta,$$

$$= \frac{1}{2\pi}\int_0^{\pi} \Big[[(1-x)\cos\theta - x]^{2n} + [(1-x)\cos\theta + x)^{2n}] \Big]\, d\theta.$$

Differentiating twice with respect to x in this last expression shows that the second derivative is positive. If we set $x = 0$ in the first derivative

$$\frac{n}{\pi}\int_0^{\pi} \Big[(1-\cos\theta)[(1-x)\cos\theta + x]^{2n-1} - (1+\cos\theta)[(1-x)\cos\theta - x]^{2n-1} \Big]\, d\theta,$$

we obtain the value

$$-\frac{2n}{\pi}\int_0^{\pi} (\cos\theta)^{2n}\, d\theta < 0,$$

so that the probability $P\{X = n\}$ attains its minimum at some $x > 0$. By setting $x = 1/2$, elementary trigonometric formulas show that the derivative is positive at that point, so that the minimum occurs at a point x_n^* in (0, 1/2). By a further analysis of this type, we show that x_n^* is decreasing in n.

Solution to Problem 1.3.16: The first term in the recurrence relation for $L(r; m, n)$ is the number of strings starting with r ones. If there are $\nu < r$ ones before the first zero, there are $L(r; m-1, n-\nu)$ strings of $m-1$ zeros and $n-\nu$ ones in which there are r or more consecutive ones. Since there must be least r ones after the first zero, $\nu \leq n - r$. The only boundary condition that is needed is:

$$L(r; 0, n) = 1, \quad \text{for } n \geq r, \qquad L(r; 0, n) = 0, \quad \text{for } n < r.$$

The recurrence relation for the probabilities $p(r; m, n)$ is obtained by dividing in the recurrence relation for the $L(r; m, n)$ by the appropriate binomial coefficient. Arrays containing the probabilities $p(r; m, n)$ are computed recursively in m. The ratio of binomial coefficients in first term and the coefficients of $p(r; m-1, n-v)$ are stored in an auxiliary array which is updated for successive m by means of easily derived recurrence relations.

Solution to Problem 1.3.19: When the given string is all zeros, the problem is trivial. Suppose that it contains $s \geq 1$ ones. From left to right, $k_1 < k_2 < \cdots < k_s \leq m+n$ are the locations of the ones in the binary representation of C. For $1 \leq r \leq s$, the set A_r consists of all strings that agree with C up to the location $k_r - 1$ and have a zero in location k_r. The sets A_r are disjoint. Their union contains all strings for which $X < C$. The set $X = C$ contains only the string C if $s = n$ and is empty otherwise.

For the strings in A_r, the positions of the first $r-1$ ones and the first $k_r - r + 1$ zeros are determined. The remaining $n - r + 1$ ones are to be placed in the $m + n - k_r$ positions to the right of location k_r. This is possible if and only if $k_r \leq m + r - 1$. The probability $P\{X < C\}$ is therefore

$$\sum \frac{\binom{m+n-k_r}{n-r+1}}{\binom{m+n}{n}},$$

where the summation is over all indices r for which $k_r \leq m + r - 1$. When $s = n$, we add

$$P\{X = C\} = \frac{1}{\binom{m+n}{n}}.$$

Solution to Problem 1.3.20: The essential ideas of the solution are outlined in the problem statement. The 168 primes less than 1,000 can be computed by a simple subroutine or are read into a separate file. In an array of 1,000 locations, we compute the exponent of the proposed form of u. For a factor $m_i!$ in the numerator, we increase each of the locations $2, \cdots, m_i$ by one. For a factor $n_j!$ in the denominator, the corresponding locations are decreased by one.

The integers $2, \cdots, M$ are tested from the top down for a prime factor. If $k = pk'$, where p is prime, we add the exponent of k to the exponents of the factors p and k' and set the exponent of k to zero. Eventually only the remaining prime factors have nonzero exponents. Using double precision, we separately sum the base-10 logarithms of the numerator and denominator. The integer parts of those sums of logarithms are subtracted to give the exponent of 10 that is reported. The antilogarithm of the difference of the fractional parts is the reported quantity U.

Cases of very large or very small u require some extra care. The exponents in the products $p_1^{a_1} p_2^{a_2} \cdots p_K^{a_K}$ for the numerator and denominator can then be very large. Loss of significance in forming the sum of the (positive) terms $a_i \log_{10} p_i$ could be appreciable. As a (partial) remedy, to balance the magnitudes of the summands, we do the following: The integer and fractional parts of the terms $a_i \log_{10} p_i$ are summed separately. Then, the integer part of the sum of fractional parts is added to the sum of the integer parts to yield the exponent R. U is computed as before.

In a test example, we computed $u = (1{,}000!)^{50}$, the largest number that could arise from the code we had written. The first procedure yielded $U = 1.70691190186099$, $R = 128380$. The second, more cautious procedure gave $U = 1.70691190190633$, $R = 128380$. The first 10 significant digits therefore appear to be correct and the trailing digits in the second result are more reliable. In less extreme examples there was no appreciable difference between the results of the two procedures. The first 15 digits were generally in agreement.

Solution to Problem 1.3.21: We write u as a quotient of factorials as in Problem 1.3.20. The arrays $\{m_i\}$ and $\{n_j\}$ are sorted and common elements are deleted. After this minor step for the sake of efficiency, we proceed as in Problem 1.3.20

Solution to Problem 1.3.25: If $N = 2n + 1$, let Joe have more than n correct predictions. His twin then cannot win. The number J of other players with more than n correct choices has a binomial density with p = 1/2. If $J = j$, then Joe is competing on an equal basis against those j players. The expected return to him (and his twin) is then $(m + 2)/(j + 1)$. Their unconditional expected return E is therefore

$$E = \sum_{j=0}^{m} \frac{m+2}{j+1} \binom{m}{j} 2^{-m} = \frac{m+2}{m+1} 2^{-m} \sum_{j=1}^{m+1} \binom{m+1}{j},$$

but the last sum is $2^{m+1} - 1$. The stated expression for E follows. When N is odd, E does not depend on n.

When $N = 2n$, we give a detailed derivation of E. It shows the type of analytic simplifications which, in the preceding case, account for the simple expression for E. Either Joe or his twin make $r > n$ correct guesses, or both have n correct guesses. The first alternative contributes the amount

$$V_1 = 2 \sum_{r=n+1}^{2n} \sum_{j=0}^{m} \binom{m}{j} \left[\binom{2n}{r} 2^{-2n} \right]^{j+1} \left[\sum_{\nu=0}^{r-1} \binom{2n}{\nu} 2^{-2n} \right]^{m-j} \frac{m+2}{j+1},$$

to E. That expression is obtained by considering the number j of the other players with the same winning score r. The factor 2 comes from the fact that Joe or his twin can hold the successful ticket. The expression successively simplifies to

$$2\frac{m+2}{m+1} \sum_{r=n+1}^{2n} \left[\left[\sum_{\nu=0}^{r} \binom{2n}{\nu} 2^{-2n} \right]^{m+1} - \left[\sum_{\nu=0}^{r-1} \binom{2n}{\nu} 2^{-2n} \right]^{m+1} \right]$$

$$= 2\frac{m+2}{m+1} \left[\left[\sum_{\nu=0}^{2n} \binom{2n}{\nu} 2^{-2n} \right]^{m+1} - \left[\sum_{\nu=0}^{n} \binom{2n}{\nu} 2^{-2n} \right]^{m+1} \right]$$

$$= 2\frac{m+2}{m+1} \left[1 - \left[\sum_{\nu=0}^{n} \binom{2n}{\nu} 2^{-2n} \right]^{m+1} \right] = 2\frac{m+2}{m+1} \left[1 - \left(\frac{1+c}{2} \right)^{m+1} \right]$$

where, for convenience, we write

$$\binom{2n}{n} 2^{-2n} = c.$$

We shall repeatedly use the equality

$$\sum_{\nu=0}^{n-1} \binom{2n}{\nu} 2^{-2n} = \frac{1-c}{2}.$$

In the second case, if $J = j$, the combined expected return to Joe and his twin is $[2(m+2)]/(j+2)$, as both now have the winning number n of correct predictions. Their expected share of the pot of $m+2$ dollars is $2/(j+2)$. The amount contributed to E is

$$V_2 = \sum_{j=0}^{m} \binom{m}{j} \left[\binom{2n}{n} 2^{-2n} \right]^{j+1} \left[\sum_{\nu=0}^{n-1} \binom{2n}{\nu} 2^{-2n} \right]^{m-j} \frac{2(m+2)}{j+2}.$$

The reduction of V_2 is a bit complicated. It may successively be written as

$$V_2 = 2(m+2)\sum_{j=0}^{m} \frac{(j+1)m!}{(j+2)!(m-j)!} c^{j+1} \left[\frac{1-c}{2}\right]^{m-j}$$

$$= \frac{2}{m+1}\sum_{j=0}^{m} (j+1)\binom{m+2}{j+2} c^{j+1} \left[\frac{1-c}{2}\right]^{m-j}$$

$$= \frac{2}{m+1}\left\{\sum_{j=1}^{m+1} (m+2)\binom{m+1}{j} c^{j} \left[\frac{1-c}{2}\right]^{m+1-j} - \frac{1}{c}\left[\left[\frac{1+c}{2}\right]^{m+2} - \left[\frac{1-c}{2}\right]^{m+2}\right] + (m+2)\left[\frac{1-c}{2}\right]^{m+1}\right\}$$

$$= \frac{2(m+2)}{m+1}\left[\left[\frac{1+c}{2}\right]^{m+1} - \left[\frac{1-c}{2}\right]^{m+1}\right] - \frac{2}{(m+1)c}\left[\left[\frac{1+c}{2}\right]^{m+2} - \left[\frac{1-c}{2}\right]^{m+2}\right] + \frac{2(m+2)}{m+1}\left[\frac{1-c}{2}\right]^{m+1}.$$

E is the sum of V_1 and V_2. It reduces to

$$E = \frac{2}{(m+1)c}\left[(m+2)c + \left[\frac{1-c}{2}\right]^{m+2} - \left[\frac{1+c}{2}\right]^{m+2}\right].$$

The expected return $E(p; m, N)$ is calculated essentially as was done for the case $N = 2n$ with $p = 1/2$. When $N = 2n+1$, we obtain

$$E(p; m, 2n+1) = 2\sum_{r=n+1}^{2n+1} \binom{2n+1}{r} 2^{-2n-1} \sum_{j=0}^{m} \binom{m}{j} \left[\binom{2n+1}{r} p^r q^{2n-r+1}\right]^j$$

$$\times \left[\sum_{\nu=0}^{r-1} \binom{2n+1}{\nu} p^{\nu} q^{2n-\nu+1}\right]^{m-j} \frac{m+2}{j+1}.$$

For its computation, it is advisable to compute and store the appropriate binomial probabilities and their sums in arrays that are updated for successive values of m. The formula for $E(p; m, 2n)$ is derived by the same argument as before. The case where Joe and his twin each make n correct guesses must again be treated separately.

Solution to Problem 1.3.28: The n items are distributed into the m containers, in

$$\binom{m+n-1}{n}$$

equally likely ways. If the number of runs of zeros and ones is $2k$, there is either an initial run of nonempty boxes and a final run of empty boxes, or vice versa. By reversing the order of the boxes, we see that the numbers of alternatives corresponding to these two cases are equal. It is therefore enough to count the number of repartitions that lead off with a run of nonempty boxes. Let r be the number of nonempty boxes, then clearly $k \le r \le m - k$. The r nonempty boxes can be assigned to the k runs in

$$\binom{r-1}{k-1}$$

ways, as each run must contain at least one nonempty box. Similarly, the $m - r$ empty boxes can be assigned to the k corresponding runs in

$$\binom{m-r-1}{k-1}$$

different ways. With the nonempty and empty boxes now chosen, the n items can be distributed in

$$\binom{n-1}{r-1}$$

ways in the nonempty boxes. Clearly, the index r must be smaller than n. That argument establishes the formula for $p(2k)$.

For $k \ge 1$, the formula for $p(2k+1)$ is proved in a similar manner. There are now either k runs of nonempty boxes with two runs of empty boxes at each end, or vice versa with the roles of empty and nonempty boxes interchanged. The formula for $p(1)$ is a simple special case.

The equality

$$p(2k+1) = \frac{m-2k}{2k} p(2k), \quad \text{for } k \geq 1,$$

can be obtained by a simple calculation in which $p(2k+1)$ is written as two separate sums and the order of summation in the second sum is reversed. There is also a direct, insightful proof of the equality, which shows that the result depends only of the 0–1 labeling of the boxes. Place the m boxes in a ring. Original labelings with $2k$ and $2k+1$ runs result in $2k$ runs in the circular arrangement. If we choose a box at random and open up the ring at that place, the resulting string of boxes has $2k$ runs if and only if we pick a box that was the first in one of the original runs. It follows that

$$\frac{p(2k)}{p(2k)+p(2k+1)} = \frac{2k}{m},$$

which is equivalent to the stated equality. By virtue of that equality, it is sufficient to compute $p(1)$ and $p(2k)$ for all allowable values of k.

To compute $p(2k)$, we need $a(k,r)$ for $1 \leq k \leq \min(m/2, n)$, and for the corresponding r satisfying $r \leq n$ and $k \leq r \leq m-k$. The recurrence relation in k for fixed r is somewhat simpler than the other. The ratio

$$\frac{a(k+1,r)}{a(k,r)} = \frac{(m-r-k)(r-k)}{k^2},$$

is at least one when

$$k \leq \left(1 - \frac{r}{m}\right) r.$$

To determine the range of k for fixed r, we distinguish between the cases $m \leq 2n$ and $m > 2n$. When $m \leq 2n$, we have

$$\text{for } 1 \leq r \leq \frac{m}{2}, \qquad 1 \leq k \leq r,$$

$$\text{for } \frac{m}{2} \leq r \leq \min(n, m-1), \qquad 1 \leq k \leq m-r.$$

For $m > 2n$, we have that $1 \le k \le r$, for $1 \le r \le n$.

For each r satisfying $1 \le r \le \min(n, m-1)$, we determine the index $k^*(r)$ for which $a(k, r)$ is largest. If the integer part k' of $(1 - r/m)r + 1$ is an allowable value for the index k, then $k^*(r) = k'$; otherwise the largest element is found at a boundary value for k. We computed that element by using the subroutine developed for Problem 1.3.20. If that maximal element is smaller than, say 10^{-13}, we skip to the next value of r to avoid computing very small quantities likely to cause underflow. The other $a(k, r)$ in the appropriate range for k are computed by the recurrence relation stated earlier. The remaining steps of the algorithm are now routine. We wrote a FORTRAN program that was tested for values of m up to 500 and n up to 400. It gives entirely consistent numerical results.

Solution to Problem 1.3.32: Store symbols for the 52 cards in some conventional order, say, as in Bridge, in an array C. That array is not modified and serves only for purposes of printout. The key insight is that we may give the successive cards randomly to one of the four players. That is probabilistically equivalent to dealing the cards in a more conventional way. We must make sure that each player receives exactly 13 cards. This is accomplished by storing 13 copies of each of the symbols 1, 2, 3, 4 in an array D. A random permutation of that array is generated. If the location i in D contains a j, $j = 1, 2, 3, 4$, the corresponding card in location i of C is given to player j. To generate a random permutation of an array of length n, exchange the content of the first location with that location J, where $J = [nU] + 1$, where U is a uniform random variate. A random element of the array is now in the first location. Repeat this for the remaining arrays of length $n - 1, \cdots, 2$.

Chapter 2

Solution to Problem 2.3.6: The mean $E(T_1)$ is given by

$$E(T_1) = 1 + \sum_{k=1}^{\infty} \prod_{r=1}^{k} \left[1 - \frac{p_1}{r}\right]^r.$$

For $x \ge 1$, the function $f(x) = (1 - x^{-1}p_1)^x$ is increasing and tends to e^{-p_1} as $x \to \infty$. If we truncate the series for $E(T_1)$ at the index N, the error R_N satisfies

$$R_N = \sum_{k=N+1}^{\infty} \prod_{r=1}^{k} \left[1 - \frac{p_1}{r}\right]^r < \sum_{k=N+1}^{\infty} e^{-kp_1} = \frac{e^{-(N+1)p_1}}{1 - e^{-p_1}}.$$

That (crude) upper bound is adequate for use in the subroutine for $E(T_1)$ needed in the search for the value of p_1 for which $E(T_1) = E(T)$. The desired p_1 increases with p. Each preceding solution can therefore serve as an initial value in computing p_1 for the next p.

Solution to Problem 2.3.7: For $0 < \alpha \le 1$, the gamma density is strictly decreasing, so that $f(a)$ is maximal at $a = 0$. For $\alpha > 1$, the mode of the density $\phi(u)$ is at $u_1 = (\alpha - 1)\lambda^{-1}$. $f'(a)$ vanishes at the unique (positive) root of the equation

$$\log\left[\frac{a}{a+1}\right] = -\frac{\lambda}{\alpha - 1}.$$

To get a lower bound for the root, we replace the left-hand side by the first two terms of the logarithmic series for $\log[1 - (1 + a)^{-1}]$ and solve the equation $y^2 + 2y - \lambda(\alpha - 1)^{-1} = 0$, where y stands for $(1 - a)^{-1}$. Using the positive root, we find that the desired value a^* of a exceeds

$$a_1 = \frac{\alpha - 1}{\lambda}\left[1 + \left[\frac{\alpha - 1 + \lambda}{\lambda}\right]^{1/2}\right] - 1.$$

If we substitute u_1 for a, we see that

$$\log\left[1 - \frac{\lambda}{\alpha - 1 + \lambda}\right] > -\frac{\lambda}{\alpha - 1 + \lambda} > -\frac{\lambda}{\alpha - 1}.$$

Therefore, $\max(0, a_1) < a^* < u_1$. The average of the lower and upper bounds is an excellent starting solution. A different starting solution is proposed in the statement of the problem. After solving for a^*, the maximum of $f(a)$ is computed by calling the incomplete gamma function.

Solution to Problem 2.3.13: Set the derivative of $\phi(\lambda; m, n)$ equal to 0 and simplify to obtain the explicit solution

$$\lambda = \left[\frac{n!}{(m-1)!}\right]^{\frac{1}{n-m+1}}.$$

Compute that value by logarithms and use a carefully written code to evaluate the corresponding value of $\phi(\lambda; m, n)$. We hail an explicit solution whenever we see one!

Solution to Problem 2.3.14: Set the derivative of

$$\left[1 - \sum_{r=0}^{k_1} e^{-\lambda} \frac{\lambda^r}{r!}\right] \times \sum_{j=0}^{k_2} e^{-\lambda} \frac{\lambda^j}{j!},$$

equal to 0 and simplify to show that λ is the root of the equation

$$\lambda = \left[\frac{k_2!}{k_1!} \frac{\sum_{j=0}^{k_2} e^{-\lambda} \frac{\lambda^j}{j!}}{1 - \sum_{r=0}^{k_1} e^{-\lambda} \frac{\lambda^r}{r!}}\right]^{\frac{1}{k_2 - k_1}}.$$

The right-hand side is decreasing in λ from $+\infty$ at 0 to 0 at $+\infty$. The equation therefore has a unique positive root. If you write a general code, compute the right-hand side accurately. Use of logarithms and a good subroutine for the Poisson probabilities.

Solution to Problem 2.3.19: The insightful proof of part *a* is as follows: Consider all 2^{2n+1} strings of $2n + 1$ strings of zeros and ones. With $p = 1/2$, the experiment is equivalent to picking such a string at random and counting the ones among the first $n + 1$ locations as wins for Player A and those in the remaining locations as wins for Player B. Take any string containing at least one zero for which A wins. By reversing that string, we obtain a string for which B wins and vice versa. The string with all ones, however, is a win for B. The probability that A wins is therefore given by

$$\frac{2^{2n}-1}{2^{2n+1}} = \frac{1}{2} - \frac{1}{2^{2n+1}}.$$

The first term in the general expression in part *b* is the probability that A scores n and B has fewer than n successes; the second term is the probability that A scores $1 \le k < n$ successes in $m + n$ trials and B scores fewer than k in his n trials. The sum

$$V_m = \sum_{k=n}^{m+n} \binom{m+n}{k} p^k (1-p)^{m+n-k}$$

is the probability that $m + n$ Bernoulli trials result in at least n successes. Clearly V_m is increasing in m for all $0 < p \leq 1$ and tends to one as $m \to \infty$. The second term is bounded above by $1 - V_m$, so that $P(p; m + n, n) \to 1 - p^n$, as $m \to \infty$. There exists a positive integer m for which $P(p; m + n, n)$ exceeds 1/2, if and only if p is smaller than the stated upper bound $p^*(n)$.

For the computation of $P(p; m + n, n)$, we store the binomial *distribution* with parameters (n, p) in one array and the binomial *density* with parameters $(m + n, p)$ in a second array. The probability $P(p; m + n, n)$ is then easily computed in terms of elements of these arrays. For successive $m \geq 0$, we update the second array by using the recurrence relations for the binomial probabilities. That permits an easy tabulation of $P(p; m + n, n)$ as a function of m and a search for the smallest m (if there is one) for which that probability exceeds 1/2. The smallest m rarely exceeds three, except for values of p that are slightly below the critical values $p^*(n)$.

Solution to Problem 2.3.37: The probability $\phi_m(k_1, k_2)$ that k_1 and k_2 successes occur during the initial two sets of m trials is given by

$$\phi_m(k_1, k_2) = \binom{m}{k_1} [p(1)]^{k_1} [q(1)]^{m-k_1} \binom{m}{k_2} [p(2)]^{k_2} [q(2)]^{m-k_2},$$

for $0 \leq k_1 \leq m$ and $0 \leq k_2 \leq m$. The expected number $S[m; N, p(1), p(2)]$ of the total number of successes is

$$\begin{aligned} S[m; N, p(1), p(2)] &= \sum_{i=1}^{m} \sum_{j=0}^{i-1} \phi_m(i, j)[i + j + (N - 2m)p(1)] \\ &= \sum_{j=1}^{m} \sum_{i=0}^{j-1} \phi_m(i, j)[i + j + (N - 2m)p(2)] \\ &+ \sum_{i=0}^{m} \phi_m(i, i)[2i + \frac{1}{2}(N - 2m)[p(1) + p(2)]]. \end{aligned}$$

That may be conveniently rewritten as

$$S[m; N, p(1), p(2)] = m[p(1) + p(2)] + (N - 2m)[p(1)\phi_1^* + p(2)\phi_2^*],$$

where $\phi_1^* = 1 - \phi_2^*$ is the probability that Coin 1 is used in the second stage trials. The probability ϕ_1^* is an interesting quantity well worth reporting in the output file. It is given by

$$\phi_1^* = \sum_{i=1}^{m} \sum_{j=0}^{i-1} \phi_m(i, j) + \frac{1}{2} \sum_{i=0}^{m} \phi(i, i).$$

We store the binomial probabilities with parameters $m, p(1)$ and $m, p(2)$ in arrays that are updated for successive values of m. For each value of m we compute $S[m; N, p(1), p(2)]$. The suggested maximum value of N is sufficiently small that the optimal value of m is most easily found by tabulating the mean number of successes. At the end, the maximizing value of m is reported.

For large N, the next observation can save much uninteresting computational effort. Depending on whether Coin 1 is best or not, the probability ϕ_1^* approaches one or zero with large m. As is to be expected, when $p(1)$ and $p(2)$ differ considerably, it is soon clear which of the two coins should be used in the remaining trials. It is therefore not always necessary to compute $S[m; N, p(1), p(2)]$ for all m up to $[N/2]$.

Solution to Problem 2.3.38: We use the notation defined in the solution to Problem 2.3.37. As before, we start by computing the binomial densities with parameters $m, p(1)$ and $m, p(2)$ in two arrays to be updated for successive values of m. Next, we compute the probabilities $\Psi_1(r; m)$ and $\Psi_2(r; m)$ that, in the first $2m$ trials, there are a total of r successes and Coin 1, respectively Coin 2, is used in the remaining trials. For $0 \le r \le 2m$, these probabilities are given by

$$\Psi_1(r; m) = \sum_{j=0}^{[r/2]-1} \phi(r-j, j) + \frac{1}{2} \delta_{r,\,even} \phi(\frac{r}{2}, \frac{r}{2}),$$

and

$$\Psi_2(r; m) = \sum_{j=[r/2]+1}^{r} \phi(r-j, j) + \frac{1}{2} \delta_{r,\,even} \phi(\frac{r}{2}, \frac{r}{2}),$$

where $\delta_{r,even} = 1$ when r is even and zero otherwise. The symbol $[x]$ stands for the integer part of x. There is a neat way of computing the arrays containing $\Psi_1(r; m)$ and $\Psi_2(r; m)$ together. Using a double DO-loop, we examine all pairs (i, j) of indices with $0 \le i, j \le m$. Depending on whether

$i < j$, $i = j$, or $i > j$, we add $\phi(i, j)$ [or $\phi(i, i)/2$] to the appropriate location (corresponding to the index r) in the proper array.

The probability $\pi_k(m, N) = P\{K = k\}$ is therefore

$$\pi_k(m, N) = \sum_{r=0}^{L_k} \Psi_1(k-r; m) \binom{N-2m}{r} [p(1)]^r [q(1)]^{N-2m-r}$$

$$+ \sum_{r=0}^{L_k} \Psi_2(k-r; m) \binom{N-2m}{r} [p(2)]^r [q(2)]^{N-2m-r},$$

for $0 \le k \le N$. $L_k = \min[k, N-2m]$. It is advisable to compute the required binomial densities with parameter $N - 2m$ starting from the modal index as discussed in Problem 1.2.5 and to save somewhat on processing time by neglecting terms smaller than 10^{-12}. This is a time-consuming algorithm as the probability density $\{\pi_k(m, N)\}$ needs to be computed for all values of $m \le N/2$. For each m, we find the index $L(m)$ for which the probability of at least $L(m)$ successes is no less than 0.75. $L(m)$ is the largest index k^* for which $\sum_{k=0}^{k^*} \pi_k(m, N) < 0.25$, or one less than the 25th percentile of the density $\{\pi_k(m, N)\}$.

Solution to Problem 2.3.40: For $j \ge 1$, any given coin of Player I will land in the box by the jth trial with probability $1 - q_1^j$, where $q_1 = 1 - p_1$. The probability that Player I succeeds before or at his jth turn is therefore $[1 - q_1^j]^m$. By the law of total probability,

$$P(m, n) = \sum_{j=1}^{\infty} [(1 - q_1^j)^m - (1 - q_1^{j-1})^m] [1 - (1 - q_2^j)^n].$$

The formula for $Q(m, n)$ is obtained by reversing the roles of the players. For $R(m, n)$, we similarly obtain

$$R(m, n) = \sum_{j=1}^{\infty} [(1 - q_1^j)^m - (1 - q_1^{j-1})^m] [(1 - q_2^j)^n - (1 - q_2^{j-1})^n].$$

Since the game always lasts an even number of turns, $T(2j + 1) = 0$, for $j \ge 0$. $T(2j)$ is the sum of the terms corresponding to the index j in the series for $P(m, n)$, $Q(m, n)$, and $Q(m, n)$. The mean number of turns until the game ends is

$$2 + 2\sum_{j=2}^{\infty} [1 - \sum_{r=1}^{j-1} T(2r)].$$

By considering all possible situations after each player has had one turn, we obtain the following equation for $P(m, n)$.

$$P(m, n) = \sum_{i=0}^{m} \sum_{j=0}^{n} \binom{m}{i} p_1^{m-i} q_1^{i} \binom{n}{j} p_2^{n-j} q_2^{j} P(i, j).$$

Similar equations hold for $Q(m, n)$ and $R(m, n)$. However, the boundary conditions are different. They are: $P(i, 0) = 0$ for $i \geq 0$, $P(0, j) = 1$ for $j \geq 1$, $Q(i, 0) = 1$ for $i \geq 1$, $Q(0, j) = 0$ for $j \geq 0$, $R(0, 0) = 1$ and $R(i, j) = 0$ if exactly one of i and j is zero.

The terms in $P(m, n)$ occur on both sides and similarly for $Q(m, n)$ and $R(m, n)$ in the corresponding equations. Solving for the term with indices m and n and taking the boundary conditions into account, we obtain the recurrence relations

$$P(m, n) = \frac{1}{1 - q_1^m q_2^n} \left[q_1^m \sum_{j=1}^{n-1} \binom{n}{j} p_2^{n-j} q_2^{j} P(m, j) + \sum_{i=0}^{m-1} \binom{m}{i} p_1^{m-i} q_1^{i} \sum_{j=1}^{n} \binom{n}{j} p_2^{n-j} q_2^{j} P(i, j) \right],$$

$$Q(m, n) = \frac{1}{1 - q_1^m q_2^n} \left[q_2^n \sum_{i=1}^{m-1} \binom{m}{i} p_1^{m-i} q_1^{i} Q(i, n) + \sum_{i=1}^{m} \binom{m}{i} p_1^{m-i} q_1^{i} \sum_{j=0}^{n-1} \binom{n}{j} p_2^{n-j} q_2^{j} Q(i, j) \right],$$

and

$$R(m, n) = \frac{1}{1 - q_1^m q_2^n} \left[p_1^m p_2^n + q_2^n \sum_{i=1}^{m-1} \binom{m}{i} p_1^{m-i} q_1^{i} R(i, n) + \sum_{i=1}^{m} \binom{m}{i} p_1^{m-i} q_1^{i} \sum_{j=1}^{n-1} \binom{n}{j} p_2^{n-j} q_2^{j} R(i, j) \right].$$

An equivalent set of recurrence relations is obtained by considering the alternatives after Player I's first turn. Let $P_1(m, n)$ be the probability the Player I wins if his turn is next and m and n coins remain. Similarly, $P_2(m, n)$ is the probability that I wins if the other player's turn is next and m and n coins remain. By the law of total probability,

$$P_1(m, n) = \sum_{i=0}^{m} \binom{m}{i} p_1^{m-i} q_1^i \; P_2(i, n),$$

and

$$P_2(m, n) = \sum_{j=0}^{n} \binom{n}{j} p_2^{n-j} q_2^j \; P_1(m, j).$$

We notice that this is just a detailed form of the first equation for $P(m, n)$. The treatment of the probabilities $Q_1(m, n)$, $Q_2(m, n)$, $R_1(m, n)$, and $R_2(m, n)$ is entirely similar.

The disadvantage of the explicit formulas is the likely loss of significance in computing the differences in each of the terms. Moreover, a criterion for the truncation of the series needs to be specified, but bounds on the truncation errors are not easily available. The recurrence relations require the computation of many binomial densities. If all these are stored, the algorithm requires much storage arrays when m and n are large. An alternative is to use two arrays, one for the binomial density in the outer sum; the other for that in the inner sum. The necessary binomial densities are then computed as often as needed. When the parameter r of a binomial density with either p_1 or p_2 is large, we neglect small tail probabilities by stopping computation of the binomial probabilities when the sum of the computed terms exceeds $1 - 10^{-12}$.

Chapter 3

Solution to Problem 3.6.16: The area Y is $1/2(X^2 + 2X)^{1/2}$. Therefore, for all $x \geq 0$,

$$P\{Y \leq x\} = P\{X^2 + 2X \leq 4x^2\} = \frac{1}{\Gamma(\alpha + 1)} \int_0^{\lambda\phi(x)} e^{-u} u^{\alpha+1} \, du,$$

where $\phi(x) = -1 + (1 + 4x^2)^{1/2}$. The probability distribution is computed by calling the library routine for the incomplete gamma ratio function.

Solution to Problem 3.6.17: The probabilities Q_{n1}, Q_{n2}, Q_{n3}, and Q_{n4} are

$$Q_{n1} = \frac{4\binom{44}{n-5}}{\binom{52}{n}}, \quad \text{for } n \geq 5, \qquad Q_{n2} = \frac{6\binom{44}{n-6}}{\binom{52}{n}}, \quad \text{for } n \geq 6,$$

$$Q_{n3} = \frac{4\binom{44}{n-7}}{\binom{52}{n}}, \quad \text{for } n \geq 7, \qquad Q_{n4} = \frac{\binom{44}{n-8}}{\binom{52}{n}}, \quad \text{for } n \geq 8.$$

These are easily computed recursively up to $n = 52$. For each n, their sum is P_n, the probability that the sample of n cards contains at least one most valuable hand of five cards. $V_{nm} = P_n R_{nm}$ is the probability that the sample of n cards contains at least one most valuable hand of five cards and a total of m points. The corresponding event has positive probability if and only if $19 \leq m \leq 40$. We compute the V_{nm} as follows. As there must be at least one King, there are 100 possible ways of choosing the number i, $1 \leq i \leq 4$, of Kings, the number j, $0 \leq j \leq 4$, of Queens, and the number r, $0 \leq r \leq 4$, of Jacks. Store the possible triplets (i, j, r) in arrays *IK*, *IQ*, and *IJ*, each of length 100. For each triplet, store the smallest value of n for which it can occur in an array *NMIN*. That number is given by $4 + i + j + r$. Also store the corresponding number of points $16 + 3i + 2j + r$ in an array *IV*.

For each triplet, we now evaluate the probability that the corresponding cards and the four Aces are found in the sample of n cards. That probability is added to the contents of the appropriate location in the array *VV* that holds the probabilities V_{nm}. For the triplet (i,j,r), that probability is

$$\frac{\binom{4}{i}\binom{4}{j}\binom{4}{r}\binom{36}{n-4-i-j-r}}{\binom{52}{n}}, \quad \text{for } n \geq 4 + i + j + r.$$

That quantity is recursively computed for successive values of n and added to the location for the probability V_{nm} with $m = 16 + 3i + 2j + r$. Finally, the desired probabilities R_{nm} are obtained by dividing V_{nm} by P_n.

Solution to Problem 3.6.18: For Player j to win by turning over the ith ace that card must be in one of the locations $vm + j$ in the deck. The sample space consists of all choices of the locations of the 4 aces in the deck.

Writing L_m for the integer part of $(52 - j)/m$, the events $A(i, j; \nu)$, $0 \le \nu \le L_m$, that the i-th ace occupies position $\nu m + j$ are disjoint. Therefore,

$$P_i(j; m) = \sum_{\nu=0}^{L_m} \frac{\binom{m\nu + j - 1}{i - 1}\binom{52 - m\nu - j}{i - 4}}{\binom{52}{4}}.$$

As $1 \le i \le 4$, there is little difficulty in a recursive computation and storage of the summands.

Solution to Problem 3.6.19: Clearly, $Y = [16(X - 1/2)^2 - 1]^2$. The polynomial $Y = P(X)$ is nonnegative and symmetric on [0, 1]. Its global maximum is 9, attained at 0 and 1. The zeros at 1/4 and 3/4 are local minima and there is a local maximum at 1/2. For $0 < u < 1$, the equation $x = P(u)$ has the four roots

$$\frac{1}{2} - \frac{1}{4}(1 + x^{1/2})^{1/2} < \frac{1}{2} - \frac{1}{4}(1 - x^{1/2})^{1/2}$$

$$< \frac{1}{2} + \frac{1}{4}(1 - x^{1/2})^{1/2} < \frac{1}{2} + \frac{1}{4}(1 + x^{1/2})^{1/2},$$

whereas for $1 < x \le 9$, it has two distinct roots at the points

$$\frac{1}{2} - \frac{1}{4}(1 + x^{1/2})^{1/2} < \frac{1}{2} + \frac{1}{4}(1 + x^{1/2})^{1/2}.$$

By considering the lengths of the intervals where $P(u) \le x$, we obtain the stated expressions for $F(\cdot)$. The mean $E(Y)$ is found by integrating $P(u)$ between 0 and 1.

Solution to Problem 3.6.20: We first write a subroutine for the Poisson probabilities. As λ may be large, we compute the modal term first by logarithms and the other terms to the left and right by the simple recurrence relation, as in Problem 1.2.5. Only terms exceeding 10^{-12} are computed and we remember the slowest and highest indices *kleft* and *kright*. The cumulative distribution is stored for the same indices between *kleft* and *kright*. The terms in the infinite sum in the formula for $P\{J_n = j\}$ are computed and summed from *kleft* to *kright*. To accommodate the fairly large exponents that may arise in each term, we computed each of the terms in the sum by logarithms. With careful coding, the computed sum of the probabilities $P\{J_n = j\}$ is very close to one.

Solution to Problem 3.6.23: With a current capital of k, the gambler bets $\min(k, N - k)$. The probabilities $P(k; N)$ therefore satisfy the equations

$$P(k; N) = pP(2k; N), \qquad \text{for } 1 \le k \le N - k,$$

$$P(k; N) = p + qP(2k - N; N), \qquad \text{for } k \ge N - k.$$

To compute $P(k; N)$ for $1 \le k \le N - 1$ we must solve a system of linear equations with a sparse coefficient matrix. For this, we recommend Gauss - Seidel iteration. We first compute the solution for $p = 0.46$, and $N = 51$. A suggested starting solution is $P(k, 51) = 0$, for $1 \le k \le 25$, $P(k, 51) = 1$ for $26 \le k \le 50$. The computed solution for that case is stored. Next, we solve the system with $N = 51$ for successively larger p values. The preceding solution becomes the starting solution for the next p.

Then we successively increase N. For $p = 0.46$, we take the solution for the preceding N and append a 1 to it. That provides a good starting solution to solve the system for N and $p = 0.46$. When that case has been solved, we solve the systems for successively larger p values, starting with the preceding solution. All systems of linear equations are so solved in a double DO-loop. The outer loop solves for successive values of N; the inner one for successively increasing p-values. The Gauss - Seidel iteration is carried out in a subroutine that takes advantage of the highly structured coefficient matrix.

Solution to Problem 3.6.34: The recurrence relation follows from the law of total probability. If ν items are marked at time j, the probability that there are r unmarked items in the sample is given by the hypergeometric probability which is the first factor in the term of the inner sum. The second factor is the binomial probability that $k - \nu$ of these r items become marked at time $j + 1$.

Clearly ϕ_j is the probability that a specific item becomes marked at time j. $P_j(k)$ is the probability that exactly $n - k$ items have not been marked by time j. Since items become marked independently, it is given by the stated binomial probabilities. To obtain an approximate solution to the equation

$F(j) = \alpha$, we write

$$\log F(j) = n \log \left[1 - \left[1 - \frac{mp}{n}\right]^j\right],$$

and we replace the right-hand side by $-n(1 - mp/n)^j$, and take logarithms again to obtain the approximate equation

$$j \log \left[1 - \frac{mp}{n}\right] = \log(-\log \alpha) - \log n.$$

In computing the probabilities $P_j(k)$, it is advisable to use logarithms.

Solution to Problem 3.6.35: There is a correct selection if the Poisson random variable with parameter λ_1 attains the smallest value among the $n + 1$ observations. Let that value be ν. If there are j more observations with value ν, then the correct population must win the lottery against these. That occurs with conditional probability $(j + 1)^{-1}$. The stated formula follows by the law of total probability.

First compute and store the Poisson densities with parameters λ and λ_1 and the tail probabilities $1 - \sum_{r=0}^{\nu} e^{-\lambda}\lambda^r (r!)^{-1}$. To write a code suitable for large λ, compute the these densities starting from the modal term. For each value of ν, compute the bound given in the hint. As long as it does not exceed, say, 10^{-9}, compute the inner sum. Its terms may be computed by a recurrence relation similar to that for the binomial density. The subroutine for $P(CS; \lambda, n, \delta)$ is called to examine that quantity as a function of λ in part *c*. Since the largest λ arising here is 5, convergence in the series is very rapid. However, we encourage the reader to explore other ranges of the parameter values.

Solution to Problem 3.6.37: If all coins have probability p_0 of showing Heads, then

$$P_{H_0}\{\max(X_1, \cdots, X_m) > r\} = 1 - \left[\sum_{k=0}^{r} \binom{n}{k} p_0^k (1 - p_0)^{n-k}\right]^m.$$

To achieve a size at least α, we choose r as the smallest integer for which

$$\sum_{k=0}^{r} \binom{n}{k} p_0^k (1 - p_0)^{n-k} > (1 - \alpha)^{1/m}.$$

For that value of r,

$$\beta = P_{H_1}\{\max(X_1, \cdots, X_m) \le r\}$$

$$= \left[\sum_{k=0}^{r} \binom{n}{k} p_1^k (1-p_1)^{n-k}\right]\left[\sum_{k=0}^{r} \binom{n}{k} p_0^k (1-p_0)^{n-k}\right]^{m-1}.$$

Notice that the sum of the binomial probabilities with parameter p_0 occurs in both formulas and should be computed only once. The tasks in parts a–c only require the computation of binomial probabilities and numerical searches for r and n. For part c we found that $r = 88$ with a sample size $n = 411$. The corresponding values of α and β are 0.0497 and 0.0481. For part e, the normal approximation to the binomial distribution with $n = 10{,}000$ and $p_0 = 1/6$ yields that for $\alpha = 0.01$, the critical value $r = 1{,}781$. To determine the values of p_1, we write the equation

$$\Phi\left[\frac{r + 0.5 - np_1}{[np_1(1-p_1)]^{1/2}}\right] = \beta\left\{\Phi\left[\frac{r + 0.5 - np_0}{[np_0(1-p_0)]^{1/2}}\right]\right\}^{1-m},$$

and substitute $r = 1{,}781$, $m = 10$, $p_0 = 1/6$, and successively the three given values for β. From the table of $\Phi(x)$ values, we determine the corresponding values of the argument of Φ on the left-hand side. Solving for p_1, we found that

$$p_1 = 0.1831, \quad \text{for } \beta = 0.01,$$

$$p_1 = 0.1846, \quad \text{for } \beta = 0.05,$$

$$p_1 = 0.1873, \quad \text{for } \beta = 0.10.$$

Solution to Problem 3.6.39: With p_ν and θ_ν defined as in the remark,

$$P_1(r, n) = \binom{m}{r} p_n^r (1-\theta_n)^{m-r}, \quad \text{for } n \ge k, \quad 0 \le r \le m,$$

$$P_2(n) = (1-\theta_{n-1})^m - (1-\theta_n)^m, \quad \text{for } n \ge k.$$

$$P_3(n) = \theta_n (1 - \theta_{n-1})^{m-1}, \quad \text{for } n \ge k.$$

and

$$P_4(n) = \theta_n (1 - \theta_n)^{m-1}, \quad \text{for } n \ge k.$$

Direct arguments can be given for each of these formulas. For the game to end at time n with the reader being the sole winner, he or she must score the kth success at time n and none of the other $m - 1$ players have been so fortunate. The truncation index N is the smallest integer for which

$$\log_{10} \left[1 - \theta_{N-1}\right] < -\frac{4}{m}.$$

The essential ingredient in the program is a subroutine to compute the negative binomial density and distribution accurately.

Solution to Problem 3.6.56: The proofs of the stated formulas are straightforward. The probability $\pi_\nu = P\{X = L_\nu\}$ is given by

$$\pi_\nu = \sum_{i=1}^{\nu} P_K(i, \nu) + \sum_{i=\nu+1}^{M} P_K^o(i, \nu).$$

We see that $P_K(i, \nu)$ and $P_K^o(i, \nu)$ are of the form $f(i, K)p_\nu$ and $g(i, K)p_\nu$, where the factors $f(i, K)$ and $g(i, K)$ do not depend on ν. This fact is exploited in an efficient computer code.

In addition to the criteria stated in parts d and e, we examined the probability that the purchaser buys at the lowest price L_1. In numerical examples $E(X)$ and $P\{X = L_M\}$ appear to be concave functions of K, while the graph of $P\{X = L_1\}$ exhibits a convex behavior in K. We leave it up to the reader to offer a qualitative explanation for the shape of these graphs.

Solution to Problem 3.6.57: There are $B(r, n - 1; k)$ subsets among the $B(r, n; k)$ that do not include n. By removing n for those that do, we obtain the $B(r - 1, n - 1; k - n)$ subsets of $r - 1$ elements chosen from among $\{1, \cdots, n - 1\}$ and with sum at most $k - n$. We see that $B(r, n; k) = 0$, for $k < r(r + 1)/2$; $B(r, n; r(r + 1)/2) = 1$ and for $k \ge n(n + 1)/2$,

$$B(r, n; k) = \binom{n}{r}.$$

We initialize the array for the $B(r, n; k)$ by setting $B(1, 1; k) = 1$, for all $1 \le k \le k^* = n(n+1)/2$. For each successive n, we also set $B(1, n; k) = \min(k, n)$, for the same range of k–values. The array for the $g(j, n; k)$ is similarly initialized. The probabilities $g(j, n; k)$ satisfy the recurrence relation

$$g(j, n; k) = \left[1 - \frac{j}{n}\right] g(j, n - 1; k) + \frac{j}{n} g(j - 1, n - 1; k - n).$$

If we are interested in only one value of n, a two-dimensional array containing the quantities $g(j, n; k)$ for $1 \le j \le n$ and $1 \le k \le n(n + 1)/2$ is computed for successive values of n. However, storing all successive $g(j, n; k)$ in a three-dimensional array preserves all earlier information on the densities of Z_k as stated in part c.

Solution to Problem 3.6.58: The following table lists, for each of the indices $1 \le i \le 10$, the values of the index n for which the probabilities $P(n; i)$, $R(n; i)$, $S(n; i)$, and $V(n; i)$ can be positive.

i	$n{:}P(n; i) > 0$	$n{:}R(n; i) > 0$	$n{:}S(n; i) > 0$	$n{:}V(n; i) > 0$
1	0 – 9	0 – 10	10 – 10	10 – 10
2	0 – 28	10 – 30	10 – 30	30 – 30
3	0 – 57	10 – 60	10 – 50	30 – 60
4	0 – 87	10 – 90	10 – 79	30 – 90
5	0 – 117	10 – 120	10 – 109	30 – 120
6	0 – 147	10 – 150	10 – 139	30 – 150
7	0 – 177	10 – 180	10 – 169	30 – 180
8	0 – 207	10 – 210	10 – 199	30 – 210
9	0 – 237	10 – 240	10 – 229	30 – 240
10	0 – 267	10 – 280	10 – 279	30 – 300

For notational convenience we set:

$$P_1(k) = \sum_{j=0}^{10-k} p(k, j), \quad \text{for } 0 \le k \le 10,$$

and

$$P_2(k) = \sum_{j=0}^{k} p(j, k-j), \quad \text{for } 0 \le k \le 9, \quad \text{and} \quad P_2(10) = \sum_{j=0}^{9} p(j, 10-j).$$

We also write $P_s(k) = p(10, 0)P_1(k)$. $P_1(k)$ is the probability that the first ball of a frame knocks over k pins. $P_2(k)$ is the probability that, in one frame, two balls are played and k pins are knocked down. $P_s(k)$ is the probability that a strike is followed by k pins falling on the first ball of the next frame. $P(n; i)$, $Q(n; i)$, $R(n; i)$, and $V(n; i)$ satisfy the following recurrence relations:

a. $P(n; 1) = P_2(n)$, and for $2 \le i \le 10$,

$$P(n; i) = \sum_{k=0}^{9} P(n-k; i-1)P_2(k) + \sum_{j=0}^{9}\sum_{k=0}^{9-j} [R(n-2j-k; i-1)$$

$$+ S(n-2j-2k; i-1) + V(n-3j-2k; i-1)]p(j, k).$$

b. $R(n; 1) = 0$, for $0 \le n \le 9$, and $R(10; 1) = P_2(10)$, and for $2 \le i \le 9$,

$$R(n; i) = \sum_{j=0}^{9} [R(n-j-10; i-1) + V(n-j-20; i-1)]p(j, 10-j)$$

$$+ [P(n-10; i-1) + S(n-20; i-1)]P_2(10),$$

and

$$R(n; 10) = \sum_{k=0}^{10} \left\{ \sum_{j=0}^{9} [R(n-j-k-10; 9) + V(n-j-k-20; 9)]p(j, 10-j) \right.$$

$$\left. + [P(n-k-10; 9) + S(n-k-20; 9)]P_2(10) \right\} P_1(k).$$

c. $S(n; 1) = 0$, for $0 \le n \le 9$, and $S(10; 1) = p(10, 0)$, and for $2 \le i \le 9$,

$$S(n; i) = [P(n-10; i-1) + R(n-20; i-1)]p(10, 0),$$

$$S(n; 10) = \left\{ \sum_{j=0}^{10} [P(n-j-20; 9) + R(n-j-30; 9)]P_s(j) \right.$$

$$\left. + \sum_{j=0}^{10} [P(n-j-10; 9) + R(n-j-20; 9)]P_2(j) \right\} p(10, 0).$$

d. $V(n; 1) = 0$, for $0 \le n \le 10$, and for $2 \le i \le 9$,

$$V(n; i) = [S(n - 20; i - 1) + V(n - 30; i - 1)]p(10, 0),$$

$$V(n; 10) = \left\{ \sum_{j=0}^{9} \sum_{k=0}^{10-j} [S(n - 2j - k - 20; 9) + V(n - 2j - k - 30; 9)]p(j, k) \right.$$
$$\left. + \sum_{k=0}^{10} [S(n - k - 40; 9) + V(n - k - 50; 9)]P_s(k) \right\} p(10, 0).$$

For each case, the expressions for $i = 1$ are clear. For $2 \le i \le 10$, the recurrence relations are obtained by conditioning on the type of the previous frame and by considering all possible ways of throwing two balls to satisfy the assumptions of the current frame.

We shall refer to a frame as being one of the following:

open: fewer than ten pins knocked down with two balls;
spare: all ten pins knocked down with two balls;
single strike: current frame is a strike but the previous frame is not;
double strike: both the previous and current frames are strikes.

The first term in the equation for $P(n; i)$ is the probability that the score in the $(i-1)$st frame is $n - k$ and both the $(i-1)$st and the ith frames are open. In the second term, the indices j and k denote the numbers of pins knocked down by the first and second balls, respectively, in the ith frame. In the case of a spare, the score at the $(i-1)$st frame had to be $n - 2j - k$; for a single strike, it had to be $n - 2j - 2k$; whereas for a double strike, it had to be $n - 3j - 2k$. The reasoning behind the equations for $R(n; i)$, $S(n; i)$, and $V(n; i)$ is similar. However, extra care is needed at the tenth frame to account for the possibility that a third ball is thrown. We note that, if the tenth frame is not open, the extra pins knocked over following the spare or strike are counted only once.

The equation for $R(n; 10)$ is similar to that for $R(n; i)$, $1 \le i \le 9$, except that there is an additional sum over the index k which indicates the number of pins knocked over by the third ball. To obtain $S(n; 10)$, we condition on whether or not the second ball of the tenth frame is a strike. In the equation for $S(n; 10)$, the first sum corresponds to the event that the second ball is a strike and in the second summation, that it is not a strike. The expression for $V(n; 10)$ is obtained in the same manner. The term with the single summation accounts for the case where the first strike of the tenth frame is followed by a second strike. The term with the double sum takes care of the alternative where the first strike is not followed by a second strike.

The probability density $\{\phi_n\}$ of the final score is given by

$$\phi_n = P(n; 10) + R(n; 10) + S(n; 10) + V(n; 10).$$

To test the code, trivial input data such as $p(3, 0) = 1$ can be used. For these, it is clear what the density of the final score has to be. In some cases, interesting multi-modal densities $\{\phi_n\}$ arise. It is worthwhile to get an intuitive understanding of why some values for the total score are favored over others. The reader may examine the results for the data:

$$p(9, 0) = 0.15, \quad p(10, 0) = 0.20, \quad p(8, 1) = 0.10,$$
$$p(9, 1) = 0.15, \quad p(7, 2) = 0.05, \quad p(8, 2) = 0.15,$$
$$p(6, 3) = 0.05, \quad p(7, 3) = 0.10, \quad p(6, 4) = 0.05.$$

and also for:

$$p(10, 0) = 0.50, \quad p(0, j) = 0.10, \quad \text{for } 6 \le j \le 10.$$

In both data sets, probabilities that are not explicitly specified are zero.

Solution to Problem 3.6.61: The probability $P_n(a, \sigma^2)$ is given by

$$P_n(a, \sigma^2) = \frac{1}{2} \int_{-1}^{1} \left[\int_{y-a}^{y+a} \frac{1}{\sigma\sqrt{2\pi}} \exp\left(-\frac{u^2}{2\sigma^2} \right) du \right]^n dy.$$

By the symmetry of the normal density, this may be rewritten as

$$P_n(a, \sigma^2) = \int_0^1 \left[\Phi\left(\frac{y+a}{\sigma} \right) - \Phi\left(\frac{y-a}{\sigma} \right) \right]^n dy.$$

For the purpose of this problem, that integral is easily computed by Simpson's rule.

Solution to Problem 3.6.62: The probability $p(n)$ is given by

$$p(n) = \int_0^\infty e^{-u}[2\Phi(u) - 1]^n \, du.$$

The integral from 7 to ∞ may be replaced by exp(−7). The integral from 0 to 7 is easily computed by Simpson's rule. As an alternative, Lagrange quadrature can be used. With Y exponential with mean 1, let N be the number of independent normal random variates that must be drawn before a value outside $(-Y, +Y)$ is observed. Then, $p(n) = P\{N > n\}$.

Solution to Problem 3.6.63: This problem is similar to the preceding one. Clearly,

$$p(\lambda) = \lambda \int_0^\infty e^{-\lambda u}[2\Phi(u) - 1]^{10} \, du.$$

To find the λ for which $p(\lambda) = 0.50$, we suggest a bisection method in conjunction with Lagrange quadrature.

Chapter 4

Solution to Problem 4.4.12: Rearrange the row and columns of P to obtain

$$P = \begin{array}{c|ccccc|} 2 & 0.10 & 0.15 & 0.15 & 0.60 & 0.0 \\ 4 & 0.25 & 0.75 & 0.0 & 0.0 & 0.0 \\ 1 & 0.20 & 0.35 & 0.25 & 0.20 & 0.0 \\ 3 & 0.0 & 0.50 & 0.0 & 0.25 & 0.25 \\ 5 & 0.10 & 0.0 & 0.15 & 0.25 & 0.50 \end{array},$$

and partition the matrix to distinguish the sets $\{2, 4\}$ and $\{1, 3, 5\}$. Call the submatrices $T(1, 1)$, $T(1, 2)$, $T(2, 1)$, and $T(2, 2)$. Write $\beta = (0.25, 0.75)$. Then the probabilities in part a are the components of the vector

$$\beta[I - T(1, 1)]^{-1}T(1, 2),$$

and the mean in part b is $\beta[I - T(1, 1)]^{-1}\mathbf{e}$. For part c, define the matrix K

by

$$K = [I - T(1, 1)]^{-1}T(1, 2)[I - T(2, 2)]^{-1}T(2, 1).$$

The required probability is the first component of βK^3.

Solution to Problem 4.4.14: For convenience, set $\phi(0) = 1$ and $\phi(K + 1) = 0$. The probabilities $\phi(i)$ then satisfy

$$\phi(i) - q(i)\phi(i - 1) - p(i)\phi(i + 1) = 0,$$

for $1 \le i \le K$. That equation may be written as

$$\phi(i - 1) - \phi(i) = \frac{p(i)}{q(i)}[\phi(i) - \phi(i + 1)], \quad \text{for } 1 \le i \le K.$$

Forming the product of left- and right-hand sides for $i \le j \le K$, we obtain

$$\phi(i - 1) - \phi(i) = \prod_{j=i}^{K} \frac{p(j)}{q(j)}\phi(K), \quad \text{for } 1 \le i \le K.$$

Summing these equations over $1 \le i \le r$, we get that

$$1 - \phi(r) = \sum_{i=1}^{r} \prod_{j=i}^{K} \frac{p(j)}{q(j)}\phi(K), \quad \text{for } 1 \le r \le K.$$

It is now easy to see that

$$\phi(r) = \frac{1 + \sum_{v=r+1}^{K} \prod_{j=v}^{K} \frac{p(j)}{q(j)}}{1 + \sum_{v=1}^{K} \prod_{j=v}^{K} \frac{p(j)}{q(j)}}, \quad \text{for } 1 \le r \le K.$$

The $\phi(r)$ are decreasing in r. The first index r for which $\phi(r) < 1/2$ is also the first for which

$$\sum_{v=r+1}^{K} \prod_{j=v}^{K} \frac{p(j)}{q(j)} - \sum_{v=1}^{r} \prod_{j=v}^{K} \frac{p(j)}{q(j)} > 1.$$

That index r° is easily found. We then examine $\phi(r^\circ)$ and $\phi(r^\circ - 1)$ to see which value is closest to 1/2.

Solution to Problem 4.4.16: The steady-state equations may be written as

$$\frac{\pi_0}{a_0} = \sum_{r=0}^{\infty} \frac{\pi_r}{A_{r+1}}, \quad \text{and} \quad \frac{\pi_i}{a_i} = \sum_{r=i-1}^{\infty} \frac{\pi_r}{A_{r+1}}, \quad \text{for } i \geq 1.$$

By subtraction we obtain that $a_1^{-1}\pi_1 = \pi_0$ and that for $i \geq 1$,

$$\frac{\pi_i}{a_i} - \frac{\pi_{i+1}}{a_{i+1}} = \frac{\pi_{i-1}}{A_i}.$$

The formula stated in part a now follows by an easy calculation. For the case in part b the steady-state probabilities are geometric. Let us write

$$\prod_{i=0}^{k-1} \frac{a_{i+1}}{A_i} = \phi_k, \quad \text{for } k \geq 0.$$

For $a_j = 1$ for $j \geq 0$, we obtain that $\phi_i = (i\,!)^{-1}$, so that the invariant probability density is Poisson with mean 1. For $a_j = j + 1$ for $j \geq 0$, we find that $\phi_i = 2^i (i\,!)^{-1}$. The invariant probability density is Poisson with mean 2.

When $a_j = (j + 1)^\alpha$, with α an integer ≤ 2, we use summation formulas for the powers of the first i integers. Specifically:

$$\sum_{j=1}^{i} j^2 = \frac{1}{6} i(i + 1)(2i + 1), \quad \text{and} \quad \sum_{j=1}^{i} j^3 = \frac{1}{4} i^2 (i + 1)^2.$$

For $\alpha = 2$,

$$\phi_k = \frac{(k + 1)!\,12^k}{(2k + 1)!},$$

and $\phi_{k+1}/\phi_k = 6(k + 2)(k + 1)^{-1}(2k + 3)^{-1}$. By the ratio test, the series converges and the Markov chain is positive recurrent. For $\alpha = 3$,

$$\phi_k = \frac{4^k(k+1)}{k!}.$$

A routine calculation shows that $\sum_{k=0}^{\infty} \phi_k = 5e^4$, so that the Markov chain is again positive recurrent. The π_i are explicitly given by

$$\pi_i = \frac{1}{5}\frac{(i+1)4^i}{i!}e^{-4}, \quad \text{for } i \geq 0.$$

For a general value of $\alpha \geq 0$, we have that

$$\frac{\phi_{k+1}}{\phi_k} = \frac{a_{k+1}}{A_k} = \frac{(k+2)^\alpha}{\sum_{r=1}^{k+1} r^\alpha},$$

$$A_k = \sum_{j=1}^{k+1} j^\alpha > \int_0^{k+1} u^\alpha \, du = \frac{1}{\alpha+1}(k+1)^{\alpha+1},$$

so that as $k \to \infty$,

$$\frac{a_{k+1}}{A_k} < (\alpha+1)\frac{(k+2)^\alpha}{(k+1)^{\alpha+1}} \to 0,$$

By the ratio test, the Markov chain is positive recurrent for all $\alpha \geq 0$.

Solution to Problem 4.4.48: At time n we need to keep track of the current state i and of the number k of reversals that have already occurred. In addition, it must be remembered whether the transition into i came from $i-1$ or from $i+1$.

Let $V_n^+(i, k)$, respectively $V_n^-(i, k)$, be the conditional probability that the state at time n was reached from $i-1$, respectively from $i+1$, with a total of k reversals. In the following recurrence relations, the initial conditions and the ranges of the indices must be specified with care. We have that for $n \geq 1$:

$$V_{n+1}^+(i, k) = p(i)V_n^+(i-1, k) + p(i)V_n^-(i-1, k-1), \quad \text{for } 2 \leq i \leq K,$$

$$V_{n+1}^-(i, k) = q(i)V_n^-(i+1, k) + q(i)V_n^+(i+1, k-1), \quad \text{for } 1 \leq i \leq K-1.$$

For $n = 1$, $V_1^+(i, k)$ and $V_1^-(i, k)$ are zero for all $1 \le i \le K$, except for

$$V_1^+(j+1, 0) = p(j), \quad \text{if and only if } 1 \le j \le K-1,$$

$$V_1^-(j-1, 0) = q(j), \quad \text{if and only if } 2 \le j \le K.$$

Furthermore, for $n \ge 1$, and $k \ge 0$, $V_n^+(1, k) = 0$, and $V_n^-(K, 1) = 0$, as it is impossible to move from 0 to 1 or from $K + 1$ to K.

For $n \ge 2$, the index k is at most $n - 1$. The probability that, at time n, the walk has not been absorbed and has experienced k reversals is given by

$$\phi_n(k) = \textstyle\sum_{i=1}^{K} [V_n^+(i, k) + V_n^-(i, k)],$$

and $\psi_n = \phi_n(1) + \cdots + \phi_n(n-1)$, is the probability that the walk has not yet been absorbed at time n. We continue recursive computation until that probability becomes negligibly small. The probability

$$P\{R = k\} = \sum_n \textstyle\sum_{i=1}^{K} [V_n^+(i, k) + V_n^-(i, k)] = \displaystyle\sum_n \phi_n(k),$$

is accumulated in the process. Notice that the algorithm requires specification of an initial state j. If an initial probability density is specified for the initial state, the recursion must be carried out for every initial state with nonzero probability and the appropriate mixture of the densities of R must be formed.

Solution to Problem 4.4.49: We discuss a method to store *only* the essential probabilities $P(i; i_1, i_2; n)$ needed in implementing a computation recursive in n. They are stored in a systematic manner in a *one-dimensional* array. That requires careful bookkeeping, but utilizes memory space very efficiently. The probabilities $P(i; i_1,i_2; n)$ for the current n will be stored in a real array *PP*. For convenience, we also use an integer array *INDX* whose contents will be explained next.

The range Y_n can assume the values $r = 0, \cdots, K$. For a given value of r, all possible values of the current state i compatible with any possible values of i_1 are store in the array *INDX*. To illustrate this, we display the contents of *INDX* for $K = 5$ and $j = 4$:

$r = 0$:	4		
$r = 1$:	3, 4	4,3	
$r = 2$:	2, 3, 4	3, 4, 5	4, 5, 6
$r = 3$:	1, 2, 3, 4	2, 3, 4, 5	3, 4, 5, 6
$r = 4$:	0, 1, 2, 3, 4	1, 2, 3, 4, 5	2, 3, 4, 5, 6
$r = 5$:	0, 1, 2, 3, 4, 5	1, 2, 3, 4, 5, 6	

We also keep track of the locations of the first token for every r. These are stored in an array LL, which in our example contains 1, 2, 6, 15, 27, 42 The total length of the array $INDX$ is $LLMAX = 53$. In general, for every r with $0 \le r \le K$, we successively store the strings of length $r + 1$ of integers $i_1, \cdots, i_1 + r$ that satisfy $\max(0, j - r) \le i_1 \le \min(j, K - r + 1)$. The number $N(r)$ of such strings is given by

$$N(r) = \min(j, K - r + 1) - \max(0, j - r) + 1,$$

and $LL(0) = 1$, $LL(r) = LL(r - 1) + rN(r - 1)$, for $1 \le r \le K$. The total length $LLMAX$ of the array $INDX$ is $L(K) + (K + 1)N(K) - 1$. For a given j with $1 \le j \le K$, the current value of $P(i; i_1, i_2; n)$ is stored in the location $LL(r) - 1 + [i_1 - \max(0, j - r)](r + 1) + i - i_1 + 1$. Use of the array $INDX$ could be avoided, but it is convenient in checking that the storage arrays have been set up correctly.

The probabilities $P(i; i_1, i_2; n)$ satisfy the recurrence relations

$$P(i; i_1, i_2; n + 1) = p(i - 1)P(i - 1; i_1, i_2; n) + q(i + 1)P(i + 1; i_1, i_2; n),$$

for $i_1 + 1 \le i \le i_2 - 1$, and

$$P(i_1; i_1, i_2; n + 1) = q(i_1 + 1)P(i_1 + 1; i_1 + 1, i_2; n),$$
$$P(i_2; i_1, i_2; n + 1) = p(i_2 - 1)P(i_2 - 1; i_1, i_2 - 1; n).$$

This, for all meaningful values of i_1, i_2, and i. These are identified in the array $INDX$. We initialize the array PP by setting

$$P(j - 1; j - 1, j; 1) = q(j), \quad \text{and} \quad P(j + 1; j, j + 1; 1) = p(j).$$

For $n = 1$, all other elements are zero. The easiest way to implement the recurrence is to use two arrays $PP1$ and $PP2$, that are used alternatingly to store the new probabilities $P(i;\, i_1, i_2;\, n)$ that are being computed.

By more belabored programming, the new $P(i;\, i_1, i_2;\, n)$ can be stored over the old ones. We see that the strings in the array *INDX* are ranked by increasing r, the index that keeps track of the length of the range. In the first of the three recurrence relations, the right-hand side involves terms with the same r as the left-hand side. For the remaining two equations we need terms for which the range is one less.

In updating the array *PP* we proceed for increasing r. For each r, there are a number of strings of the form $i_1, \cdots, i_1 + r$ to be processed. We can store the old values for the first and last elements as these are needed for $r + 1$. Using the recurrence relations, we now update the new probabilities and, with some careful bookkeeping, write them over the earlier values.

The probabilities $R(i;\, k;\, n)$ are computed by scanning the segment of *INDX* corresponding to $r = k$. For every location that contains i, we add the value of the corresponding location in *PP* to the storage location for $R(i;\, k;\, n)$. For a given n, all probabilities $R(i;\, k;\, n)$ are so computed in a single pass. Summation of these probabilities over all i produces the density of the range of the random walk with initial state j.

The probabilities $P(i;\, 0;\, i_2;\, n)$ and $P(i;\, i_1, K + 1;\, n)$ are zero except for $P(0;\, 0;\, i_2;\, n)$ and $P(K + 1;\, i_1, K + 1;\, n)$. The first is the probability that by time n, the random walk was absorbed in 0 and i_2 is the rightmost state visited. The interpretation of $P(K + 1;\, i_1, K + 1;\, n)$ is similar.

Solution to Problem 4.4.50: For $a = 6$, the transition probability matrix $P(a)$ is given

$$P(6) = \begin{array}{c|ccccccc|} 0 & 0 & 1/6 & 1/6 & 1/6 & 1/6 & 1/6 & 1/6 \\ 1 & 1/2 & 0 & 1/10 & 1/10 & 1/10 & 1/10 & 1/10 \\ 2 & 1/4 & 1/4 & 0 & 1/8 & 1/8 & 1/8 & 1/8 \\ 3 & 1/6 & 1/6 & 1/6 & 0 & 1/6 & 1/6 & 1/6 \\ 4 & 1/8 & 1/8 & 1/8 & 1/8 & 0 & 1/4 & 1/4 \\ 5 & 1/10 & 1/10 & 1/10 & 1/10 & 1/10 & 0 & 1/2 \\ 6 & 1/6 & 1/6 & 1/6 & 1/6 & 1/6 & 1/6 & 0 \end{array} .$$

In general $P_{ii}(a) = 0$, for $0 \le i \le a$, $P_{0j}(a) = a^{-1}$, for $1 \le j \le a$, $P_{aj}(a) = a^{-1}$, for $0 \le j \le a-1$, and

$$P_{ij}(a) = \frac{1}{2i}, \qquad \text{for } 1 \le i \le a-1,\ i > j,$$
$$= \frac{1}{2(a-i)}, \qquad \text{for } 1 \le i \le a-1,\ i < j.$$

By subtractions, the steady-state equations

$$\pi_0 = \sum_{r=1}^{a-1} \frac{\pi_r}{2r} + \frac{\pi_a}{a},$$

$$\pi_i = \frac{\pi_0}{a} + \sum_{r=1}^{i-1} \frac{\pi_r}{2(a-r)} + \sum_{r=i+1}^{a-1} \frac{\pi_r}{2r} + \frac{\pi_a}{a},$$

for $1 \le i \le a-1$, and

$$\pi_a = \frac{\pi_0}{a} + \sum_{r=1}^{a-1} \frac{\pi_r}{2(a-r)},$$

lead to

$$\pi_1 = \frac{2(a+1)}{3a}\pi_0, \quad \pi_{i+1} = \frac{(i+1)(2a-2i+1)}{(2i+3)(a-i)}\pi_i,$$

for $1 \le i \le a-2$, and

$$\pi_a = \frac{3a}{2(a+1)}\pi_{a-1}.$$

After some routine manipulations we obtain that for $1 \le i \le a-1$,

$$\pi_i = \frac{2(a+1)}{(2i+1)(2a-2i+1)} \frac{\binom{2a}{a}}{\binom{2i}{i}\binom{2a-2i}{a-i}}\pi_0.$$

where

$$\pi_0 = \left[2 + \sum_{i=1}^{a-1} \frac{2(a+1)}{(2i+1)(2a-2i+1)} \frac{\binom{2a}{a}}{\binom{2i}{i}\binom{2a-2i}{a-i}} \right]^{-1}.$$

The invariant density is clearly symmetric as $\pi_i = \pi_{a-i}$, for $0 \le i \le a$. By further analysis, we can show that if we place the probabilities π_i at the points i/a, their distribution function tends to the arc-sine law as $a \to \infty$.

Solution to Problem 4.4.70: By considering the eigenvalues of $\Psi(\xi)$ we see that the singularity at 0 is *removable*. The function, defined at 0 by continuity, is analytic for all (complex) ξ. Noting that $(I - e^{-\xi}P)^{-1}\mathbf{e} = (1 - e^{-\xi})^{-1}\mathbf{e}$, we see that

$$(I - e^{-\xi}P)^{-1}(I - P + \mathbf{e}\pi) = \Psi(\xi) + (1 - e^{-\xi})^{-1}\mathbf{e}\pi,$$

and therefore

$$(I - e^{-\xi}P)\Psi(\xi)Z = I - \mathbf{e}\pi.$$

The matrices P, Z, and $\Psi(\xi)$ all commute and $Z\mathbf{e} = \mathbf{e}$. Expanding the factors in the preceding equation in power series, we obtain that

$$[(I - P) - \sum_{r=1}^{\infty} \frac{(-1)^r}{r!} \xi^r P] - \sum_{k=0}^{\infty} A_k \xi^k = I - \mathbf{e}\pi.$$

By equating the coefficients of powers of ξ, we see that

$$(I - P)ZA_0 = I - \mathbf{e}\pi, \tag{1}$$

and for $n \ge 1$,

$$(I - P)ZA_n = \sum_{r=0}^{n-1} \frac{(-1)^{n-r}}{(n-r)!} ZPA_r. \tag{2}$$

Write that last equation for $n + 1$ and multiply on the left by π to obtain that for $n \ge 0$,

$$\sum_{r=0}^{n} \frac{(-1)^{n+1-r}}{(n+1-r)!}\pi A_r = 0.$$

In particular, $\pi A_0 = 0$. For $n \geq 1$, we have that

$$\pi A_n = \sum_{r=0}^{n-1} \frac{(-1)^{n+1-r}}{(n+1-r)!}\pi A_r .$$

Adding $e\pi Z A_0 = 0$ to the left-hand side of (1), we readily see that $A_0 = I - e\pi$. Similarly adding $e\pi Z A_n$ to the left-hand side of (2), it follows that for $n \geq 1$,

$$A_n = \sum_{r=0}^{n-1} \frac{(-1)^{n+1-r}}{(n+1-r)!}[e\pi - (n+1-r)ZP]A_r .$$

The stated expressions for A_1 and A_2 are obtained by direct calculation. It is also clear from the recurrence relation (2) that the matrices A_n, $n \geq 1$, are of the form $A_n = a_n e\pi + \sum_{k=1}^{n} b_{nk}(ZP)^k$, where the a_n and b_{nk} are real numbers. The recurrence relations for these coefficients are proved by induction. The first few terms have already been checked in deriving the expressions for A_1 and A_2. We successively write A_n as

$$A_n = ZP\sum_{r=0}^{n-1} \frac{(-1)^{n-r}}{(n-r)!}A_r + e\pi\sum_{r=1}^{n-1} \frac{(-1)^{n+1-r}}{(n+1-r)!}A_r$$

$$= ZP\frac{(-1)^n}{n!}(I - e\pi) + ZP\sum_{r=1}^{n-1} \frac{(-1)^{n-r}}{(n-r)!}[a_r e\pi + \sum_{k=1}^{r} b_{rk}(ZP)^k]$$

$$+ e\pi\sum_{r=1}^{n-1} \frac{(-1)^{n+1-r}}{(n+1-r)!}[a_r + \sum_{k=1}^{r} b_{rk}].$$

We interchange summations to get that

$$\sum_{r=1}^{n-1}\sum_{k=1}^{r} \frac{(-1)^{n-r}}{(n-r)!}b_{rk}(ZP)^{k+1} = \sum_{k=2}^{n}\sum_{r=k-1}^{n-1} \frac{(-1)^{n-r}}{(n-r)!}b_{r,k-1}(ZP)^k,$$

The inner sum on the right-hand side is b_{nk} for $2 \le k \le n$, and $b_{n,1} = (-1)^n/n!$, the coefficient of the term in ZP. We collect the factors of $e\pi$ to obtain

$$a_n = -\frac{(-1)^n}{n!} - \sum_{r=1}^{n-1} (n-r)\frac{(-1)^{n+1-r}}{(n+1-r)!}a_r$$
$$+ \sum_{k=1}^{n-1}\sum_{r=k}^{n-1} \frac{(-1)^{n+1-r}}{(n+1-r)!}b_{rk},$$

but the last double sum may be written as $\sum_{k=1}^{n-1} [b_{n+1,k+1} + b_{nk}]$, since

$$b_{n+1,k+1} = \sum_{r=k}^{n} \frac{(-1)^{n+1-r}}{(n+1-r)!}b_{rk}.$$

Chapter 5

Solution to Problem 5.2.12: Let there be a registered event at time zero. The alternating exponential intervals can be viewed as the sojourn times of a two-state Markov chain with state 1 corresponding to the intervals of mean $1/\theta$. At time zero, that chain is in state 1. The next registered event either occurs during the sojourn time in course at time 0; or there are $k \ge 1$ pairs of sojourn times during which no arrival is registered and then the first arrival during the next sojourn in state 1 is registered. By the law of total probability, the transform $a(s)$ is therefore

$$a(s) = \frac{\lambda}{s+\lambda+\theta} + \sum_{k=1}^{\infty}\left[\frac{\theta}{s+\lambda+\theta}\,\frac{\sigma}{s+\sigma}\right]^k \frac{\lambda}{s+\lambda+\theta},$$

and that is equivalent to the stated expression. The denominator $s^2 + (\lambda+\theta+\sigma)s + \lambda\sigma$ has two negative roots which we denote by $-\eta_1$ and $-\eta_2$. The denominator is negative at $s = -\lambda$, so that $\eta_1 > \lambda > \eta_2 > 0$. The equivalent expression in terms of p, η_1, and η_2 is obtained by a simple calculation. The quantities defined in part b are positive and therefore valid parameters for an interrupted Poisson process. If they are substituted in the expression for $a(s)$, we see that the transforms $a(s)$ and $f(s)$ agree.

An informal argument to show that an interrupted Poisson process is a renewal process proceeds as follows. At a registered event, the two-state Markov chain is in its state 1. Since the arrivals during sojourns in state 1 form a Poisson process, at registered events no information about the past of the process needs to be retained. The times between registered events are therefore independent. Their common distribution is hyperexponential.

Solution to Problem 5.2.16: The generator for the four-state Markov chain is given by

$$Q = \begin{vmatrix} -\lambda & p\lambda & q\lambda & 0 \\ \mu_1 & -\lambda-\mu_1 & 0 & \lambda \\ \mu_2 & 0 & -\lambda-\mu_2 & \lambda \\ 0 & \mu_2 & \mu_1 & -\mu_1-\mu_2 \end{vmatrix}.$$

By routine calculations we obtain that

$$\pi_0 = \mu_1\mu_2(2\lambda + \mu_1 + \mu_2)\,[\phi(\lambda, \mu_1, \mu_2, p)]^{-1},$$
$$\pi_1 = \lambda\mu_2(\lambda + p\mu_1 + p\mu_2)\,[\phi(\lambda, \mu_1, \mu_2, p)]^{-1},$$
$$\pi_2 = \lambda\mu_1(\lambda + q\mu_1 + q\mu_2)\,[\phi(\lambda, \mu_1, \mu_2, p)]^{-1},$$
$$\pi_{12} = \lambda^2(\lambda + q\mu_1 + p\mu_2)\,[\phi(\lambda, \mu_1, \mu_2, p)]^{-1},$$

where

$$\phi(\lambda, \mu_1, \mu_2, p) = \lambda^3 + \lambda^2[(1+q)\mu_1 + (1+p)\mu_2]$$

$$+ \lambda[2\mu_1\mu_2 + (\mu_1 + \mu_2)(q\mu_1 + p\mu_2)] + \mu_1\mu_2(\mu_1 + \mu_2).$$

Customers who arrive when the Markov chain is in the state 12 are lost. The fraction of lost customers is therefore given by π_{12}. For part b, we need to find the value of λ for which $\pi_{12} = 0.1$. That equation simplifies to a cubic equation in λ. It is easily solved. For fixed λ, μ_1, and μ_2, the probability π_{12} is a linear fractional function of p. Its derivative with respect to p is

$$c(\mu_2 - \mu_1)(2\lambda + \mu_1 + \mu_2)\,[\phi(\lambda, \mu_1, \mu_2, p)]^{-2},$$

where c is a positive constant. When $\mu_1 = \mu_2$ the probability π_{12} does not depend on p. Otherwise, the derivative has the same sign as the difference $\mu_2 - \mu_1$. That implies that the loss probability is smallest when an arriving customer goes to the fastest of the two idle servers.

The fraction of time that server I is busy is given by $\pi_1 + \pi_{12}$. This is again a linear fractional function in p. After some calculation we find that its derivative is given by

$$\lambda\mu_1\mu_2[2\lambda^3 + (\mu_1 + 5\mu_2)\lambda^2 + 5\mu_2(\mu_1 + \mu_2)\lambda + \mu_2(\mu_1 + \mu_2)^2]\,[\phi(\lambda, \mu_1, \mu_2, p)]^{-2},$$

which is clearly positive for all p in [0, 1]. Server I is therefore busy for the largest fraction of time when $p = 1$.

Solution to Problem 5.2.18: The distribution of $Y = (1 - \alpha)X$ is

$$F(y; \alpha) = \sum_{k=1}^{\infty} (1 - \alpha)\alpha^{k-1}F^{(k)}[(1 - \alpha)^{-1}y],$$

with Laplace - Stieltjes transform

$$\phi(s) = \sum_{k=1}^{\infty} (1 - \alpha)\alpha^{k-1}[f\,[(1 - \alpha)s\,]^k = \frac{(1 - \alpha)f\,[(1 - \alpha)s\,]}{1 - \alpha f\,[(1 - \alpha)s\,]}.$$

By l'Hospital's rule, we see that for every fixed s,

$$\lim_{\alpha \to 1} \phi(s) = \frac{1}{1 - f\,'(0)} = \frac{1}{1 + \mu'_1 s}.$$

The limit distribution is therefore exponential with mean μ'_1. The statement in part b is a classical property of geometric mixtures of PH-distributions. Use the uniformization method to compute x^*. For α close to 1,

$$F\left[\frac{x^*}{1 - \alpha}\right] \text{ is approximately } 1 - \exp\left(-\frac{x^*}{\mu'_1}\right).$$

Use the solution to the equation

$$1 - \exp\left[-\frac{x^*(1 - \alpha)}{\mu'_1}\right] = 0.25,$$

or $\alpha^\circ = 1 + \mu'_1 \log 0.75\,(x^*)^{-1}$ as a starting solution for the desired α.

Efficiently organize the solution of the equation

$$F\left[\frac{x^*}{1-\alpha}\right] = 0.25,$$

Each iteration requires evaluating the *PH*-distribution of Y by the uniformization method. For each trial value of α, that distribution is different.

Solution to Problem 5.2.20: First compute the stationary probability vector $\boldsymbol{\theta} = [\boldsymbol{\theta}(1), \boldsymbol{\theta}(2)]$ of Q. The mean time between transitions in the stationary version of the Markov chain is chosen as the unit of time. The special structure of the chain with the exponential monitor implies that the steady-state probabilities of set **2** are the same as before and that $\boldsymbol{\theta}(1^*) + \boldsymbol{\theta}(1^\circ) = \boldsymbol{\theta}(1)$. Since $\boldsymbol{\theta}(2)T(2, 1) = -\boldsymbol{\theta}(1)T(1, 1)$, we successively obtain that

$$\boldsymbol{\theta}(1^*) = \boldsymbol{\theta}(2)T(2, 1)[\gamma I - T(1, 1)]^{-1}$$

$$= -\boldsymbol{\theta}(1)T(1, 1)[\gamma I - T(1, 1)]^{-1} = \boldsymbol{\theta}(1) - \gamma\boldsymbol{\theta}(1)[\gamma I - T(1, 1)]^{-1}.$$

After routine simplifications, $K(\gamma^{-1})$ may be written as

$$K(\gamma^{-1}) = 1 - \gamma^{-1}[\boldsymbol{\theta}(1)\mathbf{e}]^{-1}\boldsymbol{\theta}(1)[\gamma^{-1}I - T(1, 1)]^{-1}\mathbf{e}.$$

To gain further insight into the meaning of that quantity, consider the conditional distribution of the remaining sojourn time τ in **1**, given that at an arbitrary epoch, the chain is in that set. It is readily seen that the Laplace - Stieltjes transform of τ is given by

$$f(s) = [\boldsymbol{\theta}(1)\mathbf{e}]^{-1}\boldsymbol{\theta}(1)[sI - T(1, 1)]^{-1}T(1, 2)\mathbf{e}.$$

Recalling that $T(1, 2)\mathbf{e} = -T(1, 1)\mathbf{e}$, we easily verify that

$$\frac{1-f(s)}{s} = [\boldsymbol{\theta}(1)\mathbf{e}]^{-1}\boldsymbol{\theta}(1)[sI - T(1, 1)]^{-1}\mathbf{e}.$$

Setting $s = \gamma^{-1}$, we find that $K(\gamma^{-1}) = f(\gamma^{-1})$.

As a measure of the *stickiness* of set 1, the function K is simply the Laplace - Stieltjes transform of the remaining sojourn time τ in that set evaluated at the mean lifetime of the monitor. The graph of K is therefore just an alternative way of visualizing the information in the distribution function of τ. If, for example, we found that $K(5) = 0.75$, we could interpret this as "if we turn on an exponential monitor with a mean lifetime of 5 each time the chain enters set 1, 75% of the time spent in that set will be recorded."

It immediately follows that $K(\gamma^{-1})$ increases from 0 to 1 as a function of its argument. A rapid increase suggests that even a short-lived monitor will record a significant portion of the time spent in set 1. That suggests that most of the sojourns in 1 are short and that the set is not sticky. This is well illustrated by the generators Q_1 and Q_2 in the problem. They only differ in the (1, 2)- and (2, 1)-elements which have been interchanged to increase the rate from a state in 1 to a state in 2. With the two Markov chains scaled to have a common time unit, the graph of K increases much more slowly for the generator Q_1.

Solution to Problem 5.2.25: We represent the state space by macro-states **i**, $1 \le i \le m$, and by two failure states I^* and II^*. The macro-state **i** consists of n states (i, j), $1 \le j \le n$, that signify that i components in module I and j in module II are functioning. The states I^* and II^* represent the two failure modes. We choose to list the macro-states in decreasing order of i; within each **i**, the (i, j) are listed in decreasing order of j. We write q for $1 - p$. We display the structure of the infinitesimal generator Q of the Markov chain:

$$
Q = \begin{array}{c|cccccccc|}
 & m & m-1 & m-2 & \cdots & 2 & 1 & I^* & II^* \\
m & A_m & B_m & 0 & \cdots & 0 & 0 & 0 & \mathbf{c}_m \\
m-1 & 0 & A_{m-1} & B_{m-1} & \cdots & 0 & 0 & 0 & \mathbf{c}_{m-1} \\
m-2 & 0 & 0 & A_{m-2} & \cdots & 0 & 0 & 0 & \mathbf{c}_{m-2} \\
\cdot & \cdot & \cdot & \cdot & & \cdot & \cdot & \cdot & \cdot \\
\cdot & \cdot & \cdot & \cdot & & \cdot & \cdot & \cdot & \cdot \\
2 & 0 & 0 & 0 & \cdots & A_2 & B_2 & 0 & \mathbf{c}_2 \\
1 & 0 & 0 & 0 & \cdots & 0 & A_1 & \mathbf{d}_1 & \mathbf{c}_1 \\
I^* & 0 & 0 & 0 & \cdots & 0 & 0 & 0 & 0 \\
II^* & 0 & 0 & 0 & \cdots & 0 & 0 & 0 & 0
\end{array},
$$

where the matrices A_k, $1 \le k \le m$, are zero except for

$$[A_k]_{rr} = -i\lambda - r\mu, \quad \text{for } 1 \le r \le n, \qquad \text{and} \qquad [A_k]_{r,r-1} = r\mu, \quad \text{for } 1 \le r < n.$$

The matrices B_k, $2 \le k \le m$, are upper triangular. The element $[B_k]_{rh}$ is, with (k,r) components functioning, the rate of a failure in module I which triggers $r - h$ shock failures in module II. It is given by

$$[B_k]_{rh} = k\lambda \binom{r}{h} p^{r-h} q^h, \qquad \text{for } 1 \le k \le m,\ 1 \le h \le r \le n.$$

For $2 \le k \le m$, $[\mathbf{c}_k]_r = k\lambda p^r$, for $1 \le r \le n$. The vector $\mathbf{d}_1$ is a column vector with all n components equal to λ and $\mathbf{c}_1$ is zero except for its last component which equals μ.

The initial state is (m, n) and the system expires when the Markov chain is absorbed in either of the states I^* or II^*. Clearly, the lifetime has a PH-distribution.

With k and r components surviving in each module, the expected time to the next component failure is $(\lambda k + \mu r)^{-1}$. That is the first term in the stated formula for $E(k, r; p)$. The probability that the first component failure occurs in module II is $\mu r(\lambda k + \mu r)^{-1}$. That failure leaves k and $r - 1$ components behind and these have an expected remaining lifetime $E(k, r-1; p)$. The third term corresponds to the case where a component in I fails and ν of the remaining components in II are destroyed by the accompanying shock. The boundary conditions are $E(k, 0; p) = 0$, for $1 \le k \le m$, and $E(0, r; p) = 0$, for $0 \le r \le n$. The ratio $R(m, n; p)$ reflects the reduction in the expected lifetime of the system due to the vulnerability of the components in II to the shock of a failure of a component in module I.

Let α be the n-vector $(1, 0, \cdots, 0)$. A Type-I failure occurs if and only if there still are some working components in module II at the mth component failure in I. The probability that ν good components remain in II when the mth failure happens in I is the component with index ν of

$$\theta = \alpha(-A_m^{-1})B_m(-A_{m-1}^{-1})B_{m-1} \cdots (-A_2^{-1})B_2(-A_1^{-1})\lambda.$$

Accounting for any failures in II induced by the last shock, we get that

$$\phi_1(0) = \sum_{\nu=1}^{n} \theta_\nu p^\nu, \quad \text{and} \quad \phi_1(r) = \sum_{\nu=r}^{n} \theta_\nu \binom{\nu}{r} p^{\nu-r} q^r, \quad \text{for } 1 \le r \le n.$$

The probabilities $\phi_2(k)$ are given by

$$\phi_2(m) = \alpha(-A_m^{-1})\mathbf{c}_m, \quad \text{and} \quad \phi_2(k) = \alpha(-A_m^{-1})B_m(-A_{m-1}^{-1})B_{m-1} \cdots (-A_k^{-1})\mathbf{c}_k,$$

for $1 \le k \le m-1$. The probabilities $\phi_1(r)$ and $\phi_2(k)$ are conveniently computed by the recursive scheme

$\mathbf{x} \leftarrow \alpha(-A_m^{-1})$.
compute $\phi_2(m)$.
if $m = 1$, go to 99
for $k = m-1, \cdots, 1$, $\mathbf{x} \leftarrow \mathbf{x}B_{k+1}(-A_k^{-1})$.
compute $\phi_2(k)$.
99 $\mathbf{x} \leftarrow \lambda\mathbf{x}$.
compute $\phi_1(r)$, $0 \le r \le n$.

Solution to Problem 5.2.36: Write $J(t)$ for the double integral, then

$$J(t)(\mathbf{e}\theta - Q) = \int_0^t \mathbf{e}\theta u \, du - \int_0^t [\exp(Qu) - I] \, du$$

$$= \frac{1}{2}t^2\mathbf{e}\theta + tI - [\mathbf{e}\theta t + [\exp(Qu) - I](\mathbf{e}\theta - Q)^{-1}],$$

from which the stated formula follows. For the asymptotic formula use $\exp(Qt) \to \mathbf{e}\theta$ as $t \to \infty$.

Solution to Problem 5.2.38: If we define its state by right-hand continuity, the system with restarts at the Poisson events is again a Markov chain. Its generator is given by $Q(\lambda, \alpha) = Q - \lambda I + \lambda\mathbf{e}\alpha$. The invariant probability vector ϕ of $Q(\lambda, \alpha)$ is

$$\phi = \frac{\alpha(\lambda I - Q)^{-1}}{\alpha(\lambda I - Q)^{-1}\mathbf{e}},$$

so that income accrues at the rate $\phi\beta = \Psi(\lambda, \alpha)$.

Next, we show that the maximum (and the minimum) of $\Psi(\lambda, \alpha)$ is attained at one of the vertices of the simplex $S = \{\alpha : \alpha \ge 0, \alpha\mathbf{e} = 1\}$. Since the matrix $(\lambda I - Q)^{-1}$ is positive, we may (by adding a positive constant to Ψ if necessary)

assume that the vector $\boldsymbol{\beta}$ is positive. The function Ψ is now of the form $\Psi(\lambda, \alpha) = (\alpha\mathbf{c})(\alpha\mathbf{d})^{-1}$, where $\mathbf{c}$ and $\mathbf{d}$ are positive vectors. If $\alpha(1)$ and $\alpha(2)$ are points in the simplex S, then $\Psi[\lambda, \theta\alpha(1) + (1 - \theta)\alpha(2)]$ is a function of θ of the form

$$f(\theta) = \frac{\theta A_1 + (1 - \theta)A_2}{\theta B_1 + (1 - \theta)B_2},$$

where A_1, A_2, B_1, and B_2 are positive constants. By evaluating the derivative, we see that $f'(\theta)$ is positive, negative, or zero depending on the sign of $A_1/B_1 - A_2/B_2$, so that the maximum of $f(\theta)$ is attained for $\theta = 0$ or $\theta = 1$. That implies that the maximum of Ψ is attained at a vertex of S. For every value of λ it therefore to sufficient compute the ratios

$$\frac{[(\lambda I - Q)^{-1}\boldsymbol{\beta}]_j}{[(\lambda I - Q)^{-1}\mathbf{e}]_j},$$

and to set $\alpha_j = 1$ for the index j corresponding to the largest ratio. If the maximum is attained for several j, then Ψ is maximum of some edge of the simplex, but that is a special case.

Solution to Problem 5.2.42: The construction of the matrix R proceeds as outlined. The eigenvalue η must be less than 1. The graph of the convex increasing function $f(\mu - \mu z)$ intersects the line $y = z$ at some point η in (0, 1) if and only if the derivative $-\mu f'(0+) = \mu\lambda'_1$ exceeds 1.

Substituting the explicit form for R in the equation $\mathbf{x}_0(T + \mu R) = \mathbf{0}$ yields

$$\mathbf{x}_0 = c\alpha(-T)^{-1}[\mu(1 - \eta)I - T]^{-1}.$$

Using the expression for $(I - R)^{-1}\mathbf{e}$, the normalizing equation leads to

$$1 = c\mu^{-1}(1 - \eta)^{-1}\alpha(-T)^{-1}[\mu(1 - \eta)I - T]^{-1}\mathbf{T}^{\circ} + c\alpha(-T)^{-1}[\mu(1 - \eta)I - T]^{-1}\mathbf{e}$$

$$= c\mu^{-1}(1 - \eta)^{-1}\lambda'_1 f^*(\mu - \mu\eta) + c\mu^{-1}(1 - \eta)^{-1}\lambda'_1[1 - f^*(\mu - \mu\eta)]$$

$$= c\mu^{-1}(1-\eta)^{-1}\lambda'_1,$$

which gives the stated value for c. The formulas for the transform of $F^*(\cdot)$ are used in this last derivation. The simplifications arise from the equality $f(c\mu - c\mu\eta) = \eta$. These same formulas are used in calculating $y_0 = \mathbf{x}_0\mathbf{e}$. Upon simplification, we obtain that $y_0 = 1 - \rho$. The same calculations yield the stated expressions for y_i with $i \geq 1$.

The term $\mathbf{x}_i\mathbf{T}^\circ \cdot$ is the elementary probability that in $(t, t+\cdot)$ there are i customers in the system and an arrival occurs. We see that $\mathbf{x}_i\mathbf{T}^\circ = (1-\eta)\eta^i(\lambda'_1)^{-1}$. The denominator in the expression for ϕ_i reduces to the arrival rate $(\lambda'_1)^{-1}$. The quantity ϕ_i is the conditional probability that there are i jobs in the system, given that an arrival occurs.

Solution to Problem 5.2.43: It is obvious that the mean of $F(\cdot)$ equals 1 when $\lambda = p + q\theta^{-1}$. The variance of the scaled distribution, which is also its squared coefficient of variation, is given by

$$\sigma^2 = \frac{2(p+q\theta^{-2})}{(p+q\theta^{-1})^2} - 1.$$

For a proper hyperexponential distribution, σ^2 is always greater than 1 and for proper choices of p and θ can assume any value in $(1, \infty)$. In the calculation of η, we shall leave the substitution for λ until later. Since the transform $f(s)$ is given by

$$f(s) = \frac{\lambda(p+q\theta)s + \lambda^2\theta}{(s+\lambda)(s+\lambda\theta)},$$

an easy calculation shows that the equation $\eta = f(\mu - \mu\eta)$ may be written as

$$\mu^2\eta(1-\eta)^2 + \lambda\mu\eta(1-\eta)(1+\theta) - \lambda\mu(1-\eta)(p+q\theta) - \lambda^2\theta(1-\eta) = 0.$$

The root $\eta = 1$ is not of interest, so we divide by $1-\eta$ to obtain the quadratic equation

$$\mu^2\eta^2 - \mu[\mu + \lambda(1+\theta)]\eta + \lambda\mu(p+q\theta) + \lambda^2\theta = 0.$$

The discriminant Δ of that equation is positive if and only if

$$\mu^{-2}\Delta = [\mu + \lambda(1 + \theta)]^2 - 4\lambda\mu(p + q\theta) - 4\lambda^2\theta$$

$$= \mu^2 + 2\lambda(1 - 2p)(1 - \theta)\mu + \lambda^2(1 - \theta)^2,$$

is positive. However, for real μ that quadratic function does not vanish and is always positive. The equation for η has two real, positive roots. The smallest root is less than one if and only if the left-hand side of the quadratic equation for η is negative at 1. That leads to the condition

$$\mu > \lambda\theta(q + \theta - q\theta)^{-1}.$$

The right-hand side is the reciprocal of the mean of $F(\cdot)$. The inequality therefore holds if and only if the queue is stable. The smallest root η is

$$\eta = \frac{\mu + \lambda(1 + \theta)}{2\mu} - \frac{1}{2\mu}[\mu^2 + 2\lambda(1 - 2p)(1 - \theta)\mu + \lambda^2(1 - \theta)^2]^{1/2}.$$

Now, we replace λ by $p + q\theta^{-1}$. The traffic intensity ρ is then μ^{-1}. Rewriting η in terms of the parameters ρ, p, and θ, we get

$$\eta = \frac{1}{2}[1 + (1 + p\theta + q\theta^{-1})\rho]$$
$$- \frac{1}{2}\{[1 + (2p - 1 - p\theta + q\theta^{-1})\rho]^2 - 4p(2p - 1 - p\theta + q\theta^{-1})\rho\}^{1/2}.$$

An interesting point is that, for any given $\rho < 1$, we can choose values of p and θ for which η is as close to 1 as we wish. That is shown by the following intuitive argument. For small θ, the dominant term in η is

$$\frac{1}{2}q\rho\theta^{-1} - \frac{1}{2}q\rho\theta^{-1}[1 - 4p\theta(q\rho)^{-1}]^{1/2} \approx p.$$

If we choose a distribution $F(\cdot)$ with p close to 1 and θ very small, then η is also close to 1, essentially independently of the traffic intensity. The means of the two exponential components of $F(\cdot)$ are then approximately p^{-1} and q^{-1}. Most interarrival times are very short, but a small fraction of very long ones keep the mean at 1. Even with a high service rate μ, most arrivals find many customers in the system. On the other hand, the steady-state probabilities $\{y_i\}$ show that most of the time the queue is empty!

With the representation $\boldsymbol{\alpha} = (p, q)$, $T = -\lambda \, diag(1, \theta)$, the components of $\mathbf{u}$ are

$$u_1 = \frac{p}{\mu(1-\eta)+\lambda}, \quad \text{and} \quad u_2 = \frac{q}{\mu(1-\eta)+\lambda\theta}.$$

The vector $\mathbf{x}_0$ is given by

$$x_{01} = \frac{\mu(1-\eta)}{p+q\theta^{-1}} \frac{p}{\mu(1-\eta)+\lambda}, \quad \text{and} \quad x_{02} = \frac{\mu(1-\eta)}{p+q\theta^{-1}} \frac{q\theta^{-1}}{\mu(1-\eta)+\lambda\theta}.$$

The steady-state distributions of the queue lengths and waiting times at arrivals and at an arbitrary time for the $H_2/M/1$ queue are now routinely calculated.

Solution to Problem 5.2.71: The arrivals of customers who find the single server occupied form a renewal process because at such epochs no further information about the past evolution of the process needs to be remembered. Let time zero correspond to such an arrival. The probability that the next arrival comes before x and the service does not end before that time is given by

$$\int_0^x e^{-\mu u} \, dF(u).$$

Alternatively, suppose that the next arrival occurs in $(u, u+du)$ and the service has ended in $(0, u)$. At time u the server becomes busy. The probability that the next arrival which finds the server busy occurs in (u, x) is therefore $F^{\circ}(x-u)$. By the law of total probability, the contribution to $F^{\circ}(x)$ of that alternative is

$$\int_0^x (1-e^{-\mu u})F^{\circ}(x-u)] \, dF(u),$$

which establishes the equation in part *a*. The Laplace - Stieltjes transform of $F^{\circ}(\cdot)$ is obtained by a routine calculation. The mean $\lambda_1^{\circ} = [f(\mu)]^{-1}\lambda_1'$.

For an insightful way of proving that $F^{\circ}(\cdot)$ is a PH-distribution, we consider the absorbing Markov chain with generator

$$R = \begin{array}{c|ccc|} 1 & T - \mu I & \mu I & \mathbf{T}^{\circ} \\ 0 & \mathbf{T}^{\circ}\boldsymbol{\alpha} & T & 0 \\ * & 0 & 0 & 0 \end{array},$$

with initial probability vector $(\boldsymbol{\alpha}, \mathbf{0}, 0)$. The phases in 0 correspond to the server being idle, those in 1 to the server being busy. We see that absorption corresponds to an arrival occurring when the server is busy. In a formal verification, we calculate the Laplace - Stieltjes transform of the absorption time distribution for the Markov chain R. After some matrix manipulations, we find that it satisfies the formula given in part *a*.

The customers who find servers $1, \cdots, i$ busy make up the i-overflow stream. By recursion on i, we see that every i-overflow stream is a renewal process and that the probability distributions $F_i^{\circ}(\cdot)$ satisfy the recurrence relations

$$F_i^{\circ}(x) = \int_0^x [e^{-\mu_i u} + (1 - e^{-\mu_i u})F_i^{\circ}(x - u)]\, dF_{i-1}^{\circ}(u),$$

where $F_0^{\circ}(x) = F(x)$. The remainder of the solution is now routine. An alternative Markov chain for the three-server system can be set up by a state description similar to that in Problem 5.2.52.

Chapter 6

Solution to Problem 6.4.5: To find the joint density of Y_1 and Y_2, consider the elementary probability $\psi(u_1, u_2)du_1du_2$ that the two points land in the intervals $(u_1, u_1 + du_1)$ and $(u_1 + u_2, u_1 + u_2 + du_2)$. If $u_1 + u_2 \leq 1/2$, then the first point must land in the leftmost interval and, conditionally, the second is then uniform in $(u_1, 1]$. Therefore,

$$\psi(u_1, u_2)\, du_1du_2 = du_1 \frac{du_2}{1 - u_1}.$$

If $u_1 > 1/2$, again the order in which the points are chosen is clear and

$$\psi(u_1, u_2)\, du_1\, du_2 = du_2 \frac{du_1}{u_1 + u_2}.$$

In the case $u_1 < 1/2 < u_1 + u_2$, the first point chosen either lands to the left or to the right of 1/2. These two cases respectively contribute

$$du_1 \frac{du_2}{1 - u_1}, \quad \text{and} \quad du_2 \frac{du_1}{u_1 + u_2},$$

to the elementary probability. The marginal densities

$$\psi_1(u_1) = 1 + \log 2, \qquad \text{for } 0 \le u_1 \le 1/2,$$

$$= -\log u_1, \qquad \text{for } 1/2 \le u_1 \le 1,$$

$$\psi_2(u_2) = 2 \log 2, \qquad \text{for } 0 \le u_2 \le 1/2,$$

$$= -2 \log u_2, \qquad \text{for } 1/2 \le u_2 \le 1,$$

of Y_2 are derived by routine calculations. Their means are $EY_1 = 5/16$, and $EY_2 = 6/16$. As is to be expected by symmetry, the length $1 - Y_1 - Y_2$ of the third interval also has mean 5/16. $P\{Y_1 < Y_2\}$ is found by integrating $\psi(u_1, u_2)$ over the subset of C where $u_1 < u_2$. After calculating elementary integrals, that probability is found to equal $5/4 - \log 2$, which is approximately 0.5569.

The differences between the random partitions of the unit interval are obvious from a plot, say, of 1,000 pairs (X_1, X_2) and 1,000 pairs (Y_1, Y_2) generated by the two methods. Summary statistics, such as the marginal histograms or the frequency of the event $\{Y_1 < Y_2\}$ give additional visual evidence.

Solution to Problem 6.4.6: The joint density of Y_1 and Y_2 is now

$$\frac{1}{2(1 - u_1)} + \frac{1}{2(u_1 + u_2)}, \qquad \text{for } u_1 + u_2 \le 1, \ u_1 \ge 0, \ u_2 \ge 0.$$

The marginal densities are $1/2 - 1/2 \log u_1$, and $-\log u_2$, on the interval [0, 1]. Their means are 3/8 and 1/4, respectively.

Solution to Problem 6.4.24: The random variables $Y_1, \cdots, Y_{m+1}$ are *exchangeable,* that is, their joint density is invariant under permutations of its variables. All $m + 1$ spacings therefore have the same marginal densities. To find the density of Y_1, notice that it is the minimum of m independent random variables, uniform on [0, 1]. Therefore,

$$\phi(u) = m(1-u)^{m-1}, \quad \text{for } 0 \le u \le 1.$$

To evaluate the expected value

$$E[-Y_1 \log Y_1] = -\int_0^1 u \log u \; m(1-u)^{m-1}\, du,$$

notice that

$$g(u) = \int_0^u mv(1-v)^{m-1}\, du = \frac{1}{m+1}[1-(1-u)^{m+1} - (m+1)u(1-u)^m].$$

An integration by parts now leads to

$$E[-Y_1 \log Y_1] = \int_0^1 \frac{g(u)}{u}\, du = \frac{1}{m+1}\left[\int_0^1 \frac{1-(1-u)^{m+1}}{u}\, du - 1\right]$$

$$= \int_0^1 \frac{(1-u)[1-(1-u)^m]}{u}\, du = \int_0^1 \frac{v(1-v^m)}{1-v}\, dv = \sum_{r=2}^{m+1} \frac{1}{r}.$$

Using the classical Euler - Mascheroni formula $\sum_{r=1}^{n} r^{-1} - \log n = \gamma + o(1)$, as $n \to \infty$, we obtain that

$$EH(m) = 1 - \frac{1-\gamma+o(1)}{\log(m+1)}.$$

where $\gamma = 0.577215665$ is the Euler - Mascheroni constant and o(1) denotes a quantity tending to 0 as $m \to \infty$. Simulation estimates of $EH(m)$ grow very slowly in m and may suggest that $EH(m)$ tends to a constant less than 1. The exact result shows that $EH(m) \to 1$ as $m \to \infty$ and also explains its slow growth.

Solution to Problem 6.4.25: Since the $Y_1, \cdots, Y_{m+1}$ are *exchangeable,* they have the identical marginal densities. The density of Y_1 is

$$\phi(u) = m(1-u)^{m-1}, \quad \text{for } 0 \le u \le 1.$$

Therefore,

$$E(L_1) = (m+1)E\left|Y_1 - \frac{1}{m+1}\right| = (m+1)\int_0^{(m+1)^{-1}} \left[\frac{1}{m+1} - u\right] m(1-u)^{m-1}\, du$$
$$+ (m+1)\int_{(m+1)^{-1}}^{1} \left[u - \frac{1}{m+1}\right] m(1-u)^{m-1}\, du = 2\left[\frac{m}{m+1}\right]^{m+1},$$

after calculation of the elementary integrals and simplification. We see that, as $m \to \infty$, $E(L_1) \to 2e^{-1}$. By a more elaborate calculation, it can be shown that the variance of L_1 is

$$\sigma^2 = \frac{8m^{m+2} + 4m(m-1)^{m+2}}{(m+2)(m+1)^{m+2}} - 4\left[\frac{m}{m+1}\right]^{2m+2},$$

which behaves as $4m^{-1}(2e^{-1} - 5e^{-2})$ as $m \to \infty$. Similarly,

$$E(L_2) = (m+1)E\left[Y_1 - \frac{1}{m+1}\right]^2$$
$$= (m+1)\int_0^1 (1-u)^2 mu^{m-1}\, du - \frac{1}{m+1} = \frac{m}{(m+1)(m+2)}.$$

Solution to Problem 6.4.41: The Pareto distribution function is given by $\Pi(x) = 0$, for $x \le \theta_0$, and $\Pi(x) = 1 - (x^{-1}\theta_0)^{\alpha}$, for $x \ge \theta_0$. Its inverse function $\Pi^{-1} = \theta_0(1-y)^{-1/\alpha}$, for $0 \le y \le 1$. Therefore, if U is a uniform variate, the random variate $Y = \theta_0 U^{-1/\alpha}$ has the given Pareto distribution. The joint density $f(x_1, \cdots, x_n \mid \theta)$ is given by

$$f(x_1, \cdots, x_n \mid \theta) = \frac{1}{\theta^n},$$

on the region where $0 < x_1 < \theta, \cdots, 0 < x_n < \theta$. Therefore the joint probability density of θ and the observations is

$$\frac{\alpha}{\theta_0}\left[\frac{\theta_0}{\theta}\right]^{\alpha+1} \frac{1}{\theta^n}.$$

To find the marginal density $m(x)$, we integrate with respect to θ over the allowable values of θ. These make up the interval from max $\{\theta_0, x_i, 1 \le i \le n\}$. Write θ° for the left endpoint, then $m(x) = \alpha\theta_0^\alpha\theta^{\circ-n-\alpha}(n+\alpha)^{-1}$, Dividing the joint density by $m(x)$ we find that

$$\pi(\theta|x) = \frac{n+\alpha}{\theta}\left[\frac{\theta^\circ}{\theta}\right]^{n+\alpha+1},$$

on $\theta > \theta^\circ$. That is the stated Pareto density. If values larger than θ_0 are observed, the support of the density is shifted to $\theta^\circ > \theta_0$. In all cases, the larger exponent $n + \alpha + 1$ causes the posterior density to drop off faster than the prior.

Solution to Problem 6.4.42: By integrating the joint density

$$\alpha e^{-\alpha(\theta-\theta_0)} \frac{1}{\theta^n},$$

over $[\theta^\circ, \infty)$, we readily find $m(x)$ and dividing the joint density by $m(x)$ yields the stated posterior density. By the change of variable $\theta^\circ u = v$, the integral in the denominator is rewritten as

$$(\theta^\circ)^{-n+1}\int_1^\infty e^{-\alpha\theta^\circ v} v^{-n}\, dv = (\theta^\circ)^{-n+1}E_n(\alpha\theta^\circ),$$

in terms of the exponential integral $E_n(z) = \int_1^\infty e^{-zv} v^{-n}\, dv$. Library routines for that function are available.

The distribution function $\Pi_1(x)$ of the present prior satisfies $1 - \Pi_1(x) = \exp[-\alpha(x - \theta_0)]$, for $x \ge \theta_0$. The corresponding tail probability for the Pareto density in Problem 6.4.41 is

$$1 - \Pi(x) = \left[\frac{\theta_0}{x}\right] = \exp[-\alpha(\log x - \log \theta_0)],$$

which tends much more slowly to 0 as $x \to \infty$.

Solution to Problem 6.4.44: The joint density of θ and of the number of successes in n Bernoulli trials with probability θ is

$$\frac{1}{B(\alpha, \beta)}\theta^{\alpha-1}(1-\theta)^{\beta-1}\left[\frac{n}{x}\right]\theta^{x}(1-\theta)^{n-x}, \quad \text{for } 0 < \theta < 1.$$

Integration over θ given the marginal density

$$m(x) = \frac{1}{B(\alpha, \beta)}\frac{1}{B(\alpha + x, \beta + n - x)}\left[\frac{n}{x}\right],$$

so that, upon diving the joint density by $m(x)$, we obtain that $\pi(\theta|x)$ is the beta density with parameters $\alpha + x$ and $\beta + n - x$, which has the stated mean. The expected a posteriori loss may now be written as

$$\int_0^1 [\theta - a(x)]^2 \pi(\theta|x)\, d\theta,$$

which is smallest if $a(x)$ is the mean of the density $\pi(\theta|x)$. We see that, as $n \to \infty$, the Bayes estimator tends to the standard empirical frequency x/n.

Solution to Problem 6.4.51: The likelihood of the n observations is

$$L(k_1, \cdots, k_n; \lambda) = e^{-\lambda n\bar{k}}\lambda^{n(\bar{k}-1)}\prod_{i=1}^{n} k_i^{k_i - 1}(k_i!)^{-1}.$$

Taking logarithms and differentiating we find the stated maximum likelihood estimator. To compute the variance, evaluate $\sigma^2 = g''(1) + g'(1) - g'^2(1)$. The explicit form of the g_k is obtained from the equation for $g(z)$ by Lagrange's expansion. It is not an easy calculation to verify directly that the generating function of $\{g_k\}$ satisfies the functional equation for $g(z)$, although that equation follows directly from a probability argument.

Solution to Problem 6.4.62: For completeness, we derive a formula for the volume $V_m(r)$ of an m-dimensional sphere of radius r. $V_m(r)$ is the integral of the indicator function of the set $\sum_{i=1}^m u_i^2 \leq r^2$. By integrating over the range of one of the variables, we get that for $m \geq 3$,

$$V_m(r) = \int_{-r}^{r} V_{m-1}[(r^2 - u^2)^{1/2}]\, du.$$

Setting $V_m(r) = C_m r^m$ and making the change of variable $ru = v$ in the integral, we obtain that

$$C_m = C_{m-1}\int_{-1}^{1} (1 - v^2)^{(m-1)/2}\, dv = C_{m-1}\int_{0}^{1} (1 - t)^{(m-1)/2} t^{-1/2} = C_{m-1}B[1/2,(m+1)/2],$$

for $m \geq 3$, with $C_2 = \pi$. It readily follows that $C_m = \pi^{m/2}[\Gamma[(m+2)/2]]^{-1}$, so that for $m = 2m'$, $V_m(r) = \pi^{m'}(m'!)^{-1} r^m$ while for $m = 2m' + 1$, $V_m(r) = 2^m (m'!)(m!)^{-1}\pi^{m'} r^m$. The probability $p(m)$ is given by $2^{-m} V_m(1)$, so that the stated formulas follow. In particular,

$$p(2) = \frac{\pi}{4}, \quad p(3) = \frac{\pi}{6}, \quad p(4) = \frac{\pi^2}{32}, \quad p(5) = \frac{\pi^2}{60}.$$

The probability that a point is accepted rapidly decreases. The simple acceptance rejection method is inefficient for larger m.

Solution to Problem 6.4.64: Integrating over the sphere $\sum_{i=2}^m u_i^2 = 1 - u^2$, we see that the marginal density of X_1 is

$$\phi(u) = \frac{1}{V_m(1)} \int \cdots \int du_2 \cdots du_m = \frac{V_{m-1}[(1 - u^2)^{1/2}]}{V_m(1)}.$$

Using the formula for $V_m(r)$ in the solution to Problem 6.4.61, we obtain the stated formula. Clearly, for $0 \leq x \leq 1$,

$$P\{X_1^2 \leq x\} = \frac{1}{B[1/2, (m+1)/2]} \int_{-\sqrt{x}}^{\sqrt{x}} (1 - u^2)^{(m-1)/2}\, du$$

$$= \frac{1}{B[1/2, (m+1)/2]} \int_{0}^{x} (1 - v)^{(m-1)/2} v^{-1/2}\, dv.$$

X_1^2 therefore has the stated beta density. The conditional joint density of $X_2, \cdots, X_m$, given that $X_1 = u$, $-1 \le u \le 1$ is

$$\frac{1}{\phi(u)} \frac{1}{V_m(1)} = \frac{1}{V_{m-1}[(1-u^2)^{1/2}]},$$

the uniform density over a sphere of radius $(1-u^2)^{1/2}$ in R^{m-1}.

Let $Y_1, \cdots, Y_m$ be independent beta variates with parameters 1/2, $(r+1)/2$, for $1 \le r \le m$. Set $X_1^2 = Y_m$. Since, given Y_m, the random variable $\sum_{i=2}^m X_i^2$ is uniform over the sphere of radius $(1-Y_m)^{1/2}$ in R^{m-1}, the random variable $(1-Y_m)^{-1}X_2^2$ has the same beta distribution as Y_{m-1}. Set $X_2^2 = (1-Y_m)Y_{m-1}$. Similarly, given X_1^2 and X_2^2, $\sum_{i=3}^m X_i^2$ is uniform over the sphere of radius

$$(1 - X_1^2 - X_2^2)^{1/2} = (1 - Y_m - Y_{m-1} + Y_m Y_{m-1})^{1/2} = (1-Y_m)^{1/2}(1-Y_{m-1})^{1/2},$$

in R^{m-2}. The random variable $X_3^2 = (1-Y_m)(1-Y_{m-1})Y_{m-2}$ therefore has the correct distribution. In general, we set

$$X_i^2 = \prod_{j=m-i+2}^{m} (1-Y_j)Y_{m-i+1}, \quad \text{for } 2 \le i \le m.$$

Each X_i has a symmetric density. The components X_i are obtained by taking the square roots of the X_i^2 and by independently assigning + or −signs with equal probabilities.

Solution to Problem 6.4.65: But for a multiplicative constant, $G_m(\cdot)$ is the standard χ_m^2-density. As the joint density of m independent normal random variables is spherically symmetric, the vector $(Y_1, \cdots, Y_m)$ has a random direction. It therefore suffices to show that its length $L_1 = [\sum_{i=1}^m Y_i^2]^{1/2}$ has the correct distribution $F_m(x) = x^m$, for $0 \le x \le 1$. However, since $L_1 = [G_m(L^2)]^{1/m}$, it follows that for $0 \le x \le 1$,

$$P\{L_1 \le x\} = P\{[G_m(L^2)]^{1/m} \le x\} = P\{G_m(L^2) \le x^m\}$$

$$= P\{L^2 \le G_m^{-1}(x^m)\} = G_m^{-1}G_m(x^m) = x^m.$$

The length of a vector uniformly distributed over the unit sphere is independent of its direction. In the procedure in part b, we generate the length from the simple probability distribution $F(\cdot)$ and we use normal variates to generate the directional cosines conveniently.

The method in part b is the most efficient *general* method that is suitable for all dimensions m. Only normal variates produce a random direction so conveniently. That follows from the fact that if m *independent,* identically distributed random variables have a spherically symmetric density, they are normal. (If you are not familiar with that property, prove it for $m = 2$).

Solution to Problem 6.4.66: For $m = 2$,

$$P_p(2) = \int_0^1 (1 - x_1^p)^{1/p}\, dx_1 = p^{-1}B(p^{-1} + 1, p^{-1}).$$

To find $P_1(m)$ explicitly, define $V_m^\circ(u)$ as the volume of the set of points with nonnegative coordinates satisfying $\sum_{i=0}^m x_i \le 1$. Then $V_2^\circ(u) = u^2/2$, and

$$V_m^\circ(u) = \int_0^u V_{m-1}^\circ(u - v)dv = \int_0^u V_{m-1}^\circ(v)\, dv,$$

for $m \ge 3$. It follows that $V_m^\circ(u) = (m!)^{-1}u^m$, so that $P_1(5) = V_5^\circ(1) = 1/120$.

Solution to Problem 6.4.73: The steady-state probabilities of the Markov chain are all equal to $1/N$. The distribution of an arbitrary interarrival time is therefore a mixture with equal weights of the distributions $G_j(\cdot)$. Its transform $f(s)$ therefore has the stated form. Clearly, for $N = 1$, $F(\cdot)$ is exponential with parameter λ. We prove by induction that the same is true for every $N \ge 1$. For $N > 1$, write

$$f(s) = \sum_{j=1}^{N-1} g_j(s)\frac{\lambda}{s + N\lambda} + \frac{\lambda}{s + N\lambda} = \left[\frac{(N-1)\lambda}{s + \lambda} + 1\right]\frac{\lambda}{s + N\lambda} = \frac{\lambda}{s + \lambda}.$$

The second equality holds by the induction assumption. The initial state j is chosen with probability $1/N$. Given j, the conditional probability that at the beginning of the $(k + 1)$st interval, the Markov chain is in state r is $[P^k]_{jr}$. That readily yields the stated expression for the joint transform $f_k(s_1, s_2)$. All X_ν, $\nu \ge 1$, have the same marginal exponential distributions. Both their mean and standard deviation are therefore λ^{-1}. The cross-moment $E(X_1 X_{k+1})$ is

$$E(X_1 X_{k+1}) = \frac{1}{N} \sum_{j=1}^{N} \sum_{r=1}^{N} g_j'(0)[P^k]_{jr} g_r'(0) = \frac{1}{N} \sum_{j=1}^{N} \sum_{r=1}^{N} \sum_{v=j}^{N} \sum_{h=r}^{N} \frac{1}{v\lambda} [P^k]_{jr} \frac{1}{h\lambda}$$

$$= \frac{1}{N} \sum_{v=1}^{N} \sum_{h=1}^{N} \frac{1}{v\lambda} \frac{1}{h\lambda} \sum_{r=1}^{v} \sum_{j=1}^{h} [P^k]_{jr}.$$

The correlation coefficient $\rho(k)$ is therefore

$$\rho(k) = \frac{1}{N} \sum_{v=1}^{N} \sum_{h=1}^{N} \frac{1}{vh} \sum_{r=1}^{v} \sum_{j=1}^{h} [P^k]_{jr} - 1.$$

It vanishes only for very special choices of the matrix P. In general, the random variables X_i are dependent. For the suggested numerical examples, the correlation coefficients are easily computed by taking successive powers of the matrix P.

For part d, plot the interarrival times as vertical lines at 1,000 equidistant points along the x-axis. A succession of long intervals corresponds to a period where there are few arrivals in the point process. For the Poisson process, the graph looks very erratic, whereas for the point processes corresponding to P_1 and P_2, you will see pronounced strings of long or short intervals. Visualizing the behavior of the three queues is particularly instructive.

Index